2/6/96

D1488681

A Practical Guide
to Compressor
Technology

A Practical Guide to Compressor Technology

Heinz P. Bloch

McGraw-Hill

New York San Francisco Washington, D.C. Auckland Bogotá
Caracas Lisbon London Madrid Mexico City Milan
Montreal New Delhi San Juan Singapore
Sydney Tokyo Toronto

Library of Congress Cataloging-in-Publication Data

Bloch, Heinz P., 1933–
    A practical guide to compressor technology / Heinz P. Bloch.
        p.   cm.
    Includes index.
    ISBN 0-07-005937-3
    1. Compressors.   I. Title.
TJ990.B546   1995
621.5'1—dc20   95-22898
                                                    CIP

McGraw-Hill

A Division of The McGraw-Hill Companies

1 2 3 4 5 6 7 8 9 0   DOC/DOC   9 0 0 9 8 7 6 5

ISBN 0-07-005937-3

*The sponsoring editor for this book was Bob Hauserman, and the
production supervisor was Pamela A. Pelton. This book was set in
Century Schoolbook by North Market Street Graphics.*

Printed and bound by R. R. Donnelley & Sons Company.

McGraw-Hill books are available at special quantity discounts to use as
premiums and sales promotions, or for use in corporate training programs. For more information, please write to the Director of Special
Sales, McGraw-Hill, 11 West 19th Street, New York, NY 10011. Or contact your local bookstore.

This book is printed on acid-free paper.

"To the making of many books there is no end, and much devotion to them is wearisome to the flesh."
—May the author of this quote always consider me his friend.

Contents

Preface   xiii

Part 1   Positive Displacement Compressor Technology   1

Chapter 1. Theory   3

1.1   Symbols   3
1.2   How a Compressor Works   5
1.3   The First Law of Thermodynamics   9
1.4   The Second Law of Thermodynamics   9
1.5   Ideal or Perfect Gas Laws   10
    1.5.1   Boyle's Law   10
    1.5.2   Charles' Law   10
    1.5.3   Amonton's Law   10
    1.5.4   Dalton's Law   11
    1.5.5   Amagat's Law   11
    1.5.6   Avogadro's Law   11
    1.5.7   The Perfect Gas Formula   11
1.6   Vapor Pressure   12
1.7   Gas and Vapor   12
1.8   Partial Pressures   13
1.9   Critical Conditions   15
1.10   Compressibility   15
1.11   Generalized Compressibility Charts   16
1.12   Gas Mixtures   17
1.13   The Mole   17
1.14   Specific Volume and Density   18
1.15   Volume Percent of Constituents   18
1.16   Molecular Weight of Mixture   19
1.17   Specific Gravity and Partial Pressure   19
1.18   Ratio of Specific Heats   20
1.19   Pseudo Critical Conditions and Compressibility   21
1.20   Weight-Basis Items   21
1.21   Compression Cycles   21
1.22   Power Requirement   23

1.23   Compressibility Correction   24
1.24   Multiple Staging   25
1.25   Volume References   27
1.26   Cylinder Clearance and Volumetric Efficiency   28
1.27   Cylinder Clearance and Compression Efficiency   31

Chapter 2. Reciprocating Process Compressor Design Overview   33

2.1   Crankshaft Design   37
2.2   Bearings and Lubrication Systems   40
2.3   Connecting Rods   43
2.4   Crossheads   43
2.5   Frames and Cylinders   43
2.6   Cooling Provisions   51
2.7   Pistons   52
2.8   Piston and Rider Rings   52
2.9   Valves   53
2.10   Piston Rods   57
2.11   Packings   60
2.12   Cylinder Lubrication   60
2.13   Distance Pieces   65

Chapter 3. Reciprocating Compressor Performance Considerations   69

3.1   Capacity Control Is Important   69
       3.1.1   Recycle/Bypass   70
       3.1.2   Suction Throttling   71
       3.1.3   Suction Valve Unloading   71
       3.1.4   Clearance Pockets   75
3.2   More About Cylinder Jacket Cooling-Heating Arrangements   78
       3.2.1   Methods of Cooling   79
3.3   Comparing Lubricated vs. Nonlubricated Conventional Cylinder Construction 82
       3.3.1   Lubricated Cylinder Designs   82
       3.3.2   Nonlubricated Cylinder Design   85
3.4   Compressor Vent and Buffer Systems Deserve Attention   86
3.5   Compressor Instrumentation Is Always Important   88
       3.5.1   Electrics vs. Pneumatics   94
       3.5.2   Switch Set Points   95
       3.5.3   Control Panels   95
       3.5.4   Valve-in-Piston Reciprocating Compressors   97

Chapter 4. Labyrinth Piston Compressors   99

4.1   Main Design Features   99
4.2   Energy Consumption   102
4.3   Sealing Problems   103

Chapter 5. Hyper Compressors   111

5.1   Introduction   111
5.2   Cylinders and Piston Seals   113
5.3   Cylinder Heads and Valves   119

5.4	Drive Mechanism	121
5.5	Miscellaneous	124
5.6	Conclusion	126

Chapter 6. Metal Diaphragm Compressors **127**

6.1	Introduction	127
6.2	Terminology	127
6.3	Description	129

Chapter 7. Lobe-Type and Sliding Vane Compressors **137**

Chapter 8. Liquid Ring Compressors **145**

Chapter 9. Rotary Screw Compressors **151**

9.1	Twin-Screw Machines	151
	9.1.1 Working Phases	151
	9.1.2 Areas of Application	154
	9.1.3 Dry vs. Liquid-Injected Machines	154
	9.1.4 Operating Principles	155
	9.1.5 Flow Power and Temperature Calculations	156
	9.1.6 Power Calculation	158
	9.1.7 Temperature Rise	159
	9.1.8 Capacity Control	160
	9.1.9 Mechanical Construction	165
	9.1.10 Industry Experience	167
	9.1.11 Maintenance History	170
	9.1.12 Performance Summary	170
9.2	Oil-Flooded Single-Screw Compressors	173

Chapter 10. Reciprocating Compressor Performance and Sizing Fundamentals **179**

10.1	Capacity and Leakage Considerations	180
	10.1.1 Theoretical Maximum Capacity	180
10.2	Capacity Losses	181
10.3	Valve Preload	182
10.4	Valve and Gas Passage Throttling	182
10.5	Piston Ring Leakage	184
10.6	Packing Leakage	185
10.7	Discharge Valve Leakage	186
10.8	Suction Valve Leakage	187
10.9	Heating Effects	187
10.10	Pulsation Effects	188
10.11	Horsepower	190
10.12	Horsepower Adders	191
10.13	Gas Properties	192
	10.13.1 Ideal Gas	192
	10.13.2 Real Gas	193

10.14   Alternative Equations of State   193
10.15   Condensation   194
10.16   Frame Loads   194
10.17   Compressor Displacement and Clearance   195
10.18   Staging   196
10.19   Fundamentals of Sizing   197
    10.19.1   Determine the Number of Stages   198
    10.19.2   Determine the Approximate Horsepower   198
    10.19.3   Determine Cylinder Bore Requirements   199
    10.19.4   Check Frame Load   200
    10.19.5   What Next?   200
10.20   Sizing Examples   200

Part 2   Dynamic Compressor Technology   209

Chapter 11. Simplified Equations for Determining Performance of Dynamic Compressors   217

11.1   Non-overloading Characteristics of Centrifugal Compressors   217
11.2   Stability   217
11.3   Speed Change   219
11.4   Compressor Drive   220
11.5   Calculations   221

Chapter 12. Design Considerations and Manufacturing Techniques   227

12.1   Axially or Radially Split?   227
12.2   Tightness   227
12.3   Material Stress   227
12.4   Nozzle Location and Maintenance   228
12.5   Design Overview   229
    12.5.1   Casings   229
    12.5.2   Flowpath   244
    12.5.3   Rotors   247
    12.5.4   Impellers   250
    12.5.5   Axial Blading   258
    12.5.6   Seals   259
12.6   Bearing Configurations   268
    12.6.1   Radial Bearings   268
    12.6.2   Thrust Bearings   270
    12.6.3   Flexure Pivot™ Tilt Pad Bearings   271
12.7   Casing Design Criteria   276
12.8   Casing Manufacturing Techniques   285
12.9   Stage Design Considerations and Manufacturing Techniques   296
    12.9.1   Stage Design Criteria   296
12.10 Impeller Manufacturing Techniques   306
12.11 Rotor Dynamic Considerations   312
12.12 Fouling Considerations and Coatings   318
    12.2.1   Polymerization/Fouling   319
    12.2.2   Fouling and Its Effect on Compressor Operation   320
    12.2.3   Coating Case Study   321

12.2.4   SermaLon Coating   322
12.2.5   Results   324

Chapter 13. Dry Gas Seal and Magnetic Bearing Systems   325

13.1   Background   325
13.2   Dry Seals   326
   13.2.1   Operating Principles   326
   13.2.2   Operating Experience   330
   13.2.3   Problems and Solutions   330
   13.2.4   Dry Seal Upgrade Developments   331
13.3   Magnetic Bearings   332
   13.3.1   Operating Principles   332
   13.3.2   Operating Experience and Benefits 335
   13.3.3   Problems and Solutions   336
13.4   Development Efforts   337
   13.4.1   Thrust-Reducing Seals   338
13.5   Integrated Designs Available   341

Chapter 14. Couplings, Torque Transmission, and Torque Sensing   347

14.1   Coupling Overview   347
   14.1.1   Low Overhung Moment   350
   14.1.2   Low Residual Unbalance Desired   351
   14.1.3   Long Life and Maintainability   353
   14.1.4   Continuous Lubrication Not a Cure-All   354
   14.1.5   Contoured Diaphragm Coupling Overview   355
14.2   Performance Optimization Through Torque Monitoring   357

Chapter 15. Lubrication, Sealing, and Control Oil Systems for Turbomachinery   367

15.1   Considerations Common to All Systems   367
15.2   Seal Oil Considerations   370

Chapter 16. Compressor Control   373

16.1   Introduction   373
16.2   Control System Objectives   373
16.3   Compressor Maps   374
   16.3.1   Invariant Coordinates   377
16.4   Performance Control   380
   16.4.1   PI and PID Control Algorithms   382
   16.4.2   Stability Considerations   384
   16.4.3   Integral or Reset Windup   385
16.5   Performance Limitations   386
   16.5.1   Surge Limit   386
   16.5.2   Stonewall   389
16.6   Preventing Surge   389
   16.6.1   Antisurge Control Variables   390
   16.6.2   Antisurge Control Algorithms   391
   16.6.3   Controlling Limiting Variables   392
16.7   Loop Decoupling   393
16.8   Conclusion   394

Chapter 17. The Head-Flow Curve Shape of Centrifugal Compressors   395

17.1   The Compressor Stage   395
17.2   Elements of the Characteristic Shape   396
       17.2.1   Basic Slope   397
       17.2.2   Blade Angle Is a Compromise   398
       17.2.3   The Fan Law Effect   400
       17.2.4   The Choke Effect   401
       17.2.5   Mach Number Considerations   402
       17.2.6   Significance of Gas Weight   403
       17.2.7   Inducer Impeller Increases Head Output   403
       17.2.8   Surge   404
       17.2.9   Vaned Diffusers   406
       17.2.10 Vaneless Diffusers   406
       17.2.11 Equivalent Tip Speeds   407
17.3   Conclusions   409

Chapter 18. Applying Multiple Inlet Compressors   411

18.1   Critical Selection Criteria   411
       18.1.1   Head Rise to Surge, Surge Margin, Overload Margin   412
       18.1.2   Head Per Section   413
       18.1.3   Compressor Parasitic Flows   414
       18.1.4   Excess Margins on the Other Process Equipment   415
       18.1.5   Representing Compressor Performance   416
       18.1.6   Practical Levels of Critical Operating Parameters   417
18.2   Design of the Sideload Compressor   419
       18.2.1   Mixing Area   419
       18.2.2   Aerodynamics   420
       18.2.3   Temperature Stratification   424
18.3   Testing   424
       18.3.1   Test Setup   424
       18.3.2   Instrumentation   425
       18.3.3   Testing Procedure   425
       18.3.4   Accuracy of Test Results   426
       18.3.5   Evaluation of Results   426

Chapter 19. Predicting Compressor Performance at New Conditions   427

19.1   How Performance Tests Are Documented   427
19.2   Design Parameters: What Affects Performance   428
       19.2.1   Thermodynamic   428
       19.2.2   Mechanical   429
19.3   What to Seek from Vendors' Documents   430
       19.3.1   Performance Test Data   430
19.4   Illustrations and Example   431

Appendix A   Properties of Common Gases   437

Appendix B   Shortcut Calculations and Graphical Compressor Selection Procedures   445

Index   501

Preface

Compressors are a vital link in the conversion of raw materials into refined products. Compressors also handle economical use and transformation of energy from one form into another. They are used for the extraction of metals and minerals in mining operations, for the conservation of energy in natural gas reinjection plants, for secondary recovery processes in oil fields, for the utilization of new energy sources such as shale oil and tar sands, for furnishing utility or reaction air, for oxygen and reaction gases in almost any process, for process chemical and petrochemical plants, and for the separation and liquefaction of gases in air separation plants and in LPG and LNG plants. And, as the reader will undoubtedly know, this listing does not even begin to describe the literally hundreds of services which use modern compression equipment.

The economy and feasibility of all these applications depend on the reliability of compressors and the capability of the selected compressors to handle a given gas at the desired capacity. It is well known that only turbocompressors made large process units, such as ammonia plants, ethylene plants, and base load LNG-plants, technically and economically feasible. Conversely, there are applications where only a judiciously designed positive displacement compressor will be feasible, or economical, or both. These compressors could take the form of piston-type reciprocating machines, or helical screw machines intended for true oil-free operation, or liquid-injected helical screw machines, and so forth. All, of course, demand the highest reliability and availability performance. These two requirements form the cornerstone of the development programs under way at the design and manufacturing facilities of the world's leading equipment producers.

Today, the petrochemical and other industries are facing intense global competition, which in turn has created a need for lower-cost equipment. Making this equipment without compromising quality, efficiency, and reliability is not easy, and only the industrial world's best manufacturers measure up to the task. Equally important, only a contemplative, informed, and discerning equipment purchaser or equipment user can be expected to spot the right combination of these two desirable and seemingly contradictory requirements: low cost and high quality.

The starting point of machinery selection is machinery know-how. From know-how we can progress to type selection, i.e., reciprocating compressor vs. centrifugal compressor, or dry vs. liquid-injected rotary screw compressor. Type selection leads to component selection—say, oil film seals vs. dry gas seals for centrifugal compressors. These could be exceedingly important considerations since both type selection and component selection will have lasting impact on maintainability, surveillability, availability, and reliability of compressors and steam turbines. Without fail, the ultimate effect will be plant profitability or even plant survival.

This text, then, is intended to provide the kind of guidance that will make it easier for the reader to make an intelligent choice. And, while it can hardly claim to be all-encompassing and complete in every detail, it is nevertheless intended to be both readable and relevant. I had planned to make the text up-to-date as regards practical, field-proven component configuration and execution of process compressors. The emphasis was to be on technology for two principal categories and their respective subgroups: positive displacement compressors and dynamic compression equipment such as centrifugal and axial turbomachines.

With experience showing machinery downtime events being linked to malfunction of auxiliaries and support equipment, we decided to include surge suppression, lubrication and sealing systems, couplings, and other relevant auxiliaries. All of these are thoroughly cross-referenced in the index and should be helpful to a wide spectrum of readers.

While compiling this information from commercially available industry source materials, I was struck by the profusion of diligent effort that some manufacturers have expended to design and manufacture more efficient and more reliable machinery. With much of this source material dispersed among the various sales, marketing, design, and manufacturing groups, I set out to collect the data and organize it into a text which first acquaints the reader with the topic by using overview and summary-type materials. The information progresses through more detailed and somewhat more design-oriented write-ups toward scoping studies and application and selection examples. Some of these are shown in both English and metric units, others were left in the method chosen by the original contributor.

The reader will note that I stayed away from an excessively mathematical treatment of the subject at hand. Instead, the focus was clearly on giving a single-source reference on all that will be needed by the widest possible spectrum of machinery users, ranging from plant operators to mechanical technical support technicians, reliability engineers, mechanical and chemical engineers, operations superintendents, project managers, or even senior plant administrators.

However, the publishers and I wish to point out that this book would never have been written without the full cooperation of a large number of highly competent equipment manufacturers in the United States and overseas. It was compiled by obtaining permission to use the direct contributions of companies and individuals listed in the Acknowledgment and Bibliography sections. These contributions were then structured into a cohesive survey of what the reader should know about compressor technology in the late 1990s. The real credit should therefore go to the various contributors and not the coordinating or compiling editor. In line with this thought, I would be most pleased if the entire effort would serve to acquaint the reader not only with the topic, but also with the names of the outstanding individuals and companies whose contributions made it all possible.

Heinz P. Bloch
Montgomery, Texas

Acknowledgments

We gratefully acknowledge the cooperation and, in many cases, intense special effort of the companies and individuals whose contributions to this text made the entire endeavor possible in the first place. Others have been timely, diligent, and kind to both review this material and to secure permission for its incorporation in the book.

Our special thanks go to A-C Compressor Corporation, Appleton, Wis.; Aerzen USA Corporation and Pierre Noack, Coatesville, Pa.; Anglo Compression, Incorporated, Mount Vernon, Ohio; The American Society of Mechanical Engineers, New York, N.Y.; Bently-Nevada Corporation, Minden, Nev.; BHS-Voith Getriebewerke, Sonthofen, Germany; Cooper Industries, Mount Vernon, Ohio; Compressor Controls Corporation, and Dr. B. W. Batson, Des Moines, Iowa; Coupling Corporation of America, York, Pa.; Demag Delaval Turbomachinery and Ken Reich, Gary Walker, and Roy Salisbury, Trenton, N.J.; Dresser Industries, Inc., Roots Division, Connersville, Indiana; Dresser-Rand Company, Engine Process Compressor Division and Ron Beyer, G. A. Lentek, and Dick Schaad, Painted Post, N.Y.; Dresser-Rand Turbo Products and Harvey Galloway, Russ Svendsen, and Art Wemmell, Olean, N.Y.; Elliott Company and Ross A. Hackel, Don Hallock, and Ken Peters, Jeannette, Pa.; Flexibox, Inc., Houston, Tex.; Indikon/Metravib Instruments, Cambridge, Mass.; KMC, Inc., West Greenwich, R.I.; Lincoln Division of McNeil Corporation, St. Louis, Mo.; Lubrication Systems Company, Houston, Tex.; Lubriquip, Inc., Cleveland, Ohio; Lucas Aerospace Corporation, Bendix Fluid Power Division, Utica, N.Y.; Nash Engineering Company, Norwalk, Conn.; Nuovo Pignone, Florence, Italy; PPI Division, The Duriron Company, Warminster, Pa.; Rotordynamics-Seal Research, North Highlands, Calif.; Sulzer-Burckhardt Engineering Works, Ltd., Winterthur, Switzerland; Sulzer Turbosystems International and Bernhard Haberthuer and Chris Rufer, New York, N.Y.; Torquetronics, Allegany, N.Y.; Revolve Technologies and Paul

Eakins, T. J. Al-Himyary, and Stan Uptigrove, Calgary, Ala., Canada; Zurn Industries, Erie, Pa.

Also, many thanks to the individual contributors, Mike Calistrat on couplings; R. Chow, B. McMordie, and R. Wiegand on coatings; and Arvind Godse for his fine paper on "Predicting Compressor Performance at New Conditions."

A Practical Guide to Compressor Technology

Positive Displacement Compressor Technology

Introduction

Positive displacement compressors comprise the first of the two principal compressor categories, the second being dynamic compressors. In all positive displacement machines, a certain inlet volume of gas is confined in a given space and subsequently compressed by reducing this confined space or volume. At this now elevated pressure, the gas is next expelled into the discharge piping or vessel system.

While positive displacement compressors include a wide spectrum of configurations and geometries, the most important process machines are piston-equipped reciprocating compressors and helical screw rotating machines. Although there are a number of others, including diaphragm and sliding vane compressors, the overwhelming majority of significant process gas-positive displacement machines are clearly reciprocating piston and twin helical screw-rotating, or rotary screw machines. For that reason, this text will focus on their operating characteristics and application ranges. The figure identifies these application ranges and allows us to compare typical flow and pressure fields for other compressor types as well.

A1   Reciprocating compressors with lubricated and nonlubricated cylinders

A2   Reciprocating compressors for high and very high pressures with lubricated cylinders

B   Helical- or spiral-lobe compressors (rotary screw compressors) with dry or oil-flooded rotors

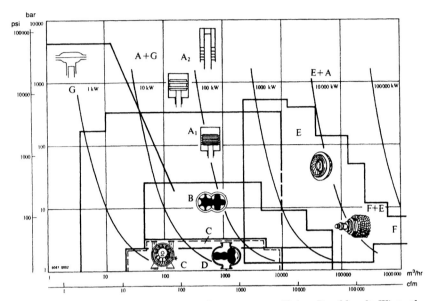

Application ranges for different types of compressors. *(Sulzer-Burckhardt, Winterthur and Basel, Switzerland)*

C   Liquid ring compressors (also used as vacuum pumps)

D   Two-impeller straight-lobe rotary compressors, oil-free (also used as vacuum pumps)

E   Centrifugal turbocompressors

F   Axial turbocompressors

G   Diaphragm compressors

The most frequently used combinations of two different compressor types are identified in three fields:

A + G   Oil-free reciprocating compressor followed by a diaphragm compressor

E + A   Centrifugal turbocompressor followed by an oil-free reciprocating compressor

F + E   Axial turbocompressor followed by a centrifugal turbocompressor

1

Theory*

This discussion of thermodynamics is limited to the processes that are involved in the compression of gases in a positive displacement compressor of the reciprocating type. A positive displacement compressor is a machine that increases the pressure of a definite initial volume of gas, accomplishing the pressure increase by volume reduction.

Only with a knowledge of basic laws and their application can one understand what is happening in a compressor and thus properly solve any given compression problem.

The definitions and units of measurement given at the end of this chapter should either be known to the reader or be thoroughly reviewed as a start.

1.1 Symbols

The following symbols (based on pounds, feet, seconds, and degrees Fahrenheit) will be used in this discussion of positive displacement compressor theory:

c	cylinder clearance, % or decimal
c_p	specific heat, constant pressure, Btu/°F/lb
c_v	specific heat, constant volume, Btu/°F/lb
CE	compression efficiency, %
k	ratio of specific heats, dimensionless
M	molecular weight (MW), dimensionless
ME	mechanical efficiency, %

---

* Developed and contributed by Dresser-Rand Company, Olean, N.Y. Based on Ingersoll-Rand Form 3519-D.

N	number of moles, dimensionless
$N_{a,b,c}$	moles of constituents, dimensionless
p	pressure, psia
$p_{a,b,c}$	partial pressure of constituents, psia
p_a	partial air pressure, psia
p_c	critical pressure (gas property), psia
p_r	reduced pressure, dimensionless
p_s	saturated vapor pressure, psia or in Hg
p_v	partial vapor pressure, psia or in Hg
psia	lb/in² absolute, psi
psig	lb/in² gauge, psi
P_t	theoretical horsepower, (work rate), hp
Q	heat, Btu
r	ratio of compression per stage, dimensionless
r_t	ratio of compression—total, dimensionless
R_0	universal or molar gas constant, ft · lb/mol °R
	(1545 when p is in lb/ft²)
R'	specific gas constant, ft · lb/lb °R
RH	relative humidity, %
s	number of stages of compression, dimensionless
S	entropy, Btu/lb/°F
SH	specific humidity, lb moisture/lb dry gas
SPT	standard pressure and temperature, 14.696 psia and 60°F
T	absolute temperature, °R
T_c	critical temperature, °R
T_r	reduced temperature, dimensionless
v	specific volume, ft³/lb
$v_{a,b,c}$	partial volume of constituents, ft³/lb
v_r	pseudo-specific reduced volume, ft³/lb
V	total volume, ft³
VE	volumetric efficiency, %
W	weight, lb
W_a	weight of dry air in a mixture, lb
W_v	weight of vapor in a mixture, lb
$W_{a,b,c}$	weight of constituents in a mixture, lb
Z	compressibility factor, dimensionless
η_v	volumetric efficiency, %

1.2 How a Compressor Works

Every compressor is made up of one or more *basic elements*. A single element, or a group of elements in parallel, comprises a *single-stage* compressor.

Many compression problems involve conditions beyond the practical capability of a single compression stage. Too great a compression ratio (absolute discharge pressure divided by absolute intake pressure) causes excessive discharge temperature and other design problems. It therefore may become necessary to combine elements or groups of elements in series to form a *multistage* unit, in which there will be two or more steps of compression. The gas is frequently cooled between stages to reduce the temperature and volume entering the following stage. Let's assume the gas is air.

Note that each stage is an individual basic compressor within itself. It is sized to operate in series with one or more additional basic compressors, and even though they may all operate from one power source, each is still a separate compressor.

The basic reciprocating compression element is a single cylinder compressing on only one side of the piston (*single-acting*). A unit compressing on both sides of the piston (*double-acting*) consists of two basic single-acting elements operating in parallel in one casting.

The reciprocating compressor uses automatic spring-loaded valves that open only when the proper differential pressure exists across the valve. Inlet valves open when the pressure in the cylinder is slightly below the intake pressure. Discharge valves open when the pressure in the cylinder is slightly above the discharge pressure.

Figure 1.1 shows the basic element with the cylinder full of a gas, say, atmospheric air. On the theoretical pV diagram (indicator card), Point 1 is the start of compression. Both valves are closed.

Figure 1.2 shows the compression stroke, the piston having moved to the left, reducing the original volume of air with an accompanying rise in pressure. Valves remain closed. The pV diagram shows compression from Point 1 to Point 2, and the pressure inside the cylinder has reached that in the receiver.

Figure 1.3 shows the piston completing the delivery stroke. The discharge valves opened just beyond Point 2. Compressed air is flowing out through the discharge valves to the receiver. After the piston reaches Point 3, the discharge valves will close, leaving the clearance space filled with air at discharge pressure.

During the expansion stroke, Fig. 1.4, both the inlet and discharge valves remain closed, and air trapped in the clearance space increases in volume, causing a reduction in pressure. This continues as the piston moves to the right until the cylinder pressure drops below the inlet pressure at Point 4.

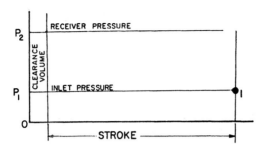

Figure 1.1   Basic compressor element with the cylinder full of gas. On the theoretical pV diagram (indicator card), Point 1 is the start of compression. Both valves are closed. *(Dresser-Rand Company, Painted Post, N.Y.)*

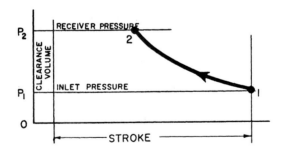

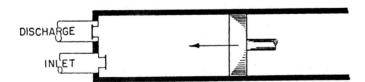

Figure 1.2   Compression stroke. The piston has moved to the left, reducing the original volume of gas with an accompanying rise in pressure. Valves remain closed. The pV diagram shows compression from Point 1 to Point 2 and the pressure inside the cylinder has reached that in the receiver. *(Dresser-Rand Company, Painted Post, N.Y.)*

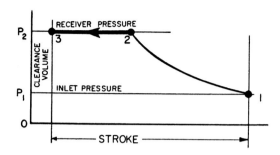

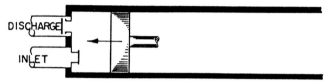

Figure 1.3   The piston is shown completing the delivery stroke. The discharge valves opened just beyond Point 2. Compressed air is flowing out through the discharge valves to the receiver. *(Dresser-Rand Company, Painted Post, N.Y.)*

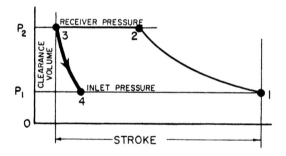

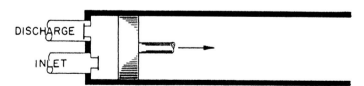

Figure 1.4   During the expansion stroke shown, both the inlet and discharge valves remain closed and gas trapped in the clearance space increases in volume, causing a reduction in pressure. *(Dresser-Rand Company, Painted Post, N.Y.)*

The inlet valves now will open, and air will flow into the cylinder until the end of the reverse stroke at Point 1. This is the intake or suction stroke, illustrated by Fig. 1.5. At Point 1 on the pV diagram the inlet valves will close, and the cycle will repeat on the next revolution of the crank.

In an elemental two-stage reciprocating compressor the cylinders are proportioned according to the total compression ratio, the second stage being smaller because the gas, having already been partially compressed and cooled, occupies less volume than at the first-stage inlet. Looking at the pV diagram (Fig. 1.6), the conditions before starting compression are Points 1 and 5 for the first and second stages, respectively; after compression, conditions are Points 2 and 6, and after delivery, Points 3 and 7. Expansion of gas trapped in the clearance space as the piston reverses brings Points 4 and 8, and on the intake stroke the cylinders are again filled at Points 1 and 5 and the cycle is set for repetition. Multiple staging of any positive displacement compressor follows this pattern.

Certain laws that govern the changes of state of gases must be thoroughly understood. Symbols were listed previously in this chapter.

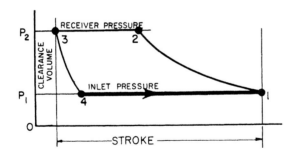

Figure 1.5   At Point 4, the inlet valves will open and gas will flow into the cylinder until the end of the reverse stroke at Point 1. *(Dresser-Rand Company, Painted Post, N.Y.)*

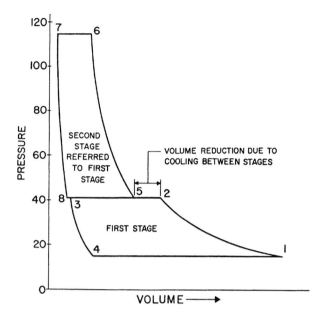

Figure 1.6 The *pV* diagram for a two-stage compressor. *(Dresser-Rand Company, Painted Post, N.Y.)*

1.3   The First Law of Thermodynamics

This states that energy cannot be created or destroyed during a process (such as compression and delivery of a gas), although it may change from one form of energy to another. In other words, whenever a quantity of one kind of energy disappears, an exactly equivalent total of other kinds of energy must be produced.

1.4   The Second Law of Thermodynamics

This is more abstract and can be stated several ways.

1. Heat cannot, of itself, pass from a colder to a hotter body.
2. Heat can be made to go from a body at lower temperature to one at higher temperature *only* if external work is done.
3. The available energy of the isolated system decreases in all real processes.
4. Heat or energy (or water), of itself, will flow only downhill.

Basically, these statements say that energy exists at various levels and is *available for use* only if it can move from a higher to a lower level.

In thermodynamics a measure of the *unavailability* of energy has been devised and is known as *entropy*. It is defined by the differential equation:

$$dS = d\frac{Q}{T} \tag{1.1}$$

Note that entropy (as a measure of unavailability) increases as a system loses heat but remains constant when there is no gain or loss of heat (as in an adiabatic process).

1.5 Ideal or Perfect Gas Laws

An ideal or perfect gas is one to which the laws of Boyle, Charles, and Amonton apply. There are no truly perfect gases, but these laws are used and corrected by compressibility factors based on experimental data.

1.5.1 Boyle's law

At constant temperature, the volume of an ideal gas varies inversely with the pressure. In symbols:

$$\frac{V_2}{V_1} = \frac{p_1}{p_2} \tag{1.2}$$

$$p_2 V_2 = p_1 V_1 = \text{constant} \tag{1.3}$$

This is the *isothermal law.*

1.5.2 Charles' law

The volume of an ideal gas at constant pressure varies directly as the absolute temperature.

$$\frac{V_2}{V_1} = \frac{T_2}{T_1} \tag{1.4}$$

$$\frac{V_2}{T_2} = \frac{V_1}{T_1} = \text{constant} \tag{1.5}$$

1.5.3 Amonton's law

At constant volume, the pressure of an ideal gas will vary directly with the absolute temperature.

$$\frac{p_2}{p_1} = \frac{T_2}{T_1} \tag{1.6}$$

$$\frac{p_2}{T_2} = \frac{p_1}{T_1} = \text{constant} \tag{1.7}$$

1.5.4 Dalton's law

This states that the total pressure of a mixture of ideal gases is equal to the sum of the partial pressures of the constituent gases.

The *partial pressure* is defined as the pressure each gas would exert if it alone occupied the volume of the mixture at the mixture temperature.

Dalton's law has been proven experimentally to be somewhat inaccurate, the total pressure often being higher than the sum of the partial pressures, particularly as pressures increase. However, for engineering purposes it is the best rule available and the error is minor.

This can be expressed as follows, all being at the same temperature and volume.

$$p = p_a + p_b + p_c + \ldots \tag{1.8}$$

1.5.5 Amagat's law

This is similar to Dalton's law but states that the volume of a mixture of ideal gases is equal to the sum of the partial volumes that the constituent gases would occupy if each existed alone at the *total* pressure and temperature of the mixture. As a formula this becomes:

$$V = V_a + V_b + V_c + \ldots \tag{1.9}$$

Note: Dalton's and Amagat's laws are discussed further under Partial Pressures, Sec. 1.8.

1.5.6 Avogadro's law

Avogadro states that equal volumes of all gases, under the same conditions of pressure and temperature, contain the same number of molecules. This law is very important and is applied in many compressor calculations. Further discussion is given under The Mole, Sec. 1.13.

1.5.7 The perfect gas formula

Starting with Charles' and Boyle's laws, it is possible to develop the formula for a given weight of gas.

$$pV = WR'T \qquad (1.10)$$

where W is weight and R' is a specific constant for the gas involved. This is the perfect gas equation.

Going one step further, by making W in pounds equal to the molecular weight of the gas (1 mol), the formula becomes

$$PV = R_0 T \qquad (1.11)$$

This is very useful. R_0 is known as the *universal gas constant,* has a value of 1545, and is the same for all gases. Note, however, that R_0 is 1545 only when p is lb/ft^2; V is ft^3/lb mol; and T is °R (°F + 460). When p is lb/in^2, R_0 becomes 10.729. The *specific* gas constant (R') for any gas can be obtained by dividing 1545 by the molecular weight.

1.6   Vapor Pressure

As liquids physically change into a gas (as during a temperature rise), their molecules travel with greater velocity, and some break out of the liquid to form a vapor above the liquid. These molecules create a vapor pressure, which (at a specified temperature) is the only pressure at which a pure liquid and its vapor can exist in equilibrium.

If, in a closed liquid-vapor system, the volume is reduced at constant temperature, the pressure will increase imperceptibly until condensation of part of the vapor into liquid has lowered the pressure to the original vapor pressure corresponding to the temperature. Conversely, increasing the volume at constant temperature will reduce the pressure imperceptibly, and molecules will move from the liquid phase to the vapor phase until the original vapor pressure has been restored. Temperatures and vapor pressures for a given gas always move together.

The temperature corresponding to any given vapor pressure is obviously the *boiling point* of the liquid and also the *dew point* of the vapor. Addition of heat will cause the liquid to boil, and removal of heat will start condensation. The three terms, saturation temperature, boiling point, and dew point, all mean the same physical temperature at a given vapor pressure. Their use depends on the context in which they appear.

Typical vapor pressure curves for common pure gases are shown in App. A. Tables of the properties of *saturated* steam show its temperature-vapor pressure relationship.

1.7   Gas and Vapor

By definition, a *gas* is a fluid having neither independent shape nor form, tending to expand indefinitely. A *vapor* is a gasified liquid or solid; a substance in gaseous form. These definitions are in general use today.

All *gases* can be liquified under suitable pressure and temperature conditions and therefore could also be called *vapors*. The term *gas* is more generally used when conditions are such that a return to the liquid state (condensation) would be difficult within the scope of the operations being considered. However, a gas under such conditions is actually a superheated vapor.

The terms *gas* and *vapor* will be used rather interchangeably, with emphasis on closer approach to the liquid phase when using the word vapor.

1.8   Partial Pressures

Vapor pressure created by one pure liquid will not affect the vapor pressure of a second pure liquid, when the liquids are insoluble and nonreacting, and the liquids and/or vapors are mixed within the same system. There is complete indifference on the part of each component to the existence of all others. The total vapor pressure for mixtures is the sum of the vapor pressures of the individual components. This is Dalton's law, and each individual vapor has what is called a *partial pressure* as differentiated from the total pressure of the mixture.

During compression of any gas other than a pure and dry gas, the principles of partial pressure are at work. This is true even in normal 100 psig air compression for power purposes, because there is always some water vapor mixed with the intake air and the compressor must handle both components. Actually, air is itself a mixture of a number of components, including oxygen, nitrogen, argon, etc., and its total pressure is the sum of the partial pressures of each component. However, because of the negligible variation in the composition of *dry* air throughout the world, it is considered and will hereafter be treated as a single gas with specific properties of its own.

After compression, partial pressures are used to determine moisture condensation and removal in intercoolers and aftercoolers. Partial pressures are also involved in many vacuum pump applications and are encountered widely in the compression of many mixtures.

Dalton's and Amagat's laws have been referred to in Secs. 1.5.4 and 1.5.5. See Eqs. (1.8) and (1.9), which apply here.

Since water vapor is by far the most prevalent constituent involved in partial pressure problems in compressing gases, it is usually the only one considered in subsequent discussions.

In a mixture, when the dew-point temperature of any component is reached, the space occupied is said to be saturated by that component. A volume is sometimes specified as being partially saturated with water vapor at a certain temperature. This means that the vapor is actually superheated and the dew point is lower than the actual temperature. If the moles (see definition, Sec. 1.13) of each component are known, the partial pressure of the component in question can be determined. Otherwise, it is customary to multiply the vapor pressure of the component at the existing mixture temperature by the relative humidity to obtain the partial pressure.

The terms *saturated gas* or *partially saturated gas* are incorrect and give the wrong impression. It is *not* the gas that is saturated with vapor; it is the volume or space occupied. The vapor and gas exist independently throughout the volume or space. Understanding of this true concept is helpful when working with partial pressures and gas mixtures.

Relative humidity is a term frequently used to represent the quantity of moisture present in a mixture, although it uses partial pressures in so doing. It is expressed as follows:

$$\text{RH } (\%) = \frac{\text{actual partial vapor pressure} \times 100}{\text{saturated vapor pressure at existing mixture temperature}}$$

$$= \frac{p_v \times 100}{p_s} \tag{1.12}$$

Relative humidity is usually considered only in connection with atmospheric air, but since it is unconcerned with the nature of any other components or the total mixture pressure, the term is applicable to vapor content in any problem, no matter what the conditions.

The saturated water vapor pressure at a given temperature is always known from steam tables or charts. It is the existing partial vapor pressure that is desired and is therefore calculable when the relative humidity is stated.

Specific humidity, used in calculations on certain types of compressors, is a totally different term. It is the ratio of the weight of water vapor to the weight of *dry* air and is usually expressed as pounds (or grains) of moisture per pound of dry air

$$\text{SH} = \frac{W_v}{W_a} \tag{1.13}$$

also

$$\text{SH} = \frac{0.622 p_v}{p - p_v} = \frac{0.622 p_v}{p_a} \tag{1.14}$$

where p_a is partial air pressure.

The *degree of saturation* denotes the actual relation between the weight of moisture existing in a space and the weight that would exist if the space were saturated.

$$\text{Degree of saturation \%} = \frac{\text{SH actual} \times 100}{\text{SH saturated}} \qquad (1.15)$$

$$= \text{RH} \times \frac{p - p_s}{p - p_v} \qquad (1.16)$$

Usually p_s and p_v are quite small as compared to p; therefore, the degree of saturation closely approximates the relative humidity. The latter term is commonly used in psychrometric work involving air-water vapor mixtures while degree of saturation is applied mainly to gas-vapor mixtures having components other than air and water vapor.

The practical application of partial pressures in compression problems centers to a large degree around the determination of mixture volumes or weights to be handled at the intake of each stage of compression, the determination of mixture molecular weight, specific gravity, and the proportional or actual weight of components.

1.9   Critical Conditions

There is one temperature above which a gas will not liquefy with pressure increases no matter how great. This point is called the *critical temperature*. It is experimentally determined.

The pressure required to compress and condense a gas at this critical temperature is called the *critical pressure*.

The critical constants of many gases are given in App. A.

1.10   Compressibility

All gases deviate from the perfect or ideal gas laws to some degree, and in some cases the deviation is rather extreme. It is necessary that these deviations be taken into account in many compressor calculations to prevent cylinder volumes and driver sizes being sadly in error.

Compressibility is experimentally derived from data on the actual behavior of a particular gas under p-V-T changes. The compressibility factor Z is a multiplier in the basic formula. It becomes the ratio of the actual volume at a given p-T condition to the ideal volume at the same p-T condition.

The ideal gas Eq. (1.11) is therefore modified to:

$$pV = ZR_0T \qquad (1.17)$$

or

$$Z = \frac{pV}{R_0 T} \tag{1.18}$$

In these equations, R_0 is 1545 and p is lb/ft^2.

A series of compressibility and temperature-entropy charts has been drafted to cover all gases on which reliable information could be found. These will be found in specialized texts or handbooks. In some cases, they represent consolidation and correlation of data from several sources, usually with a variance of less than 1 percent from the basic data. These charts may be considered authoritative.

The temperature-entropy charts are useful in the determination of theoretical discharge temperatures that are not always consistent with ideal gas laws. Discharge temperatures are required to obtain the compressibility factor at discharge conditions as involved in some calculations.

These specific Z and T-S charts will provide the necessary correction factors for most compression problems involving the gases covered.

1.11   Generalized Compressibility Charts

Because experimental data over complete ranges of temperature and pressure are not available for all gases, scientists have developed what are known as *generalized compressibility charts*. There are a number of these. One set has been selected as being suitable for screening calculations and is included in App. A.

These charts are based on what are called *reduced* conditions. Reduced pressure p_r is the ratio of the absolute pressure in lb/in^2 at a particular condition to the absolute critical pressure. Similarly, reduced temperature T_r is the ratio of the absolute temperature at the particular condition to the absolute critical temperature. The formulas are:

$$p_r = \frac{p}{p_c} \tag{1.19}$$

$$T_r = \frac{T}{T_c} \tag{1.20}$$

It has been found that compressibility curves on the reduced basis for a large number of gases fall together with but small divergence. There are only a few gases that are too individualistic to be included.

Some charts show a reduced volume $v_r{}'$ also, but this is really a pseudo (pretended) reduced condition obtained by use of the following formula. Reduced volumes are not shown on the charts included here.

$$v_r' = \frac{vp_c}{R_0 T_c} \qquad (1.21)$$

From this we can also write:

$$v = \frac{v_r' R_0 T_c}{P_c} \qquad (1.22)$$

In these formulas, v and v_r' are the specific volumes of 1 mol of gas.

Critical pressures and temperatures for many gases are given in App. A.

1.12 Gas Mixtures

Mixtures can be considered as equivalent ideal gases. Although this is not strictly true, it is satisfactory for present purposes.

Many mixtures handled by compressors contain from 2 to 10 separate components. It is necessary to determine as closely as possible many properties of these as *equivalent gases*. Chief among these properties are:

Specific volume

Density

Volume and mole percent

Molecular weight

Specific gravity

Partial pressure

Ratio of specific heats (k)

Pseudo reduced pressure

Pseudo reduced temperature

Compressibility

Gas constant

Specific heats

1.13 The Mole

The *mole* is particularly useful when working with gas mixtures. It is based on Avogadro's law that equal volumes of gases at given pT conditions contain equal numbers of molecules. Since this is so, then the *weight* of these equal volumes will be proportional to their molecular weights.

The volume of 1 mol at any desired condition can be found by the use of the perfect gas law:

$$pV = R_0T \quad \text{or} \quad pV = 1545T \tag{1.23}$$

Choosing standard pressure and temperature (SPT) conditions we solve for V in the previous formula (p is lb/ft^2 and T is °R). This turns out to be 379.4 ft^3. For simplicity, use 379 ft^3/mol.

To repeat, this is the volume of a weight (expressed in pounds) of any gas at 14.696 psia and 60°F—the weight being the same number as the molecular weight.

Thus, a mole of hydrogen occupies a volume of 379 ft^3 at standard conditions and weighs 2.016 lb. A mole of air occupies 379 ft^3 at the same conditions but weighs 28.97 lb. A mole of isobutane, still 379 ft^3, weighs 58.12 lb. This, of course, assumes that they act as perfect or ideal gases, which most of them do at or near standard conditions (SPT) or 14.696 psia and 60°F. Most mole calculations involve these or similar conditions.

Note, however, that a mole is a *weight* of gas. It is *not* a volume.

In spite of the deviation from a perfect gas being sometimes in question, the following methods of obtaining mixture pseudo properties are of great value, and in some cases are the only approach.

1.14   Specific Volume and Density

Since the volume and the weight of a mole of any gas is known from the defined relations, it follows that the specific volume in ft^3/lb or density in lb/ft^3 is obtained by simple division.

Gas	Specific vol., ft^3/mol	lb/mol and molecular weight	Specific vol., ft^3/lb	Density, lb/ft^3
Hydrogen	379	2.016	188.3	.00531
Air	379	28.97	13.1	.0763
Isobutane	379	58.12	6.51	.153

Note that these data are on the basis of perfect gas laws. Some gases—isobutane is one—deviate even at SPT conditions. The actual figures, for example, on isobutane are 6.339 ft^3/lb and 0.1578 lb/ft^3.

1.15   Volume Percent of Constituents

Mole percent is the ratio of the number of moles of one constituent to the total number of moles of mixture. Mole percent also happens to be percent by volume. This statement should be questioned since a mole is defined as a weight. Look at the following table for proof. The gas analysis in these and following tables is that of a typical raw ammonia synthesis gas.

Gas	Mol %	Mol/mol of mixture	Vol (SPT) of 1 mol	Vol/mol of mixture	Vol %
H_2	61.4	0.614	379	232.7	61.4
N_2	19.7	0.197	379	74.7	19.7
CO_2	17.5	0.175	379	66.3	17.5
CO	1.4	0.014	379	5.3	1.4
	100.0	1.000		379.0	

1.16   Molecular Weight of Mixture

The average molecular weight of the mixture is often needed. It is obtained by multiplying the molecular weight of each component by its mole fraction (mol %/100) and adding these values as shown in the following.

Gas	Mol % or vol. %	Mol. wt.	Proportional mol. wt.
H_2	61.4	2	1.228
N_2	19.7	28	5.516
CO_2	17.5	44	7.700
CO	1.4	28	.392
	100.0		14.84

Therefore, the average (or pseudo) molecular weight of the mixture is 14.84.

1.17   Specific Gravity and Partial Pressure

Normally, specific gravity for gases is a ratio of the lb/ft³ of the gas involved to the lb/ft³ of air, both at SPT conditions. Considering a mole of each gas, the volumes are the same and the weight of each volume is the same as the molecular weight. Therefore specific gravity is figured as the ratio of these molecular weights and becomes (for the previous example) 14.84 divided by 28.97 or 0.512.

It can be stated that the fraction of the total pressure in a gas mixture due to a given component is equal to the fraction which that component represents of the total moles of gas present.

$$p_a = \frac{pN_a}{N} \quad \text{and} \quad p_b = \frac{pN_b}{N} \quad \text{and} \quad p_c = \frac{pN_c}{N} \quad (1.24)$$

Thus, in a mixture of 15 mol at 15 psia total pressure containing 2 mol of hydrogen, the partial pressure of the hydrogen would be 2/15 of 15 psia, or 2 psia. Volume fractions, if available, may be used in place of mole fractions here.

1.18   Ratio of Specific Heats

The value of k enters into many calculations. A definite relationship exists between the specific heat at constant volume and the specific heat at constant pressure. If we take a mole of gas and determine its heat capacity:

$$Mc_p = Mc_v + 1.99 \tag{1.25}$$

$$Mc_v = Mc_p - 1.99 \tag{1.26}$$

In these formulas, M is the weight of a mole of gas (the molecular weight). These are easily resolved into:

$$k = \frac{Mc_p}{Mc_v} = \frac{Mc_p}{Mc_p - 1.99} \tag{1.27}$$

Remembering the unit of specific heat as Btu/lb/°F temperature rise, we can calculate the heat required to increase the temperature of each component gas 1°F and add them to get the total for the mixture. Mc_p is the heat requirement for 1 mol. For compressor work, it is usual to use this molar heat capacity at 150°F, which is considered an average temperature. A calculation table follows:

Gas	Mol %	Mol gas/ mol mixture	Mc_p at 150°F of component	Product
H_2	61.4	0.614	6.94	4.26
N_2	19.7	0.197	6.98	1.38
CO_2	17.5	0.175	9.37	1.64
CO	1.4	0.014	6.97	0.10
	100.0	1.000		7.38

Note: For convenience, the molar heat capacity at 150°F (Mc_p) is given in App. A for most gases.

The molar specific heat (Mc_p) of the mixture is therefore 7.38. Entering this in the formula:

$$k = \frac{7.38}{7.38 - 1.99} = 1.369 \text{ (say 1.37)} \tag{1.28}$$

1.19   Pseudo-Critical Conditions and Compressibility

Mention has been made of reduced pressure and reduced temperature under the discussion of compressibility. Generalized compressibility curves on this basis are in App. A. They are also applicable to mixtures—for approximations, at least.

It is necessary to figure mixture pseudo-critical pressure and temperature conditions to be used in calculating the pseudo-reduced conditions to be used in entering the charts. Pressures and temperatures must be in absolute values.

Gas	Mol %	Individual critical temp., °R	Pseudo T_c, °R	Individual critical press., psia	Pseudo p_c, psia
H_2*	61.4	83	51.0	327	201.0
N_2	19.7	227	44.7	492	96.9
CO_2	17.5	548	95.9	1073	187.8
CO	1.4	242	3.4	507	7.1
Mixture pseudo-criticals			195		493

* Must use effective critical conditions (see App. A.)

Using these values the pseudo-reduced conditions can be calculated, and probable Z factors obtained from generalized charts.

1.20   Weight-Basis Items

To certain gas properties of a mixture, each component contributes a share of its own property in proportion to its fraction of the total *weight*. Thus, the following are obtained, the weight factors being fractions of the whole.

$$R' = \frac{W_a R'_a + W_b R'_b + W_c R'_c + \ldots}{W} \tag{1.29}$$

$$c_p = \frac{W_a c_{pa} + W_b c_{pb} + W_c c_{pc} + \ldots}{W} \tag{1.30}$$

$$c_v = \frac{W_a c_{va} + W_b c_{cb} + W_c c_{rc} + \ldots}{W} \tag{1.31}$$

1.21   Compression Cycles

Two theoretical compression cycles are applicable to positive displacement compressors. Although neither cycle is commercially attainable, they are used as a basis for calculations and comparisons.

Isothermal compression occurs when the temperature is kept constant as the pressure increases. This requires continuous removal of the heat of compression. Compression follows the formula:

$$p_1V_1 = p_2V_2 = \text{constant} \qquad (1.32)$$

Adiabatic (isentropic) compression is obtained when there is no heat added to or removed from the gas during compression. Compression follows the formula:

$$p_1V_1^{k} = p_2V_2^{k} \qquad (1.33)$$

where k is the ratio of the specific heats.

Figure 1.7 shows the theoretical no clearance isothermal and adiabatic cycles on a pV basis for a compression ratio of 4. Area ADEF represents the work required when operating on the isothermal basis, and ABEF the work required on the adiabatic basis. Obviously, the isothermal area is considerably less than the adiabatic and would be the cycle for greatest compression economy. However, the isothermal cycle is not commercially approachable, although compressors are usually designed for as much heat removal as possible.

Adiabatic compression is likewise never exactly obtained since there is always some heat rejected or added. Actual compression therefore takes place along a *polytropic cycle* where the relationship is:

$$p_1V_1^{n} = p_2V_2^{n} \qquad (1.34)$$

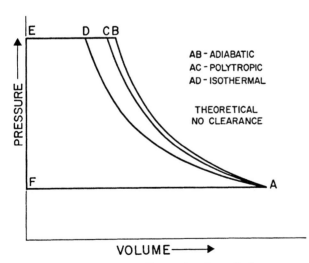

Figure 1.7   The pV diagram illustrating theoretical compression cycles. *(Dresser-Rand Company, Painted Post, N.Y.)*

The exponent n is experimentally determined for a given type of machine and may be lower or higher than the adiabatic exponent k. In positive displacement compressors n is usually less than k. Figure 1.7 shows a typical polytropic compression curve for a reciprocating water-jacketed compressor cylinder.

Thermodynamically, it should be noted that the isentropic or adiabatic process is reversible, while the polytropic process is irreversible. Also, all compressors operate on a *steady-flow* process.

Either n or $(n-1)/n$ can also be experimentally calculated from test data if inlet and discharge pressures and temperatures are known. The following formula may be used:

$$\frac{T_2}{T_1} = \left(\frac{p_2}{p_1}\right)^{(n-1)/n} = r^{(n-1)/n} \tag{1.35}$$

This formula can also be used to estimate discharge temperatures when n or $(n-1)/n$ is known.

It is obvious that k and n can have quite different values. In certain engineering circles, there has been a past tendency to use these symbols interchangeably to represent the ratio of specific heats. This is incorrect, and the differentiation between them should be carefully observed.

1.22   Power Requirement

The power requirement of any compressor is the prime basis for sizing the driver and for selection and design of compressor components. The actual power requirement is related to a theoretical cycle through a *compression efficiency,* which has been determined by test on prior machines. Compression efficiency is the ratio of the theoretical to the actual gas horsepower and, as used by the industry, does not include mechanical friction losses. These are added later either through the use of a mechanical efficiency or by adding actual mechanical losses previously determined. Positive displacement compressors commonly use mechanical efficiencies ranging from 88 to 95 percent, depending on the size and type of unit.

Historically, the isothermal cycle was the basis used for many years. It is used today in only a few cases. Positive displacement machines are now compared to the isentropic or adiabatic cycle, which more nearly represents what actually occurs in the compressor.

In calculating horsepower, the compressibility factor Z must be considered since its influence is considerable with many gases, particularly at high pressure.

An inlet volume basis is universal with positive displacement compressors. It is important to differentiate between an inlet volume on a perfect gas basis (V_{p1}) and one on a real gas basis (V_{r1}). Volumes are at inlet pressure and temperature (p_1 and T_1).

$$V_{r1} = V_{p1}Z_1$$

The basic theoretical adiabatic single stage horsepower formula follows:

$$P_T(\text{ad}) = \frac{p_1 V_{r1}}{229} \times \frac{k}{k-1} \times (r^{(k-1)/k} - 1) \times \frac{Z_1 + Z_2}{2Z_1} \qquad (1.36)$$

This represents the area of a theoretical adiabatic pV diagram for the volume per minute (V_1) being handled.

A frequently used basis for V_1 is 100 cfm (real) at inlet conditions, in which case the formula becomes:

$$P_T(\text{ad})/100 = \frac{P_1}{2.29} \times \frac{k}{k-1} \times (r^{(k-1)/k} - 1) \times \frac{Z_1 + Z_2}{2Z_1} \qquad (1.37)$$

Another form current in the industry is the basis for frequently used charts. In this, a volume of 1 million ft³/d or MMCFD is used. *In this case only,* V_1 is measured as *perfect gas* at *14.4 psia* and intake temperature, and the actual compressor capacity must be referred to these conditions before computing the final horsepower.

$$P_T(\text{ad})/MMCFD = 43.67 \times \frac{k}{k-1} \times (r^{(k-1)/k} - 1) \times \frac{Z_1 + Z_2}{2} \qquad (1.38)$$

Since the isothermal cycle is based on no temperature change during compression, heat is removed continuously as generated, and there is theoretically no gain by multiple staging. Therefore, Eq. (1.39) holds for any number of stages as long as r is the overall or total compression ratio.

$$P_T(\text{iso}) = \frac{p_1 V_{r1} \ln r}{229} \times \frac{Z_1 + Z_2}{2Z_1} \qquad (1.39)$$

1.23   Compressibility Correction

In the preceding equations a correction is indicated for deviation from the perfect gas laws—the compressibility. This involves the determination of compressibility at both intake and discharge conditions. Intake pressure and temperature are known, and compressibility at these conditions can be obtained directly from specific gas charts or by the reduced condition method using generalized charts. To obtain Z at discharge conditions it is necessary to determine the discharge temperature. Discharge pressure is known.

On the adiabatic cycle as applied to positive displacement units, it is customary to use the *theoretical* discharge temperature in calculations. In an actual compressor there are many factors acting to cause variation from the theoretical but, *on an average,* the theoretical temperature is closely approached, and any error introduced is slight.

Adiabatic compression is isentropic, i.e., the entropy remains constant. If temperature-entropy diagrams are available for the gas involved, the theoretical discharge temperature can be read directly. Otherwise, it is necessary to calculate it by the following relationships:

$$\frac{T_2}{T_1} = \left(\frac{p_2}{p_1}\right)^{(k-1)/k} = r^{(k-1)/k} \tag{1.40}$$

Note that all pressures and temperatures are absolute.

Equations (1.36), (1.37), (1.38), and (1.39) are theoretical and are not affected by gas characteristics such as molecular weight, specific gravity, and actual density at operating conditions. These all have an effect on actual power requirements, however, and proper allowances are made by designers.

1.24   Multiple Staging

All *basic* compressor elements, regardless of type have certain limiting operating conditions. Basic elements are single stage, i.e., the compression and delivery of gas is accomplished in a single element, or group of elements arranged in parallel. The most important limitations include the following:

1. Discharge temperature

2. Pressure differential

3. Effect of clearance (ties in also with compression ratio)

4. Desirability of saving power

There are other reasons for multiple staging than these, but they are largely for the designer of the specific unit to keep in mind. No ready reference rules can be given.

When any limitation is involved, it becomes necessary to multiple stage the compression process, i.e., do it in two or more steps. Each step will use at least one *basic element* designed to operate in series with the other elements of the machine.

A reciprocating compressor usually requires a separate cylinder for each stage with intercooling of the gas between stages. Figure 1.8 shows the *pV* combined diagram of a two-stage 100 psig air compressor. Further stages are added in the same manner. In a reciprocating unit, all stages are commonly combined into one unit assembly.

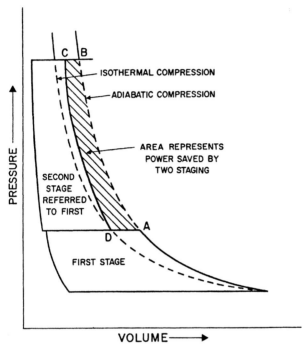

Figure 1.8   Combined pV diagram for a two-stage air compressor. *(Dresser-Rand Company, Painted Post, N.Y.)*

It was previously noted that the isothermal cycle (constant temperature) is the more economical of power. Cooling the gas after partial compression to a temperature equal to original intake temperature (back to the isothermal) obviously should reduce the power required in the second stage. Area ABCD represents the work saved over single-stage adiabatic compression in this particular case.

For minimum power *with perfect intercooling* between stages, there is a theoretically best relation between the intake pressures of succeeding stages. This is obtained by making the ratio of compression the same in each stage and assumes intake temperature to be the same in all stages. The formula used is based on the *overall* compression ratio

where r_s = compression ratio per stage
r_t = overall compression ratio ($p_{final}/p_{initial}$)
s = number of stages

$$r_s = \sqrt[s]{r_t} \tag{1.41}$$

For example

$$\text{Two-stage:}\quad r_s = \sqrt[2]{r_t}$$

$$\text{Three-stage:} \qquad r_s = \sqrt[3]{r_t}$$

$$\text{Four-stage:} \qquad r_s = \sqrt[4]{r_t}$$

Commercial displacement compressors are, as a rule, initially sized on the preceding basis. The final machine, however, usually operates at compression ratios varying slightly from these to allow for other factors the designer must consider. Each stage is figured as a separate compressor, the capacity (V_1) of each stage being separately calculated from the first-stage real intake volume, and corrected to the actual pressure and temperature conditions existing at the higher-stage cylinder inlet and also for any change in the moisture content if there is condensation between stages in an intercooler. The theoretical power per stage can then be calculated and the total horsepower obtained.

On the basis of perfect intercooling and equal compression ratios per stage, Eqs. (1.36), (1.37), and (1.38) can be altered to obtain *total* theoretical power by multiplying the first term by the number of stages s and dividing the exponent of r by s. The compression ratio r must be the total ratio. However, since compression ratios seldom are equal and perfect intercooling is seldom attained, it is believed that the best general method of figuring is to use one stage at a time.

1.25 Volume References

Since the most generally required quantities are original inlet volume and inlet volume to subsequent stages (both on a per-minute basis), a summary of equations follows in which the word *dry* means there is no water vapor in the quantity of gas or gas mixture involved:
From SCFM (cfm measured at 14.7 psia, 60°F, dry)

$$V_1 = \text{SCFM} \times \frac{14.7}{p_1} \times \frac{T_1}{520} \times Z_1 \qquad (1.42)$$

From weight flow (W lb/min, dry)

$$V_1 = \frac{W \times 1545 \times T_1}{144 \times p_1 \times M} \times Z_1 \qquad (1.43)$$

From mole flow N mol/min, dry)

$$V_1 = \frac{N \times 379 \times 14.7 \times T_1}{p_1 \times 520} \times Z_1 \qquad (1.44)$$

From cfm measured at conditions other than those at intake CFM_g at p_g, T_g, Z_g, dry

$$V_1 = \text{CFM}_g \times \frac{p_g}{p_1} \times \frac{T_1}{T_g} \times \frac{Z_1}{Z_g} \qquad (1.45)$$

In all the preceding, pressure is lb/in^2 absolute.

If water vapor is a component in the gas analysis and the total analysis percentage amounts to 100, the previous equations may be applied to the wet gas. Use the proper value for M in Eq. (1.43), however. Often water vapor is segregated, and the space it occupies must be included separately. This is a partial pressure problem (see Partial Pressures, Sec. 1.8). Multiplying any of the preceding volume equations by the following will apply the correction.

$$\frac{p_1}{p_1 - p_v} \qquad (1.46)$$

where p_v is the actual vapor pressure of the contained moisture.

1.26   Cylinder Clearance and Volumetric Efficiency

Cylinder clearance cannot be completely eliminated. *Normal* clearance will vary approximately between 4 and 16 percent for most standard cylinders. There are special low-compression-ratio cylinders where normal clearance will be much greater. Normal clearance does not include clearance volume that may have been added for other purposes, such as capacity control.

Although the amount of clearance in a given cylinder is of little importance to the average user (guarantees being made on actually delivered capacity), its effect on capacity should be understood because of the wide application of a variation in clearance volume for capacity control and other purposes. Normal clearance variations have no effect on power requirements.

When a piston has completed the compression and delivery stroke and is ready to reverse its movement, gas at discharge pressure is trapped in the clearance space. This gas expands on the return stroke until its pressure is sufficiently below intake pressure to open the suction valves. On a pV diagram (Fig. 1.9) is shown the effect of this reexpansion on the quantity of fresh gas drawn in. The actual capacity is materially affected.

The theoretical formula for volumetric efficiency as a percentage is:

$$\eta_v = 100 - C(r^{1/k} - 1) \qquad (1.47)$$

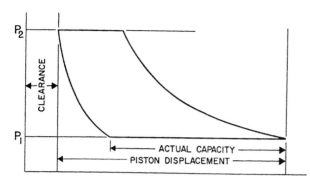

Figure 1.9 Work done on a volume of gas trapped in cylinder clearances (clearance volume) represents an inefficiency. *(Dresser-Rand Company, Painted Post, N.Y.)*

As a practical matter, there are factors that modify this, and an accepted formula for rough estimates follows:

$$\eta_v = 100 - C(r^{1/k} - 1) - L \tag{1.48}$$

Here, term L is introduced to allow for the effect of variables such as internal leakage, gas friction, pressure drop through valves, and inlet gas preheating. The term L is difficult to generalize, but it might be 5 percent for a moderate-pressure oil-lubricated air compressor. A higher value of L will be necessary with a light gas than with a heavy gas because of increased leakage.

Inspection of the equations shows that the *VE decreases:* (1) as the clearance *increases,* (2) as the compression ratio *increases,* and (3) as k *decreases.*

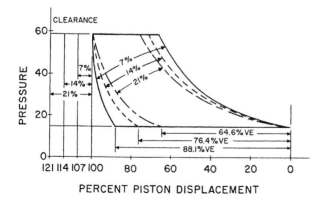

Figure 1.10 Theoretical pV diagrams based on a compression ratio of 4.0, k of 1.40, and clearances of 7, 14 and 21 percent. *(Dresser-Rand Company, Painted Post, N.Y.)*

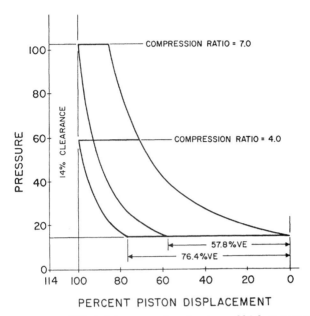

Figure 1.11   Effect of clearance at moderate- and high-compression ratio conditions. A pV diagram for a ratio of 7 is superimposed on a diagram for a ratio of 4, all else being the same. *(Dresser-Rand Company, Painted Post, N.Y.)*

Figure 1.10 shows a series of theoretical pV diagrams based on an r of 4.0, k of 1.40, and clearances of 7, 14, and 21 percent. The effect of clearance is clearly indicated. Rather wide clearance ranges have been used for illustrative purposes.

Figure 1.11 illustrates the effect of clearance at moderate- and high-compression ratio conditions. A pV diagram for a ratio of 7 is superim-

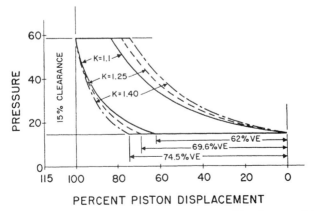

Figure 1.12   Effect of k on volumetric efficiency. The clearance is high for illustrative purposes. *(Dresser-Rand Company, Painted Post, N.Y.)*

posed on a diagram for a ratio of 4, all else being the same. A relatively high clearance (14 percent) is used for illustrative purposes. The clearance for any commercial compressor designed for a ratio of 7 would be much less than 14 percent.

Figure 1.12 illustrates the effect of k on volumetric efficiency. The clearance is high for illustrative purposes.

Clearance obviously concerns the designer more at the higher compression ratios and when handling gases with low specific heat ratios, although he will always endeavor to maintain clearance at the lowest value consistent with adequate valving and running clearances.

1.27 Cylinder Clearance and Compression Efficiency

Just as clearance in a cylinder has predominant control over volumetric efficiency VE, so does the valve area in a cylinder have predominant control over compression efficiency CE. However, to obtain low clearance and a high VE, the designer finds it necessary to limit the size and number of valves. This, however, may tend to lower the efficiency of compression and raise the horsepower. The designer must therefore evaluate both factors and arrive at a compromise, quite a common engineering procedure.

As a general rule, high VE and high CE (low power requirement) do not go together. One cannot attain both. There are, however, four rough divisions the designer uses, the type of application determining to a considerable degree whether one or the other factor takes precedence or whether they are balanced. They might be classified as follows:

	Compression ratio	More important factor
Very high	10–30 (vacuum pumps)	Clearance
High	8–10 max	Clearance principally
		Valving somewhat
Moderate	5 max	Balanced
Low	2 or less	Valving

Chapter

2

Reciprocating Process
Compressor Design Overview*

The reader may best be introduced to the subject by way of a component or construction feature review. We may note that today's reciprocating process compressors are the result of many years of development and experience. As of 1992, one U.S. manufacturer alone had manufactured reciprocating engines and compressors for over 100 years with over 50,000 process units shipped.

Reciprocating process compressors are a very efficient and reliable method to compress almost any gas mixture from vacuum to over 3000 atm. They have numerous applications in refining, chemical, and petrochemical plants. Power ratings vary up to 18,000 kW with capacities up to about 35,000 m^3/h at compressor inlet conditions.

Reciprocating compressors have great flexibility. Being positive displacement compressors, reciprocating units can easily compress a wide range of gas densities from hydrogen having a molecular weight of 2 to gases like chlorine with a molecular weight of 70.

Reciprocating compressors can quickly adjust to varying pressure conditions with stage compression ratios ranging from 1.1 on recycle services to over 5 on gases with low k values or low ratios of specific heat. Typical compression ratios are about 3 per stage to limit discharge temperatures to perhaps 150 to 175°C (300 to 350°F).

Some reciprocating compressors have as many as six stages to provide a total compression ratio over 300.

Conservative rotative and piston speeds are used for process compressors as most units operate continuously for many years with only

* Developed and contributed by G. A. Lentek, Ronald W. Beyer, and Richard G. Schaad, Sr. Product Specialists, Dresser-Rand Company, Engine Process Compressor Division, Painted Post, N.Y.

occasional shutdowns for maintenance. With many applications the gases can cause problems because of being corrosive, containing entrained liquid, and/or foreign abrasive particles. For these reasons low- to medium-speed compressors are used, having rotative speeds from 275 to 600 r/min with piston speeds varying from 3 to 5 m/s (600 to 1000 ft/min) and compressor strokes from 150 to 460 mm (6 to 18 in). Normally, for larger kW rated units, longer strokes and slower speeds are used.

Also for nonlubricated applications, lower rotative and lower piston speeds are recommended to obtain improved piston and packing ring life.

The most common reciprocating compressor in use today is the balanced-opposed design (Figs. 2.1 through 2.3). This design maximizes the operating life of larger reciprocating units by minimizing unbalanced forces and moments. Two to ten cylinders are used with the reciprocating and rotating weights balanced as closely as possible. Also, single-cylinder units can be built with opposed balance weight crossheads used as necessary. Figure Y and similar vertically arranged reciprocating compressors are shown in Figs. 2.4 through 2.8.

Unbalanced forces are produced by reciprocating and rotating masses. Reciprocating forces occur in all compressors from the acceleration and deceleration of the reciprocating weights (piston and rod, crosshead, and a portion of the connecting rod). A compressor designer tries to equalize the reciprocating weights on each crankthrow to balance the forces. Rotating forces result from the centrifugal force pro-

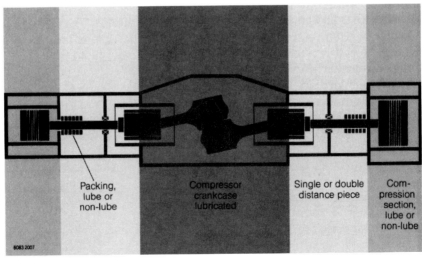

Figure 2.1 Principle of balanced-opposed reciprocating process compressor. *(Sulzer-Burckhardt, Winterthur and Basel, Switzerland)*

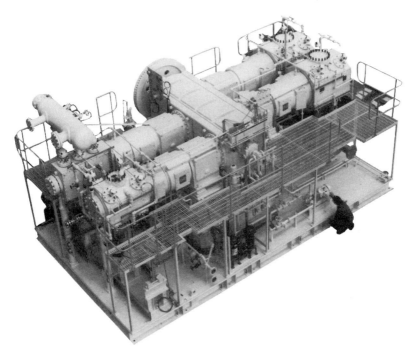

Figure 2.2 Balanced-opposed reciprocating compressor package. *(Dresser-Rand Company, Painted Post, N.Y.)*

duced from the unbalanced weights of the crankthrow and part of the connecting rod.

Only primary unbalanced forces occurring at the compressor speed and secondary unbalanced forces occurring at twice the compressor speed are considered significant in compressor foundation design (Refer to Fig. 2.9). Unbalanced primary and secondary moments also exist in most compressors. With a two-cylinder unit having equal reciprocating weights on crankthrows set at 180° to each other, all primary and secondary forces cancel each other. Only couples or moments are transmitted to the foundation. With good foundation design, these moments are not harmful.

Only six crankthrow units can be perfectly balanced with all unbalanced forces and moments zero. However, perfect balance is normally required only for offshore platform installations or for foundations installed without use of piles on extremely poor (swamplike) soil conditions.

Compressors installed on well-designed foundations over compacted soil or rock will withstand the normal unbalanced forces and moments of a reciprocating compressor. Multiple compressors should be installed on foundations tied together with a common concrete mat to spread out the area resisting any unbalanced forces and moments. Compressors are designed to withstand these forces and moments and

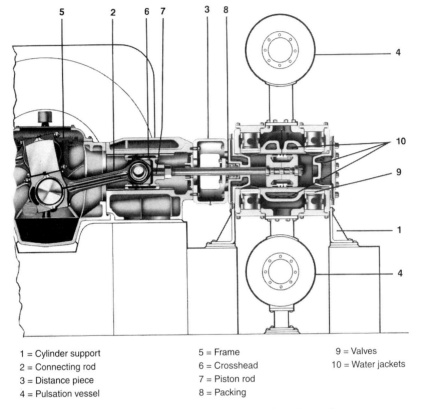

1 = Cylinder support   5 = Frame   9 = Valves
2 = Connecting rod   6 = Crosshead   10 = Water jackets
3 = Distance piece   7 = Piston rod
4 = Pulsation vessel   8 = Packing

Figure 2.3 Right-hand portion of balanced-opposed reciprocating compressor. *(Dresser-Rand Company, Painted Post, N.Y.)*

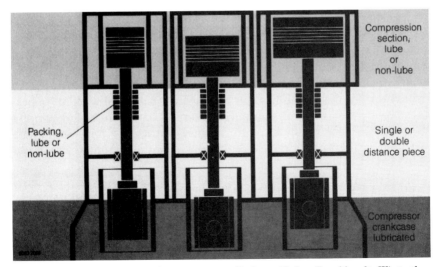

Figure 2.4 Vertically arranged compressor cylinders. *(Sulzer-Burckhardt, Winterthur and Basel, Switzerland)*

Figure 2.5 Vertically oriented reciprocating compressor. *(Cooper Cameron Corporation, Cooper-Bessemer Reciprocating Products Division, Grove City, Pa.)*

merely transmit them to the foundation. However, it is very important that a strong grout bond be obtained between the compressor frame and the foundation. Epoxy-type grouts are highly recommended.

2.1 Crankshaft Design

Up to 10 crankthrow units have been supplied for large compressors. The cranks are arranged with equal angles between each crank to provide optimum unbalanced forces and the smoothest overall crank effort torque (see Fig. 2.10). Even-number crankthrow units are arranged with 180° opposed pairs of cranks to cancel out inertia forces; odd-number crankthrow units require special crank-angle layout or dummy crossheads, as shown in Fig. 2.11.

Cylinders create pulsating compression forces and vibratory torque on the crankshaft with peaks that can exceed the average compressor horsepower torque by up to five times. The crankshaft design must be

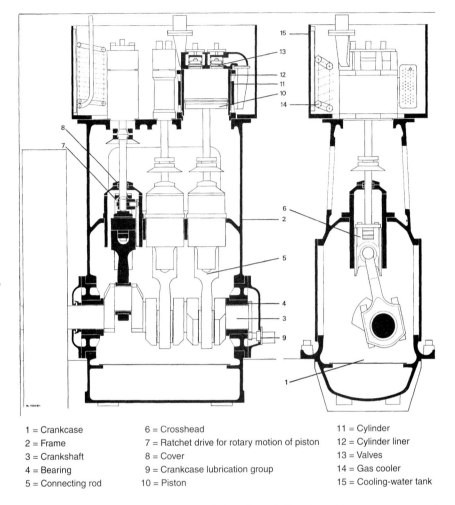

1 = Crankcase	6 = Crosshead
2 = Frame	7 = Ratchet drive for rotary motion of piston
3 = Crankshaft	8 = Cover
4 = Bearing	9 = Crankcase lubrication group
5 = Connecting rod	10 = Piston

11 = Cylinder
12 = Cylinder liner
13 = Valves
14 = Gas cooler
15 = Cooling-water tank

Figure 2.6   Sectional drawing of water-lubricated vertical reciprocating compressor used in oxygen service. *(Sulzer-Burckhardt, Winterthur and Basel, Switzerland)*

conservative to withstand these crank effort and vibratory stresses. For compressors over a small size of about 150 kW per crank, the crankshafts should be forged steel.

A customarily applied American Petroleum Institute Specification (API 618) requires all crankshafts to be forged steel and heat-treated with ground bearing surfaces. Experienced manufacturers further require the cranks to be upset forged from the steel billet to provide stronger grain flow through the crank webs and cranks instead of machining the cranks from a billet. Materials used are alloy steel AISI 1045 or AISI 4140 with ultrasonic inspection by the crankshaft supplier. Crankshafts are purchased completely finished from forging sup-

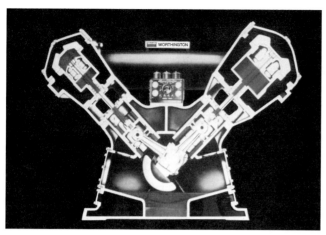

Figure 2.7   Reciprocating compressor with Y arrangement of cylinders. *(Dresser-Rand Company, Painted Post, N.Y.)*

pliers who have special facilities to upset forge and grind the journals and crankpins. Also, special attention must be given to providing polished radii between the cranks and crank webs. Oil passages are drilled to permit oil flow from the journals to the crankpins. The intersections of these holes must be radiused and polished to prevent stress concentration points. Separate bolted-on or integral counterweights are used to help offset unbalanced forces and moments. (Figs. 2.12 and 2.13.)

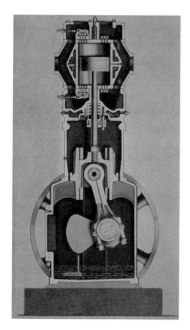

Figure 2.8 Reciprocating compressor with vertically arranged cylinders. *(Dresser-Rand Company, Painted Post, N.Y.)*

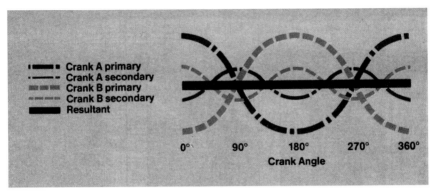

Figure 2.9 Crank-angle diagram demonstrates how primary and secondary forces balance each other out by acting in opposite directions. *(Dresser-Rand Company, Painted Post, N.Y.)*

2.2 Bearings and Lubrication Systems

Most units have replaceable, precision-bored, sleeve-type aluminum alloy crankpin and main bearings. No field fitup or adjustment is necessary. Aluminum alloy has a high bearing load capability and is not likely to score the crankshaft surface should a bearing failure occur. Other bearing materials used are steel-backed aluminum, steel- or bronze-backed, babbitt-lined, and trimetal (steel-bronze-babbitt). With any of these bearings systems it is very important to maintain clean oil piping and filters and specified lubrication pressures and temperatures.

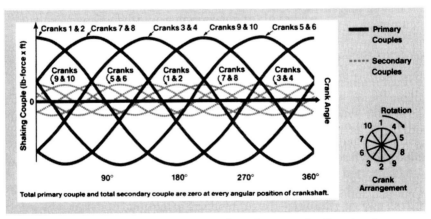

Figure 2.10 Preferred crank arrangement and resultant couples for a ten-throw compressor. Pairs of cranks are uniformly displaced at 36°. Therefore, reciprocating and rotating weights in opposing cylinders will not normally be equal. A variety of techniques are used to add weights to reciprocating parts to achieve the desired balance. *(Dresser-Rand Company, Painted Post, N.Y.)*

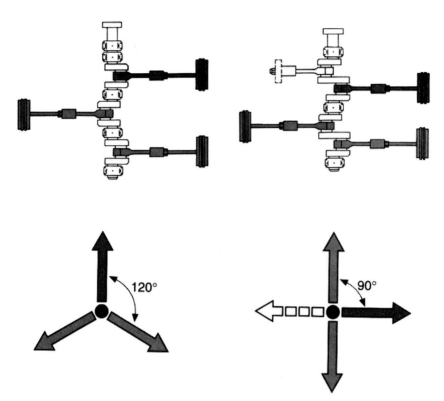

Figure 2.11   Three-throw crank angles at 120° (left) vs. 90° (right). Note dummy cross-head required on right. *(Dresser-Rand Company, Painted Post, N.Y.)*

Figure 2.12   Two-throw crankshaft. *(Dresser-Rand Company, Painted Post, N.Y.)*

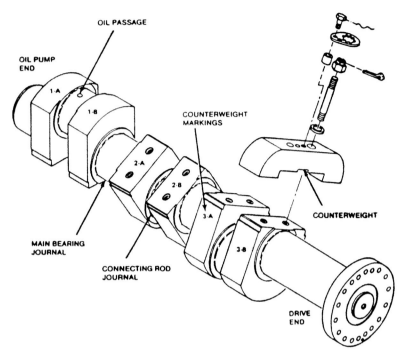

Figure 2.13 Three-throw crankshaft showing oil passages and counterweights. *(Dresser-Rand Company, Painted Post, N.Y.)*

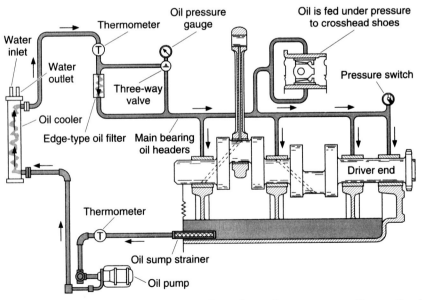

Figure 2.14 Force-feed lubrication system for reciprocating compressor. *(Dresser-Rand Company, Painted Post, N.Y.)*

All large process compressors require forced-feed lubrication with a minimum scope of supply shown in Fig. 2.14, including oil pump, oil cooler, and oil filter. Redundancy and instrumentation requirements are governed by the criticality of a given process, and API 618 covers available options.

2.3 Connecting Rods

(In Figs. 2.15 and 2.16). Connecting rods on process reciprocating machines are typically made of forged steel and manufactured with a closed die to provide good grain flow throughout the piece. Forced lubrication oil passages are drilled the length of the rod to permit oil flow from the crankpin to the crosshead pin bushing. Crosshead bushings are replaceable bronze. Connecting rod bolts are special forgings, and larger sizes have rolled threads for maximum strength.

2.4 Crossheads

This sliding component is typically manufactured of cast steel, or cast or ductile iron, with options for cast steel to meet API 618. For units over 150 kW, replaceable shim-adjusted top and bottom crosshead shoes are supplied. Most crossheads are a floating-pin design (Figs. 2.17 and 2.18); however, some larger units use a fixed-pin design. Either type is acceptable for long-term, reliable operation.

2.5 Frames and Cylinders

Although a few manufacturers have offered fabricated, or welded, equipment frames, the majority of compressor frames are cast iron (Fig.

Figure 2.15 Typical connecting rod. *(Dresser-Rand Company, Painted Post, N.Y.)*

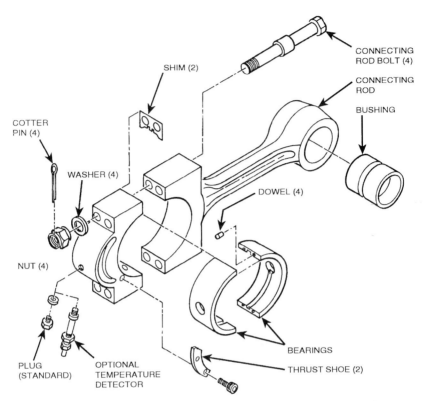

SHIM (2)

CONNECTING ROD BOLT (4)

CONNECTING ROD

BUSHING

COTTER PIN (4)

WASHER (4)

DOWEL (4)

NUT (4)

BEARINGS

PLUG (STANDARD)

OPTIONAL TEMPERATURE DETECTOR

THRUST SHOE (2)

Figure 2.16 Component parts of typical connecting rod. *(Dresser-Rand Company, Painted Post, N.Y.)*

Figure 2.17 Crosshead for reciprocating compressor. *(Dresser-Rand Company, Painted Post, N.Y.)*

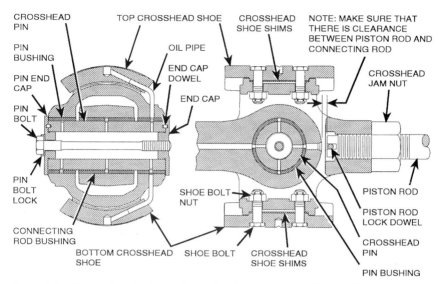

CROSSHEAD PIN

PIN BUSHING

PIN END CAP

PIN BOLT

PIN BOLT LOCK

CONNECTING ROD BUSHING

TOP CROSSHEAD SHOE

OIL PIPE

END CAP DOWEL

END CAP

CROSSHEAD SHOE SHIMS

SHOE BOLT NUT

BOTTOM CROSSHEAD SHOE

SHOE BOLT

CROSSHEAD SHOE SHIMS

NOTE: MAKE SURE THAT THERE IS CLEARANCE BETWEEN PISTON ROD AND CONNECTING ROD

CROSSHEAD JAM NUT

PISTON ROD

PISTON ROD LOCK DOWEL

CROSSHEAD PIN

PIN BUSHING

Figure 2.18 Crossectional views of crosshead components. *(Dresser-Rand Company, Painted Post, N.Y.)*

2.19). Frames have suitable ribbed bearing supports to eliminate frame deflection and maintain crankshaft alignment under all operating conditions. Larger frames over 750 kW have either a tie-rod or tie-bar over each main bearing to prevent deflections from the inherent high horizontal gas and inertia forces. Frames are totally enclosed to withstand outdoor conditions and have large maintenance access covers.

Figure 2.19 Cast-iron compressor frame assembly showing double bearings at the drive end. *(Dresser-Rand Company, Painted Post, N.Y.)*

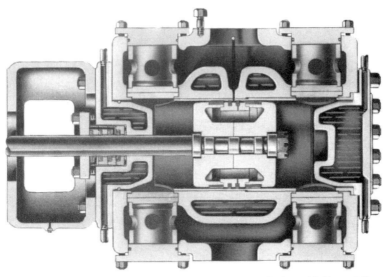

Figure 2.20 Low- or medium-pressure double-acting cylinder with flanged liner and three-piece piston. Liberally sized jackets reduce thermal stresses and aid in heat dissipation. *(Dresser-Rand Company, Painted Post, N.Y.)*

Each cylinder must be designed for the capacity, pressure, temperature, and gas properties of a specific project. Available cylinder materials include cast iron (Figs. 2.20 and 2.21), ductile or nodular iron (Fig. 2.22), cast steel, and forged steel (Figs. 2.23 and 2.24). Manufacturers such as Dresser-Rand have also built large numbers of fabricated (welded) compressor cylinders (Fig. 2.25), with carbon and stainless steel being the primary materials.

Tandem cylinders, Fig. 2.26, are supplied where space and cost savings are important. Similar considerations may lead to the selection of step or truncated cylinders, Fig. 2.27.

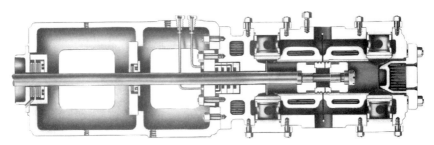

Figure 2.21 Medium- or high-pressure double-acting cylinder with flanged liner. The liner is locked in place by a flange between head and cylinder barrel. A two-compartment distance piece designed to contain flammable, hazardous, or toxic gases is illustrated. *(Dresser-Rand Company, Painted Post, N.Y.)*

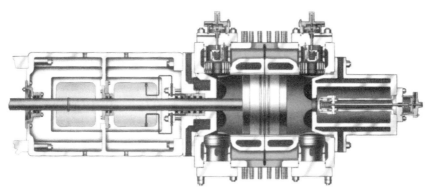

Figure 2.22   Cast-iron or nodular iron cylinders shown with two-compartment distance piece and frame extension. Pressures to 1500 psi are typical. *(Dresser-Rand Company, Painted Post, N.Y.)*

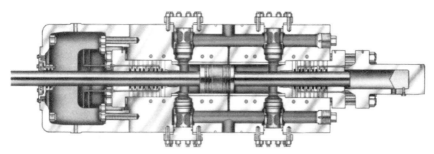

Figure 2.23   Forged steel cylinder with tailrod design for pressures to 7500 psi. Tailrod construction is used to pressure-balance a piston or to achieve rod load reversals. This load reversal may be needed to properly lubricate the crosshead pin bearing. *(Dresser-Rand Company, Painted Post, N.Y.)*

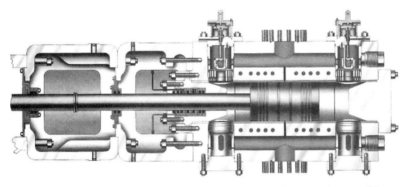

Figure 2.24   Forged steel cylinder with two-compartment distance pieces and frame extension for 3000 psi refinery service. *(Dresser-Rand Company, Painted Post, N.Y.)*

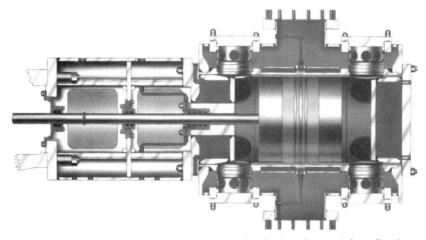

Figure 2.25   Fabricated carbon or stainless steel cylinders for special applications. *(Dresser-Rand Company, Painted Post, N.Y.)*

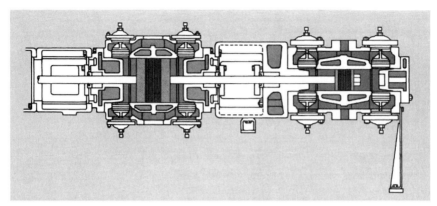

Figure 2.26 Tandem cylinders are furnished with a second piston connected in-line with the first piston. *(Dresser-Rand Company, Painted Post, N.Y.)*

Cylinders normally have separate force-feed lubrication systems using special oils. However, many services can be supplied nonlubricated. Nonlubricated service requires very clean gas by suction filtration to 1 μm if necessary and piston speeds reduced to below 4 m/s (700 ft/min) for acceptable piston, rider, and packing ring life. (Labyrinth piston compressors represent a special and important subcategory of nonlubricated process gas compressors and will be covered later in this text).

Truncated Cylinders

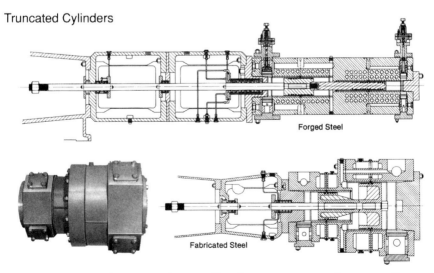

Forged Steel

Fabricated Steel

Figure 2.27 Truncated, or step cylinders allow for space-saving multistaging. *(Cooper Cameron Corporation, Cooper-Bessemer Reciprocating Products Division, Grove City, Pa.)*

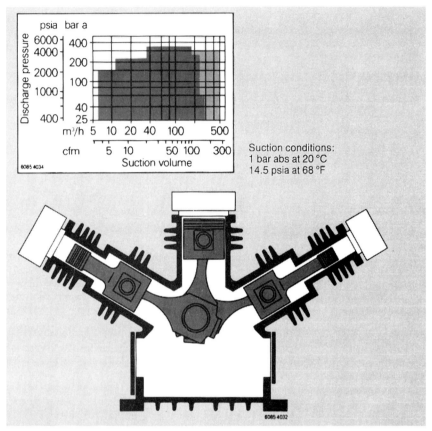

Figure 2.28 Trunk piston compressor with conventional Stage 1 and step-type higher stage pistons. *(Sulzer-Burckhardt, Winterthur and Basel, Switzerland)*

Most cylinders are double-acting, i.e., compression takes place in one half of the cylinder as the piston moves toward the cylinder head, and also as the piston moves toward the crank end of the machine. However, cylinders can be made single-acting for special applications such as for high pressures where only a small displacement is required. Both conventional and tandem cylinders are shown in Fig. 2.28. This illustration depicts a trunk piston compressor. Trunk piston machines resemble automobile engines in that they do not incorporate crossheads.

Most process gas cylinders on large double-acting compressors are equipped with replaceable full-length liners that are held in place to prevent end movement or rotation. Liners, Fig. 2.29, are always used in steel cylinders. Standard liners are centrifugal cast iron that provides a good dense bearing surface. Other materials such as Niresist are available for special applications.

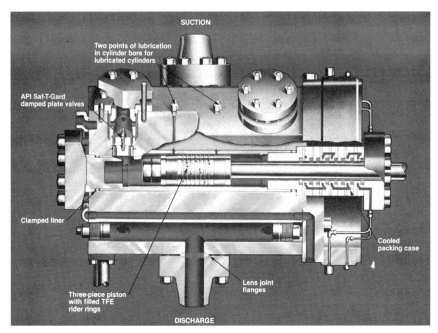

Figure 2.29   Compressor cylinder with clamped liner, cooling and lubricating provisions. *(Dresser-Rand Company, Painted Post, N.Y.)*

2.6   Cooling Provisions

For large process gas compressors, forced cooling through the cylinder barrel and heads is most common (Fig. 2.30). If water is used, it is very important that clean treated water be used. Untreated river water is not acceptable because excessive deposits and fouling buildup will occur in the cylinder jackets, creating serious damage from cylinder overheating. A closed cooling system with a tempered water glycol mixture is highly recommended to minimize deposits and also prevent liquid dropout from saturated gases in the cylinders. A typical cooling system as shown in the schematic from API 618 (Fig. 2.31) may be used for one or more units.

The purpose of cylinder cooling is to equalize cylinder temperatures and prevent heat buildup. This cooling removes only the frictional heat. The heat of compression is removed by the inter- or aftercoolers.

Thermosyphon or static-filled cooling (Fig. 2.32) can be used for cylinders having discharge temperatures below 88°C (190°F).

The coolant supply temperatures should be at least 6°C (10°F) above the gas inlet temperature to prevent formation of liquid in the cylinder gas passages, which can cause serious valve and piston problems.

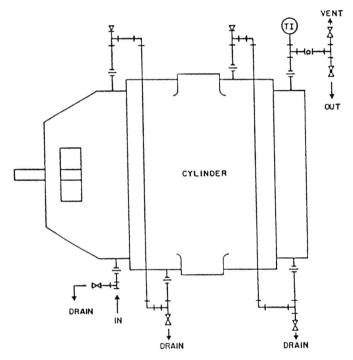

Figure 2.30 Forced cooling arrangement for large compressor cylinders. *(Dresser-Rand Company, Painted Post, N.Y.)*

2.7 Pistons

Cast iron is the piston material of choice for most applications. Aluminum is used for large pistons and on higher-speed units to reduce and balance inertia forces. For some high-pressure applications, over 150 atmospheres absolute pressure (ata), one-piece integral steel piston and rod construction is used for higher piston strengths.

2.8 Piston and Rider Rings

Most process units today are equipped with *Teflon* (PTFE) piston rings. Normally two or three single-piece, diagonal cut rings without expanders are used. For some high-pressure applications (over 300 ata) three-piece bronze segmental rings are used. Also for some nonlubricated applications other special plastics or high-performance polymers have been used. One typical assembly is illustrated in Fig. 2.33.

For many lubricated and all nonlubricated applications PTFE rider rings are used. The rider rings support the weight of the piston and piston rod. Rider rings may be split type located in the center of the piston (Fig. 2.34) or band type stretched onto the piston. The bearing pressure

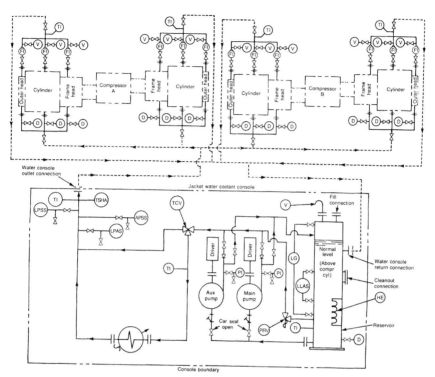

Figure 2.31 Closed cooling water system per API 618. *(American Petroleum Institute, Washington, D.C.)*

on rider rings is normally below 0.7 kg/cm^2 (10 lb/in^2). As noted earlier, it is critical to have clean gas for long piston, rider, and packing ring life. Dirt or piping rust and scale carryover into cylinders will cause very rapid ring, cylinder bore, and valve wear.

2.9 Valves

Virtually all process gas and moderate-to-large size air compressors use spring-loaded, gas-actuated valves. Two of the many basic valve configurations are depicted in Figs. 2.35 through 2.39.

Although certain claims and counterclaims are made by the various manufacturers, they share in the desire to provide durable configurations compatible with gas composition and pressure. Also, valves are almost always symmetrically placed around the outer circumference of the cylinder and can normally be removed and serviced from the outside.

Good specifications mandate configurations and arrangements that will preclude installation errors. Reversing a suction valve could make

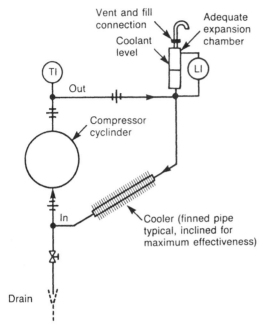

Figure 2.32   Thermosyphon cooling arrangement per API 618. *(American Petroleum Institute, Washington, D.C.)*

it function as a discharge valve and vice versa. Similarly, a bad valve design might risk deteriorating components falling into the compression space of a cylinder. Quite obviously, catastrophic damage and safety incidents could be the end result.

To ensure against structural failure of the guard or seat, API-compliant valve designs feature the use of a center bolt. The bolt is designed so that even in the event of its failure, it cannot drop into the compression chamber. The center bolt provides a very important part in valve fixed-clearance and physical strength. Without the bolt, all of the differential pressure would be sustained by the valve seat alone. The center bolt allows the designer to use the physical strength available in the guard since the center bolt ties the guard and seat together. The result is smaller clearance volumes, which directly result from thinner seats and guards than would be possible with designs not using a center bolt.

Poorly designed valves can also cause noticeable decreases in compression efficiency; valve lift and valve area affect gas velocity and must be properly dimensioned.

Figure 2.35 depicts a plate valve. Enhanced versions of plate valves will sometimes apply the principle of pneumatic cushioning by allowing a small amount of gas to be trapped, as shown in Fig. 2.36. Deck-

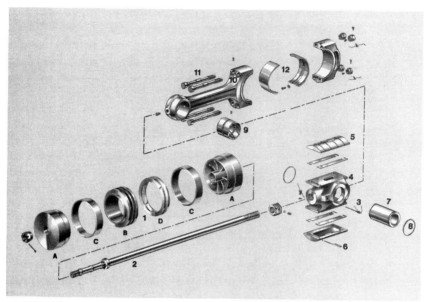

1. **3-piece piston** allows for two solid slip-on Rider Bands.
   A. Piston end bells
   B. Ring carrier
   C. Solid rider bands
   D. Piston rings
2. **Piston rod**

3. **Dowel pin** so piston rod can not turn
4. **Box-section crosshead**
5. **Babbitt-faced cast iron cross-head slippers**
6. **Crosshead slipper key**
7. **Full-floating crosshead pin**

8. **Crosshead-pin retainers**
9. **Renewable crosshead pin bearing**
10. **High-strength connecting rod**
11. **Connecting bolts**
12. **Split crankpin bearing**

Figure 2.33 Reciprocating components of typical compressor. *(Dresser-Rand Company, Painted Post, N.Y.)*

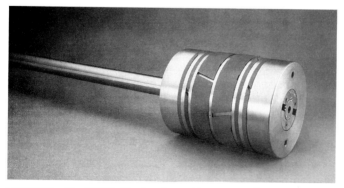

Figure 2.34 The PTFE rider bands used to support pistons of contact-type nonlubricated compressors. *(Dresser-Rand Company, Painted Post, N.Y.)*

Balanced spring system delivers
equivalent motion to the valve plates.

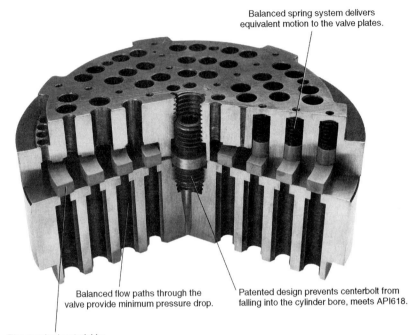

Balanced flow paths through the
valve provide minimum pressure drop.

Patented design prevents centerbolt from
falling into the cylinder bore, meets API618.

The standard material for
plates and buttons is "Hi-Temp".

Figure 2.35   Plate valve. *(Dresser-Rand Company, Painted Post, N.Y.)*

and-one-half and double-deck valves, Figs. 2.37 and 2.38, are designed
to incorporate larger flow areas and, thus, improved efficiency.

While the description of valves has emphasized the plate-type
design, circular channel ring-type valves, as well as poppet designs are
available. Straight channel, circular channel ring, and poppet designs
were primarily created for high-pressure and medium-pressure, low-
ratio applications, respectively. In most high-pressure applications the
damped valve has replaced the channel design. The lower plate mass,
the greater damping and a plate with fewer stress concentrations has
led to the success of the damped valve over the channel design. The
poppet valve has primarily been applied to low-ratio, slow-to-medium
speed, gas transmission service. Because of the alignment problems of
valve seat to valve guard, maintenance has occasionally been a problem. Nevertheless, well-designed poppet valves, Fig. 2.39, are widely
used in the application range illustrated in Fig. 2.40.

The basis for a valve or compressor manufacturer's dynamic calculations is depicted in Fig. 2.41. Poor valve designs are revealed in lift
vs. crank-angle diagrams (Fig. 2.42) and pV diagrams such as shown
in Fig. 2.43.

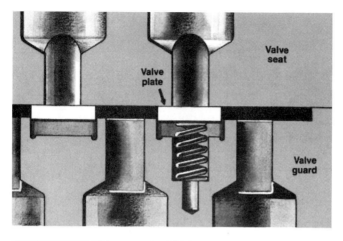

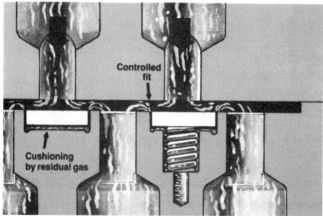

Figure 2.36   Cutaway of closed and open positions of damped plate valve illustrating pneumatic damping feature. *(Dresser-Rand Company, Painted Post, N.Y.)*

2.10   Piston Rods

The standard recommended piston rod material is AISI 4142 alloy steel, induction hardened throughout the packing travel to Rockwell Hardness 50 C minimum. Other materials such as stainless steel are available for increased corrosion resistance properties. However, these materials cannot be hardened above Rockwell Hardness 40 C. Special hard coatings, including tungsten carbide and ceramic materials, are available as well. High-quality process compressors incorporate piston rods furnished with precision-controlled rolled threads (Fig. 2.44) that offer much greater fatigue strength over cut threads. API 618 also specifies use of rolled threads.

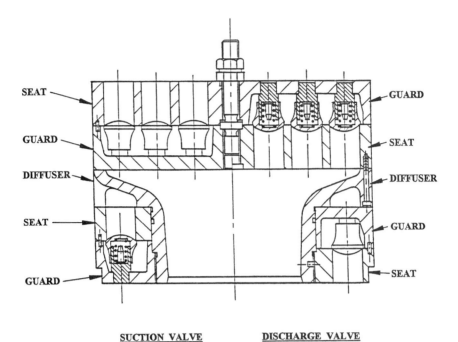

<u>SUCTION VALVE</u>   <u>DISCHARGE VALVE</u>

Figure 2.37   Deck-and-one-half compressor valve. *(Anglo Compression, Inc., Mount Vernon, Ohio)* (Alternative Selection: Deck-and-one-half Accu-Guide poppet valve)

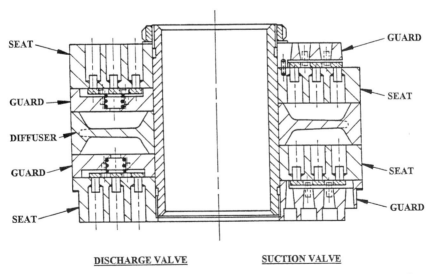

<u>DISCHARGE VALVE</u>   <u>SUCTION VALVE</u>

Figure 2.38   Double-deck valve, ported plate, opposed flow type. *(Anglo Compression, Inc., Mount Vernon, Ohio)*

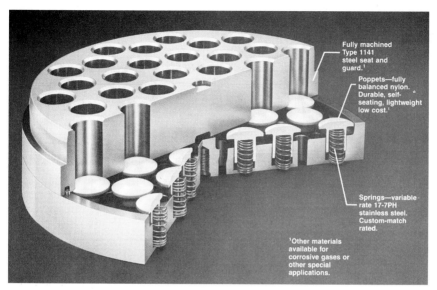

Figure 2.39   Poppet valve. *(Cooper Cameron Corporation, Cooper-Bessemer Reciprocating Products Division, Grove City, Pa.)*

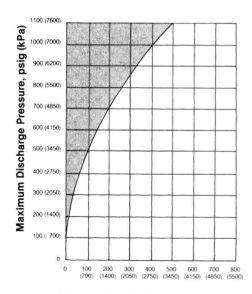

Present application range is represented below the curve.

Figure 2.40   Poppet valve application range. *(Cooper Cameron Corporation, Cooper-Bessemer Reciprocating Products Division, Grove City, Pa.)*

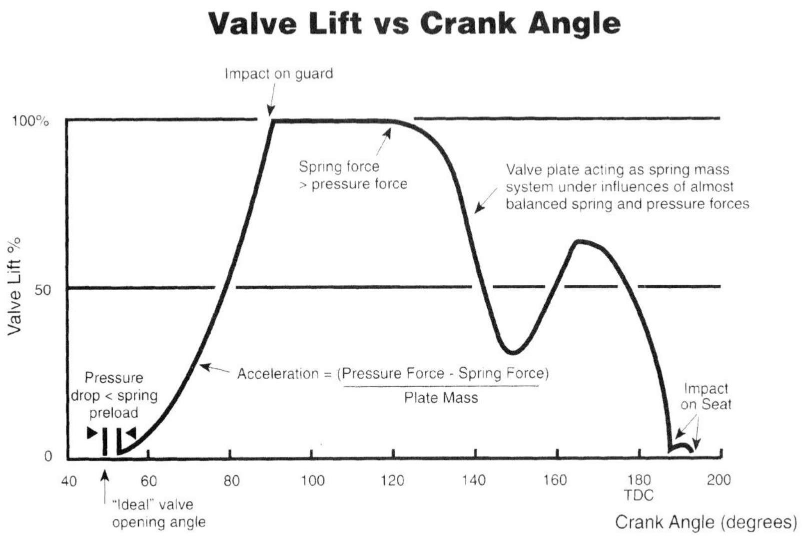

Valve Lift vs Crank Angle

Figure 2.41   Basis of valve dynamics calculation. *(Dresser-Rand Company, Painted Post, N.Y.)*

2.11   Packings

Packings (Fig. 2.45) are required wherever piston rods protrude through compressor cylinders and distance pieces. Vented, full-floating, self-lubricating PTFE packing is standard and provides long-lasting operation with a minimum of gas leakage. One lubrication feed is common except high pressures over 150 ata may have two feeds (Fig. 2.46). Many packing cases are also equipped with internal cooling passages. A typical self-contained cooling system for piston rod pressure packing is shown in Fig. 2.47.

2.12   Cylinder Lubrication

Proper cylinder lubrication may greatly reduce compressor maintenance requirements. Cylinders are typically lubricated by a forced-feed lubricator system that is separate from the crankcase system. The lubricator is normally crankshaft driven; however, motor drive is also available. Cylinders have at least one lubricator feed each in the top of the bore and in the packing. Large-diameter and high-pressure cylinders may have additional feeds in the bottom of the bore and packing. Each lubricator pumping unit delivers from 10 to 50 drops/min and has individual flow adjustment. Required flow rate can only be determined by experience and cylinder inspection on each application by checking that an adequate oil film exists in the cylinder.

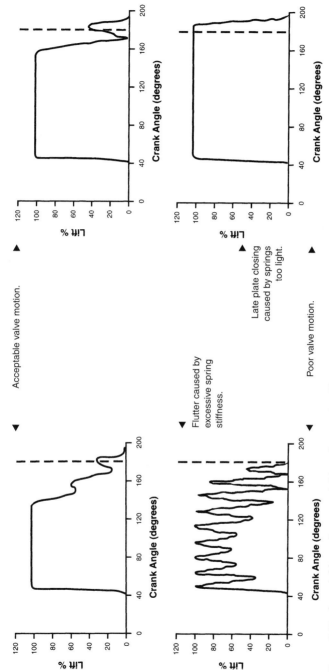

Figure 2.42 Acceptable and unacceptable valve motion illustrated on lift vs. crank-angle diagrams. *(Dresser-Rand Company, Painted Post, N.Y.)*

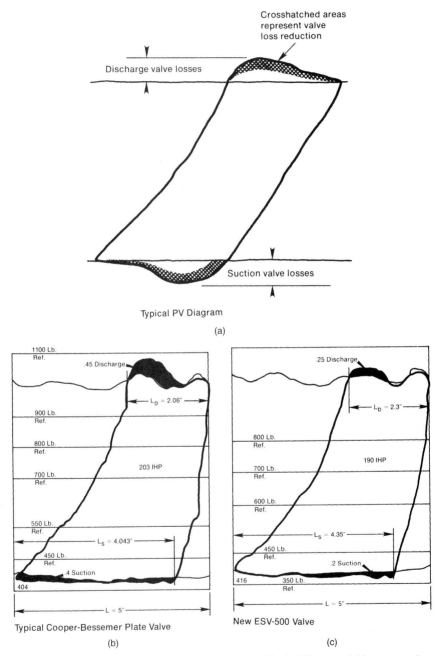

Figure 2.43   The pV diagrams can reveal valve losses. Typical diagram (a) is compared to traditional plate valve (b) and enhanced design (c). *(Cooper Cameron Corporation, Cooper-Bessemer Reciprocating Products Division, Grove City, Pa.)*

Figure 2.44 Rolled thread on piston rod. *(Dresser-Rand Company, Painted Post, N.Y.)*

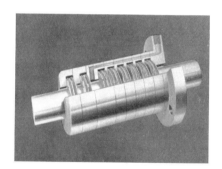
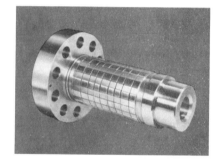
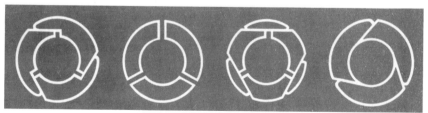

Figure 2.45 Packing cartridges and available arrangements. Single-, double-, radial-, or tangential-cut rings with passages for lubrication, coolant and venting are provided, as required by the application. Surfaces are usually lapped. *(Dresser-Rand Company, Painted Post, N.Y.)*

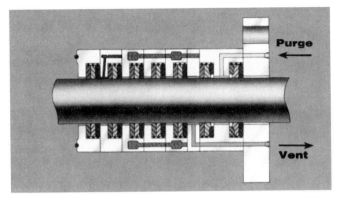

Figure 2.46 Lubrication and cooling passages on rod packing. *(Dresser-Rand Company, Painted Post, N.Y.)*

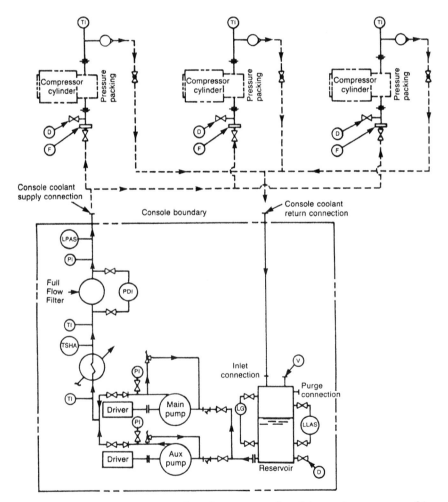

Figure 2.47 Typical self-contained cooling system for piston rod pressure packing schematic, per API 618. *(American Petroleum Institute, Washington, D.C.)*

The lubricant required for cylinders is a heavy, well-refined oil and compounded with 5 to 10 percent animal fat if the gas is saturated. Diester synthetic lubricants are also extremely well suited for cylinder lubrication.

2.13   Distance Pieces

These components are usually furnished as steel or cast-iron castings, or steel weldments. Distance piece geometry can vary to meet the application. The standard distance piece is a single compartment with vent and drains (Fig. 2.48). The pressure packing is vented separately. All distance pieces typically have large openings with gasketed, gastight covers for ease of packing maintenance.

For nonlubricated or other services requiring an oil slinger, an extended-length distance piece is used so that no portion of the piston rod enters both the crankcase oil wipers and the cylinder packing (Fig. 2.49). The primary function of an oil wiper is to prevent loss of crankcase oil; it will not prevent gas from entering the crankcase.

For gases that could contaminate the crankcase oil or are hazardous, a two-compartment distance piece is recommended having an interme-

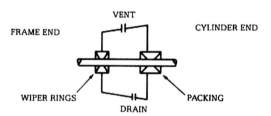

Figure 2.48   Single-compartment, API standard distance piece for general compression applications. *(Dresser-Rand Company, Painted Post, N.Y.)*

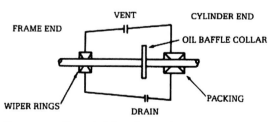

Figure 2.49   Extended length, single-compartment distance piece. No portion of the piston rod that enters the packing will enter the frame oil wiper rings. *(Dresser-Rand Company, Painted Post, N.Y.)*

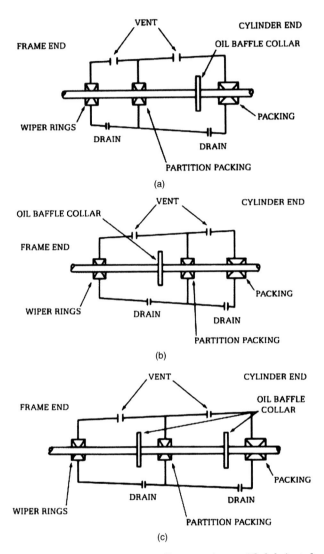

Figure 2.50 Two-compartment distance pieces with lubricated partition packing. Baffle collars at cylinder end (*a*), frame end (*b*), or both ends (*c*) prevent oil migration. Venting or purging of each compartment is possible. (*Dresser-Rand Company, Painted Post, N.Y.*)

diate packing (Fig. 2.50). The cylinder side compartment is vented to a safe area, and the crankcase side compartment is normally vented to atmosphere or purged with nitrogen. An oil slinger may be located in this compartment. Packing vent and distance piece vent and drain manifolds should always be kept separate.

We are now ready to consider other aspects of reciprocating compressor technology in greater detail.

3

Reciprocating Compressor Performance Considerations

3.1 Capacity Control Is Important

As already discussed, a reciprocatiᵣg compressor is a positive displacement device. During normal operation it will take in a quantity of gas from its suction line and compress the gas as required to move it through its discharge line. Unlike centrifugal pumps, the reciprocating compressor cannot self-regulate its capacity against a given discharge pressure; it will simply keep displacing gas until told not to. This would not be a problem if we had an unlimited supply of gas to draw from and an infinite capacity downstream to discharge into; however, in the real world of refineries, chemical plants, and gas transmission lines, we find that we have specific parameters within which to work, and that capacity is a unique quantity at any point in time. Thus, we have a real need to control the capacity of the reciprocating compressor.

We also find that in most instances a reciprocating compressor needs to be unloaded for start-up. Simplistically, it would seem that starting a positive displacement machine fully loaded would require a driver with 100 percent starting torque. However, what we find is that a reciprocating compressor can typically have a 3:1 peak to mean torque ratio. This peak torque requirement coupled with breakaway friction means that the driver now must have as much as a 350 percent starting torque capability. Again looking at the real world, we find that motors are designed to have 40 to 60 percent starting torque capability, thus necessitating an unloaded start. Additionally, we can convince ourselves that it is healthier for the compressor to start and continue running unloaded to give the compressor time to warm up.

From basic thermodynamics and compression theory we know that there are equations describing the performance of a reciprocating compressor cylinder. The equations relate such parameters as compression ratios, cylinder clearances, volumetric efficiency, and thus cylinder flow capacity. From our earlier discussion, we recall

$$VE = \eta_v = 100 - C(r^{1/k} - 1) \tag{3.1}$$

We can also draw a pressure-volume (pV) diagram that visually gives us a feel for how the cylinder is operating. This was shown earlier in Sec. 1.2. Referring back to this information allows us to understand how some of the capacity control methods work.

3.1.1 Recycle/bypass

One of the simplest methods of controlling capacity is to recycle, or bypass, the compressed gas back to the compressor suction. This is physically accomplished by piping from the compressor discharge line through some type of control valve and going back to the compressor suction line. To reduce the flow to process one simply opens up the bypass line and diverts the excess flow back to the compressor suction. In addition to being simple, this system also has the advantage of being infinitely controllable (within the limitation of the size of the bypass line). Many, if not most, process-type compressors have some sort of recycle line so that operators can fine-tune the flow to process.

A recycle system does, however, have shortcomings, the greatest being its inefficiency. Consider that we are taking gas at an elevated discharge pressure after having invested considerable horsepower in compressing it to that discharge pressure. Now we are going to expand it back down to the lower suction pressure simply so that we can invest more horsepower by compressing it again. As far as the compressor is concerned, it is always running at 100 percent load and is consuming 100 percent horsepower, even though the flow actually delivered to process could be a low percentage, or even zero. Another problem is in the actual design of the bypass line. The greater the percentage of bypass, the larger the piping has to be. In addition, depending on the particular gas characteristics, it may be necessary to include a cooler in the bypass line to dissipate the heat generated in compressing and continually recycling the gas. Also, consider the case of multistage compression where multiple bypass lines may be needed. The cost of installing an extensive bypass line could become prohibitive.

The most practical application for the bypass line is for small degrees of fine capacity control or for limited duration start-up unloading where a simple loop around the compressor can be opened for a short period of time to relieve the initial compression load.

3.1.2   Suction throttling

Although not very widely used, suction throttling is another method of controlling the capacity of a reciprocating compressor. The technique is to reduce the suction pressure to the compressor by limiting or throttling the flow into the cylinder. By referring back to our pV diagram we see that if all else is held constant, reducing the suction pressure creates a narrow card indicating a lower volumetric efficiency and thus less flow.

Additionally, the density of the gas will be reduced at this lower pressure, thus helping to reduce the delivered mass flow.

Suction throttling has its limitations. It takes a fairly dramatic reduction in suction pressure to give any sizeable reduction in capacity. Additionally, as the suction pressure is reduced and the discharge pressure held constant, the compression ratio is increased. This causes higher discharge temperatures and also higher rod loads.

3.1.3   Suction valve unloading

Probably the most common method of controlling compressor capacity is via suction valve unloading. The technique here is to physically keep the cylinder from compressing gas by maintaining an open flow path between the cylinder bore and the cylinder suction chamber.

The cylinder will take in gas normally; however, instead of completing the normal cycle of compression and discharge, the cylinder will simply pump the gas still at suction pressure back into the suction chamber via this open pathway. There is absolutely no gas discharged to process. Additionally, since there is no compression occurring, there is virtually no horsepower consumed other than passageway losses.

There are three basic types of suction valve unloading. The first, and oldest, is finger-type unloaders, shown in Fig. 3.1. These unloaders consist of a series of small fingers that are housed in the valve crab assembly and actuated via a push rod from an outside actuator. To unload the valve, the fingers are lowered so that they depress the valve-sealing components and thus hold the valve in the open position. The pathway then between the cylinder bore and the gas passage is through these open suction valves. Finger-type unloaders will typically be mounted on each suction valve so that the flow area of the unloaded pathway is maximized. Also, since the fingers are simply holding open the existing suction valves, no special valve design is required. Actuation of finger-type unloaders can be manual using a handwheel and screw or lever arrangement to lower the fingers, or by using a small air cylinder on the top of the unloader stem.

One of the biggest problems associated with finger-type unloaders is the potential for damaging the valve-sealing elements with the fingers.

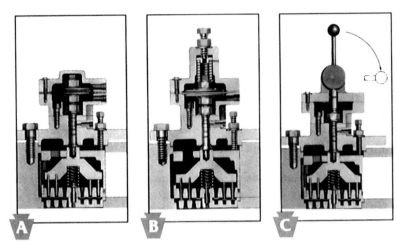

Figure 3.1 Finger-type unloaders, pneumatically operated: (*a*) direct-acting (air-to-unload); (*b*) reverse-acting or fail-safe (air-to-unload) which automatically unloads the compressor in the event of control air failure; (*c*) manual operation. (*Cooper Cameron Corporation, Cooper-Bessemer Reciprocating Products Division, Grove City, Pa.*)

If one visualizes the force generated by pneumatically actuated fingers as they are driven down against the valve-sealing components, one can see how this could contribute to premature valve failure.

An alternative to finger-type unloaders is plug-type unloaders, Fig. 3.2. Here, instead of acting on the valve-sealing components themselves, we have a passageway bored through the middle of the valve. In normal operation this passageway is sealed with a plug. To unload the cylinder we simply remove the plug and allow the gas to flow in and out of this passageway with no compression taking place.

The plug-type unloader offers a major benefit over the finger-type unloader because it does not work on the valve-sealing components as did the finger-type unloader. However, since there is now a passageway through the center of the valve, we have reduced the normal effective flow area of the valve. In the case of a heavy gas, this could mean higher consumed horsepower because of the reduced valve flow area and increased pressure drop through the valve. Again as with the finger-type unloaders, each suction valve would typically require a plug-type unloader.

The third type of unloader is the port, or passage-type unloader, Fig. 3.3. This type of device also uses a plug to seal the unloader passageway. However, instead of working in a passageway through the valve, the unloader uses a separate port between the cylinder bore and the gas passage. The port is created by removing one suction valve per cylinder end and replacing it with a large plug assembly. There are

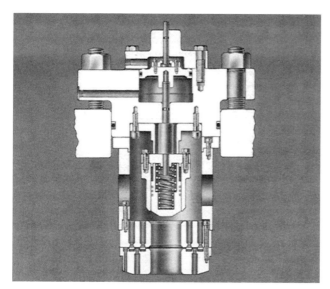

Figure 3.2 Outside operated plug-type unloader. Actuating air cannot mix with the gas being compressed. *(Dresser-Rand Company, Painted Post, N.Y.)*

many advantages to the port-type unloader. The flow area of the passageway is large, since the plug is the full diameter of a normal valve rather than just a portion of the diameter. Moreover, when opened, the plug will lift from 1 to 2 in off its seat. With, thus, only one unloader needed per cylinder end, the compressor incorporates fewer devices for maintenance purposes. Additionally, since the port unloader does not work on an active valve, it does not need to be removed for regular valve maintenance. Port unloaders are ideal for low molecular weight gas applications where the total number of suction valves is typically reduced to improve pressure drop across the active valves.

Again, all valve unloader types can be manually operated or actuated by a pneumatic cylinder. When pneumatically actuated, these devices can be designed to load or unload upon either application or removal of air pressure. The advantages of pneumatic operation are the ability to remotely control the capacity of the compressor or even automate this control.

Since suction valve unloaders keep compression from occurring, they control capacity in discrete steps. For example, a double-acting cylinder can be operated 100 percent loaded, 50 percent unloaded by unloading one end of the cylinder, or fully unloaded by unloading both ends of the cylinder.

An interesting alternative to these discrete steps of unloading is the stepless capacity control system offered by Hoerbiger compressor con-

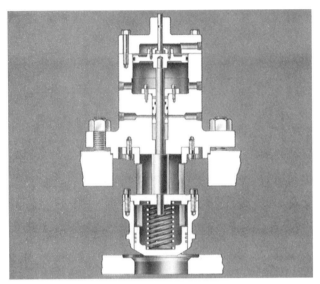

Figure 3.3 Outside operated port-type unloader requires the use of only one unloading device per cylinder end and is typically used for lower molecular weight gases. Air cannot mix with the gas being compressed. *(Dresser-Rand Company, Painted Post, N.Y.)*

trols. This system uses finger-type unloaders that are pneumatically actuated. However, rather than keeping the valve unloaded continuously, the stepless system actually unloads the valve for only a portion of its stroke.

By allowing compression to occur during only part of the stroke, one obtains partial flow instead of full flow or no flow for a fully unloaded cylinder. Actuation of these unloaders is by a specially designed control panel that monitors process flow requirements and unloads the compressor as necessary. Because of its principles of design, this system is limited in turndown, and applications must be reviewed individually to confirm their suitability.

In general, suction valve unloading is an excellent method of controlling capacity. The devices are simple and easy to maintain and operate. They are efficient and are very good for start-up unloading so that starting torque requirements are extremely low. Nevertheless, suction valve unloading does have some drawbacks.

If a cylinder is operated fully unloaded for an extended period of time, gas temperatures in the inlet passage will rise, since the same gas is being worked back and forth through the unloader passageway. This can become a problem, especially where the gas k-value is high. The solution is to periodically load the cylinder so that the heated gas will be pumped to process, and the cooler inlet gas stream will normalize the temperatures. Typically, this cyclic loading should occur for 10 min out of every hour of unloaded operation. This heating problem does not occur at 50 percent loads, since the active end of the cylinder will scavenge the heated gas and keep the stream of cool suction gas in the suction chambers.

Another area to be looked at when operating suction valve unloaders is the potential for a nonreversing crosshead pin load. For a given cylinder description, there may be conditions where unloading one end of the cylinder, typically the frame end, can cause a nonreversing load. This would prevent lubricant from flowing into the pin clearance. To guard against this, each mode of unloading should be studied for pin load reversal and any unacceptable operating modes highlighted to the compressor operator, or, in the case of an automated system, locked out of the unloader logic.

3.1.4 Clearance pockets

Clearance pockets, Fig. 3.4, are also a common way of reducing the capacity of a compressor. From compressor theory, we know that the volumetric efficiency of a cylinder is dependent on its clearance or nondisplaced volume. We again recall:

$$VE = \eta_v = 100 - C(r^{1/k} - 1) \tag{3.2}$$

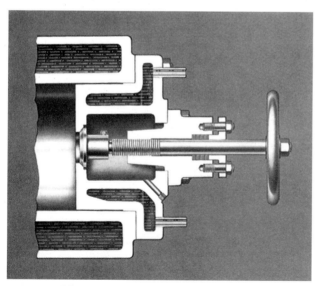

Figure 3.4 Manual fixed-clearance pocket valve is generally located in the outer head of a cylinder, as shown. This type of control is used for applications that require limited and infrequent capacity changes. *(Dresser-Rand Company, Painted Post, N.Y.)*

This equation illustrates that, as clearance is increased, the volumetric efficiency decreases, thus reducing the amount of flow to process. It should be noted here that the degree to which the volumetric efficiency is affected by increase in clearance is governed by the compression ratio of the service. In practical terms we see that when the compression ratio is 2.5 or higher, an increase in clearance will actively reduce the volumetric efficiency. However, for ratios of 2.0 or less it takes a fairly major increase in clearance for any reasonable reduction in volumetric efficiency. As a generalization, clearance pockets are ineffective and not useful with low compression ratios (1.5 or less).

The design of a clearance pocket is simple and can be visualized from Fig. 3.4 and Fig. 2.22, earlier. It is essentially an empty volume, typically in the outer head of the cylinder, with a valved passage to the cylinder bore. During normal operation, the valve is closed and the cylinder operates at full capacity. For reduced capacity operation the valve is opened, and the cylinder capacity is reduced by the effect of this added clearance on the volumetric efficiency. Typically, clearance pockets are of a fixed volume and sized to reduce flow precisely to a predetermined level. It is not uncommon to use multiple fixed-volume clearance pockets to allow for numerous discrete reduced capacity steps of control.

For example, a large cylinder may be executed with two fixed-volume clearance pockets, one small and one large. Using the small and large independently and then together, one can obtain three reduced flow modes of operation from two clearance pockets. It is also possible to put clearance pockets on the frame end of a cylinder so that many different clearance volume combinations are feasible.

Fixed-volume clearance pockets, just as valve unloaders, can be manually or pneumatically actuated.

Figure 3.5 depicts an option available with clearance pockets. Instead of executing them with a fixed volume, they can be configured for a variable volume.

A variable-volume clearance pocket is typically mounted on the outer head of a cylinder and is basically a cylinder itself with a piston moved back and forth to increase or reduce the volume of the pocket. An advantage to the variable-volume clearance pocket is that instead of having fixed steps of unloading, stepless turndown is achieved by simply adjusting the pocket. This is particularly useful in meeting ever-changing process conditions since the pocket can be adjusted in the field as required. Historically, variable-volume clearance pockets have been manually actuated by means of a very large handwheel to move the piston back and forth. When the correct volume is achieved, the pocket can be locked in place by means of a second locking wheel.

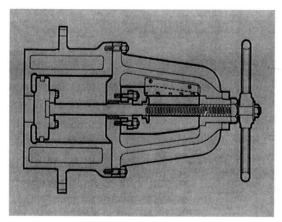

Figure 3.5 Manual, variable-volume clearance pocket. This clearance pocket provides capacity reduction in an infinite number of steps over a given range. This pocket can also be automatically actuated by a hydraulic system that varies the position of the piston in the pocket. *(Dresser-Rand Company, Painted Post, N.Y.)*

The handwheel method sounds fine until one considers the force against the pocket piston created by the process gas pressure inside the cylinder. At times it might take the strength of two operators to turn the handwheel and adjust the pocket. Also, as corrosion or deposits are formed from the process gas, the piston may tend to stick in one position and may no longer be useful for adjustment.

An improvement over the manually controlled pocket is the hydraulically controlled variable-volume clearance pocket. Here, the manual handwheel has been replaced with a hydraulic cylinder. The cylinder is directly coupled to the sliding piston and uses hydraulic pressure from a support package to move the piston back and forth. Not only can the hydraulic cylinder create a higher force than the typical field operator, but it also allows the pocket to be automated so that the pocket will adjust itself to field conditions.

All of the capacity control methods described so far have assumed that the compressor is running at a constant speed. Since the reciprocating compressor is a positive displacement device and does not rely on sophisticated gas dynamics for its compression, it is possible to vary the speed to directly adjust the throughput. One typical variable-speed driver is the steam turbine. Whenever steam is economically available in plants, it is not uncommon to have a large steam turbine coupled to a reciprocating compressor through a gear speed reducer. By varying the speed of the turbine, the capacity of the compressor can be directly controlled. However, because of the complexity of the drive train (large flywheel, soft couplings, multireduction gear boxes) the system is limited in its turndown by torsional problems. Typically, a turbine gear drive arrangement is limited to a ±10 percent speed variation. Operation outside of the approved speed range could lead to massive failure of the drive system. New technology makes it feasible to use variable-speed AC motors for running compressors. (DC drive systems have been available for some time but were limited to low-horsepower applications.) Modern control technology allows AC motors of all sizes to operate with fully variable speeds. However, since this technology is relatively new to the reciprocating compressor industry, there are still a number of questions to be addressed: torsional response; valve action at reduced rotative speeds; feedback into the electrical grid due to torque pulsations from the compressor; economic feasibility; whether variable speed should be employed analogous to bypasses in the past, i.e., to simply trim flow between major steps of operation. All these areas need to be considered before variable-speed AC motors are specified as drivers for reciprocating compressors.

3.2   More About Cylinder Jacket Cooling-Heating Arrangements

During their normal compression cycle, reciprocating compressor cylinders typically generate considerable amounts of heat. The heat comes from the work of compression plus the friction of the piston rings against the cylinder wall. Unless some of this heat is dissipated, undesirably high operating temperatures will occur. Most cylinders intended for process gas operation are designed with a jacket, shown earlier in Figs. 2.20 through 2.22, to allow this heat to be removed by some cooling medium.

There are a number of advantages in dissipating this heat.

1. By lowering the cylinder wall operating temperatures, one can reduce losses in capacity and horsepower due to the suction gas being preheated by warm cylinder gas passages. The cooler the inlet gas, the denser it is, and greater mass flow per unit volume will result. Removing heat from the gas during compression lowers its final discharge temperature and reduces the power required for compression.

2. Dissipating the heat from the cylinder and reducing the inlet gas temperature creates a better operating climate for the compressor valves, yielding longer valve life and reduced formation of deposits.

3. A jacketed cylinder filled with coolant will maintain a more even temperature throughout the cylinder and reduce hot spots, which

could cause uneven thermal expansion and undesirable deformation of the cylinder.

4. A lower cylinder wall temperature leads to better bore lubrication. Lubricants will break down less on a cool wall than they would on a hot wall, and better lubrication leads to extended ring life and less maintenance.

However, care is needed to not reduce the cylinder operating temperatures too much. Consider the problems created by introducing a warm saturated gas into a cylinder with cold metal sections. Condensation will occur in the bore; thus, washing the lubricant from the cylinder walls will cause accelerated wear of the piston and rider rings. Even worse, a large quantity of condensed liquid could collect in the inlet gas passage and be introduced into the cylinder as a slug of liquid. This could lead to at least broken valves and maybe a broken cylinder. To avoid this condensation problem, it is considered good practice to use a cylinder coolant temperature approximately 6°C (10°F) warmer than the inlet gas temperature.

3.2.1 Methods of cooling

When evaluating a process compressor for cooling, a cylinder will fall into one of four general categories.

Noncooled. For a cylinder operating in cryogenic service where gas temperatures are typically below –60°C (–75°F), no cooling is required. In fact, no cooling medium is suitable for providing uniform, acceptable cylinder temperatures. For applications like these, cylinders are often designed with no cooling jacket at all and will simply be insulated from the ambient air in an attempt to avoid severe temperature differentials or frost formation on the cylinder exterior.

Static cooling. Static cooling is used for applications where gas discharge temperatures are below 88°C (190°F) and mean temperatures are low (below 60°C [140°F]). This type of cooling is also applied where there will be no unloaded cylinder operation that could create abnormally high temperatures.

In a static system the cylinder water jacket is simply filled with a cooling medium such as a water-glycol mixture. No attempt is made to circulate the mixture. A small reservoir vented to atmosphere is provided to allow for thermal expansion.

Thermosyphon cooling. A cylinder may be thermosyphon cooled where discharge temperatures are moderate (88 to 90°C [190 to 210°F]), mean temperatures are in the range of 60 to 66°C (140 to 150°F), and where there will be no extended periods of fully unloaded operation that could increase operating temperature. This cooling method was illustrated earlier in Fig. 2.32.

A thermosyphon system is similar to the static system; however, there is now a small section of pipe connecting the top cooling medium outlet to the bottom of the cylinder. The idea here is that as the warm water in the radiative sections cools, it will flow to the bottom of the cylinder, creating a slight circulation through the cylinder jackets.

Full-circulation cooling. For applications where gas mean and discharge temperatures are in excess of the previously stated limits or where extended periods of fully unloaded operation are anticipated, the cylinder requires that a coolant be circulated through its jacket to dissipate the heat buildup. This was discussed previously (see Fig. 2.31).

For large process cylinders, it is common to have water-cooled cylinder heads as well as a jacket around the cylinder bore. These sections are connected by external jumper pipe. As mentioned earlier, the temperature of the coolant should be controlled so that it is maintained approximately 6°C (10°F) higher than the gas inlet temperature. In addition, the flow should be controlled so that the temperature rise across the cylinder circuit is between 3 and 11°C (5 and 20°F). Flow through a cylinder is controlled by means of a globe valve on the discharge of the jumper piping around the cylinder. It is customary to throttle the coolant outlet to ensure that the cylinder is constantly flooded with coolant. A thermometer and sight flow indicator are located immediately upstream of the discharge globe valve to assist the operator in adjusting the coolant flow to maintain the correct temperature rise across each cylinder.

In calculating coolant flow required for a given cylinder, we find that the coolant temperature rise across the cylinder and the coolant flow rate are inversely proportional. In other words, a high flow rate of water will pick up only a slight increase in temperature. Conversely, a slow trickle of water will rise appreciably in temperature. In optimizing the balance of flow rate and temperature rise, it is important to remember to keep temperature rises moderate (3 to 11°C [5 to 20°F]). This will ensure a fairly even cylinder temperature and will keep the flow rate within certain limits. Too low a flow rate will allow silt or other entrained particulates to drop out in the water jacket, eventually leading to a blockage of flow. On the other hand, too high a flow rate can create a prohibitively high pressure drop through the coolant circuit. Typically, velocities between 4 and 8 ft/s based on the size of the coolant jumper pipe have been used as guidelines.

Actual calculation of coolant flow is, at best, an estimate based on empirical formulas and, at worst, a black art. An old rule of thumb for determining coolant flow in GPM is to divide cylinder horsepower by the allowable temperature rise. This method is simple but neglects accounting for most of the critical parameters. Coolant temperatures, gas temperatures, cylinder size, and frictional factors are all important parameters in calculating cylinder coolant flow requirements.

Calculation methods today are based on empirical formulas that use these parameters and have been found to have good field correlations.

Untreated water can be used, provided the temperatures are acceptable and that the water is filtered before it goes through the cylinder jackets. Nevertheless, refineries typically use cooling tower water, which is temperature controlled, filtered, and treated. Coolant discharging from a shell and tube intercooler is a good source of coolant since it has been preheated by the compressed gas stream. This should reduce the risk of incurring moisture condensation and liquid slugging.

The ultimate source of coolant is from a closed cylinder jacket coolant console dedicated specifically to an individual compressor. Figure 3.6 depicts such a console. The system can be tailored specifically to the compressor in question. Motor-driven pumps (main and backup) provide circulation. Coolant temperature can be kept high enough by the use of a reservoir heater, kept cool enough by a shell and tube or radiative style cooler, and controlled by an automated temperature control circuit. Instrumentation is added to monitor and protect the system.

Figure 3.6 Packaged jacket water cooling system. *(Dresser-Rand Company, Painted Post, N.Y.)*

3.3 Comparing Lubricated vs. Nonlubricated Conventional Cylinder Construction

One of the major areas of motion, and thus wear, in a reciprocating compressor is in the cylinder. Considering that during a year's normal operation a piston travels nearly 100,000 miles in the cylinder bore, it can be seen that cylinder lubrication merits closer investigation.

3.3.1 Lubricated cylinder designs

Probably 80 percent of all process reciprocating compressor cylinders are lubricated. Lubricating the cylinder bore makes sense. It reduces friction between the piston rings and cylinder bore, thus reducing frictional heat and wear of both cylinder bore and piston rings. It lubricates the cylinder valves, helping them to survive the 100 million+ cycles they go through in a year's operation. A film of lubricant in the cylinder also helps protect the cylinder components from the effects of corrosive gases.

When using lubricated construction, the cylinder designer uses the lubricating film to his best advantage. Because the sliding surfaces will be lubricated, harder piston and rider ring materials can be used. Typical materials of construction would be glass and/or molybdenum-filled Teflon (PTFE). Because of its relative hardness, this material has excellent durability and, when lubricated, the wear characteristics of both rings and bore contacted are excellent. Carbon or graphite-filled PTFE, which has become a rather universal ring material, is also frequently used in lubricated service. Since the piston will be riding on a film of lubricant, the piston can be relatively heavy. It should be noted that, although lubricated construction allows a piston to be run directly in the cylinder bore, it has become common practice in the process compressor industry over the past decades to design pistons with rider bands, shown earlier in Figs. 2.33 and 2.34, supporting the pistons in the cylinder bore.

Rider bands can be considered as an expendable support shoe for the piston. As wear occurs, the ring could be readily replaced. For lubricated designs, rider band bearing loads are typically in the 8- to 10-psi range when considering a contact area of 120° of arc.

Piston rod packing rings would also be made of glass and/or molybdenum-filled PTFE for lubricated service. Again, this relatively hard compound shows excellent durability and wear characteristics in a lubricated service, without being excessively abrasive to the piston rod surface. As with the piston and rider rings, carbon or graphite-filled PTFE is also commonly used.

Figure 3.7 Pump-to-point lubricator mounted at center of frame supplies different amounts of lubricant to packing and cylinders. *(Dresser-Rand Company, Painted Post, N.Y.)*

Because the compressor cylinders are operating at elevated pressure, the lubricant must be pumped into the cylinder in a controlled manner. There are a variety of cylinder lubricators designed precisely for this purpose. The most common style of cylinder lubricator is the pump-to-point lubricator, Fig. 3.7.

Pump-to-point lubricators are designed with a fabricated steel box serving not only as the main body of the system, but also as the lubricator sump. Through this box runs a multicammed shaft that is driven either by the compressor crankshaft or by an electric motor. The actual pumping units are located on top of the box. The units are equipped with a suction straw that drops into the sump and a follower that rides on the cams to actuate the pumping plunger. On top of the pumping units is a transparent cap that allows observing the lubricant as it is drawn up through the suction straw and drips down to be pumped to the cylinder. Counting the frequency of drops facilitates monitoring of lubrication flow rates (Fig. 3.8).

A cylinder typically has multiple points of lubrication. As illustrated earlier in Fig. 2.29, there may be several feeds in the main bore, depending on size and pressure. There could also be several feeds in the packing case, and sometimes a feed in a partition packing. Each point of lubrication is fed by an individual pump from the lubricator which is why this style is called a *pump-to-point* lubricator. The advantage of using the pump-to-point lubricator is that each pump is individually adjustable. Distribution of lubricant to different areas in the cylinders can thus be tailored to best suit the lubrication requirements.

The other major style of lubricator is the divider block lubricator, Fig. 3.9. Divider block lubricators incorporate a single high-pressure injection pump feeding a number of divider blocks, also referred to as *splitter blocks,* or *divider valves.* The function of the divider block is to divide the flow of lubricant from the pump into multiple streams of various predetermined proportions so that each point can be fed an

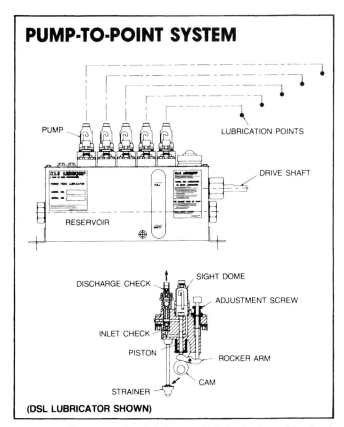

Figure 3.8 Pump-to-point lubricator. *(Lubriquip, Inc., Cleveland, Ohio)*

appropriate amount of lubricant. This style of lubricator is popular because, once constructed, the proportion of flow from the divider blocks is fixed and cannot get out of adjustment in the field. Moreover, divider block lubricator arrangements are easily instrumented to annunciate deficiencies in any point being lubricated. This must be attributed to the cascading motion of the divider blocks that are interconnected through ports and thus require each block to deliver its portion of lubricant before the next block functions.

Should any individual point become clogged or blocked, all oil flow would stop. A single no-flow switch can thus be used to monitor all lubrication points. Additionally, there are cycle monitors that can be used to monitor the rate of cycling of the blocks. If for some reason the blocks take too long to cycle, indicating a reduced flow rate to the cylinder, this device can sound an alarm. One of the disadvantages of divider block–type lubricators is the difficulty to field-adjust the differ-

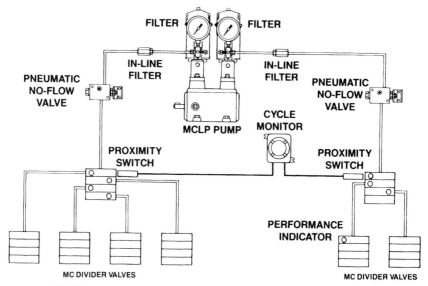

Figure 3.9   Divider block lubricator system. *(Lincoln Division of McNeil Corporation, St. Louis, Mo.)*

ently proportioned flow amounts. Such adjustments may become desirable whenever substantial differences exist in cylinder pressure levels.

3.3.2   Nonlubricated cylinder design

There are some processes that will not tolerate oil entrained in the gas stream. Oil separators can be installed in the compressor discharge lines; however, sometimes these are not effective enough for the level of cleanliness required, or there may be safety problems associated with a particular gas contacting the lubricant. In such cases, the only alternative is to use nonlubricated cylinder designs. Probably 20 percent of all process gas compressors are designed for nonlubricated operation because of process demands. Oil in the gas stream could lead to catastrophic problems in an oxygen compressor or even in a high-pressure air compressor. Also, many chemical processes cannot tolerate the existence of lubricant in their catalysts.

Special consideration must be given to nonlubrication applications. Without an oil film, piston rings do not seal in the cylinder bores as efficiently, thus causing blowby in the cylinder, resulting in a lower delivered gas flow from the cylinder. There are certain operational limits that must be addressed. At higher operating gas pressure, piston rings have much greater force against cylinder walls, thus creating greater wear problems. Except for special applications, nonlubricated construction is limited to pressures below 2000 psi.

Without the lubricating film to reduce wear, we must change out piston ring materials. Typically a nonlubricated piston or rider ring is made from carbon- or graphite-filled PTFE. The carbon or graphite filling gives it a measure of lubricity. Carbon- or graphite-filled PTFE is also a little softer and less abrasive in cylinder bore contact than glass and/or molybdenum-filled PTFE. For applications with very dry gases Morganite Graflon compounds have proven to be very good. For oxygen applications, copper-filled PTFE is used, replacing lead-filled rings due to the environmental problems associated with lead. The oxygen atmosphere causes the copper-filled rings to form a copper oxide layer that gives them a natural lubricity similar to the carbon or graphite filling. Copper filling is preferred over carbon filling because it is less active in the oxygen atmosphere. Of course, this is where noncontacting, labyrinth piston design excels. This will be discussed later.

For conventional nonlubricated designs, the cylinder designer must address bearing load in the cylinder. Without the advantage of the lubricating film, the designer must reduce the bearing load on the rider bands typically to 3 to 5 psi, by reducing the piston weight or increasing rider band area.

Piston rod packing designs for nonlubricated construction will also differ from those used in lubricated construction. Typical ring materials are again a carbon-filled PTFE material. Because of the additional frictional heat generated between nonlubricated rings and piston rod, it is wise to have a cored packing case and circulate coolant through it at pressures above 250 psi. In some cases, it may also be necessary to include bronze backup rings in the packing case to aid in heat transfer from the rod into packing case and cooling media.

3.4 Compressor Vent and Buffer Systems Deserve Attention

Whenever a moving surface must be sealed against a pressure differential, one must expect leakage. In a reciprocating compressor, this leakage typically occurs at the piston rods. Even with state-of-the-art multiring packing cases, it is likely that process gas operating at hundreds or thousands of psi will leak from cylinders past the rod packings. Since this process gas is typically hazardous, flammable, or even toxic, it is certainly undesirable and often illegal to simply allow the gas to leak into the atmosphere. It is thus necessary to capture this leakage gas.

Normally a packing case will have a vent connection piped from between the last two rings in the packing case, Fig. 2.46. This connection not only serves as a vent to remove gas before it is leaked into the cylinder distance piece but also serves as a drain for excess cylinder lubricating oils. This is why packing vents should always be piped to a point below the piston rod. These vent connections can be piped to a flare stack or gas recovery system or back to gas suction if the pressure is low enough.

Venting the packing case will not, however, give complete assurance against the leakage of gas into the cylinder distance piece.

To prevent this gas from leaking into the compressor frame where it can cause damage to the frame end running gear, cause an explosion, or simply leak to the atmosphere through the frame breather, it is customary to vent the distance piece to a flare stack or recovery system. A more conservative approach is to use a two-compartment distance piece where the cylinder end compartment is buffered with an inert gas and the frame end compartment is vented to flare (Fig. 2.50). By pressurizing the cylinder end compartment to a higher pressure than the flare stack, any leakage past the rod packing will be by the inert gas leaking toward the packing vent rather than the process gas leaking out of the packing case into the distance piece. The vented frame end compartment ensures that no gas will leak into the frame.

In cases where it is imperative to capture as much leaking gas as possible, one can use the same type purge and vent system in the packing case itself. This was illustrated earlier in Fig. 2.46. Here a buffer of inert gas is introduced between the last two rings in the packing case and vented between the second and third-to-last packing rings. Since the last several rings in the packing case are now at a very reduced pressure level, one can no longer rely on the gas pressure to adequately seal the rings against the rod and packing case. This problem can be alleviated by using mechanically loaded ring sets in the last two packing cups.

The question of where to vent the leakage is best answered by the end user. Nonhazardous, nonflammable gases may be allowed to leak directly to atmosphere. Some compressors will have a simple gooseneck on top of the distance piece vent connection or use louvered covers on the distance piece doors to allow ambient air to purge any leakage. The most obvious vent destination for a flammable process gas is the flare stack. Flare systems typically run at pressures between 5 and 15 psig. For higher-pressure systems it may be necessary to put an additional sealing ring at the end of the packing case beyond the vent to help encourage the gas leakage to run into the high-pressure flare header instead of into the distance piece and then to the atmosphere.

The most economical place to vent to is the first-stage suction of the compressor. The process gas leakage is thus recovered and put back into the process. Discretion must be used in taking a packing vent to suction. If suction pressures are much above 25 or 30 psig, it may create a major problem for the packing case. This case would now possibly experience pressurization of the midportion with 30 psig gas and this might compromise its sealing capabilities.

3.5 Compressor Instrumentation Is Always Important

The area of compressor design that offers the greatest degree of freedom and personal input from the equipment specifier is instrumentation. Since compressor instrumentation does not directly contribute to the pumping duty of the machine, it is often neglected or overlooked. The amount of instrumentation on a given unit can range from a few simple pressure and temperature gauges to sophisticated electronic or computer-based systems.

As we analyze this subject we find that instrumentation serves three basic purposes: to monitor, to protect, and to diagnose. Monitoring instruments are the most basic. They are the gauges, or readout devices, that allow operators to examine the compressor and its process and to determine whether the unit is functioning properly. It is important to install monitoring instruments for all parameters critical to the operation of a compressor.

Protective devices are those that alert operators to an upset condition or keep the compressor from destroying itself. Should operator oversights occur, the compressor protective devices will shut the unit down before the problem reaches disastrous dimensions.

Diagnostic instruments are those that monitor various parameters, integrate their findings, and make a diagnosis as to the health of the compressor. Diagnostics determine not only whether the compressor is running properly or has failed; they also predict an approaching failure or operational problem. Used properly, diagnostic instrumentation helps schedule maintenance shutdowns rather than having the compressor trip unexpectedly.

Reciprocating compressor instrumentation most often covers the following parameters.

Pressure. Since the basic function of a reciprocating compressor is to elevate gas pressure from one level to another, pressure would appear to be the most basic parameter to look at. For monitoring purposes, the simple pressure gauge tells us at a glance exactly what we need know. Pressure gauges should have large diameter dials so that they are easily read. As with any instrument, the scale on the gauge should be selected so that, under normal operating conditions, the pointer is approximately midrange. As with all pressure instruments, pressure gauges should be provided with an isolation valve to facilitate replacement or servicing.

Pressure switches are an important component of any instrumentation system. Whenever pressures go beyond the normal operating limits, the pressure switch can activate either an alarm, or protective shutdown, or both.

There are many different designs of pressure switches available: single-pole, double-throw; double-pole, double-throw; single-level switches; multilevel switches; dual switches in a single housing; internally adjusted; externally adjusted; factory adjusted. A typical switch should be ruggedly constructed of materials suitable for the application and should be listed and approved for operation in the area classification for its installed environment. An internal adjustment is convenient to allow for field readjustment in case of changing pressure conditions and to prevent accidental misadjustment from inadvertent outside contact.

Conventional switch logic is to have the contacts normally closed in operation and open for alarm or shutdown actuation. This way, if the field lines are ever accidentally cut, the circuit opens; the machine shuts down and the unit will not run without protection. Pressure switches should be installed with suitable block and bleed valves. This allows the switch to be blocked out and bled down so that its set point can be verified. Normally, a pressure gauge or connection for a gauge will be piped next to the pressure switch to assist in calibration.

Temperature. Temperature is as important as pressure in most processes. A 4½-in dial-type thermometer is the common temperature indicator for the process industry. When measuring a fluid temperature it should be installed in a thermowell of suitable material. Thermometers with flexible head mechanisms (every angle design) are convenient because they allow the thermometer face to be adjusted for ease of viewing.

Another method of monitoring temperatures is via a thermocouple, or *resistance temperature detector* (RTD). This allows the temperature to be monitored at a remote location on a readout device. Thermocouples and RTDs can also be useful as sensors for protective circuitry. Instead of a monitoring instrument, or perhaps included in the monitoring instrument, is a comparative circuit that compares the actual input from the thermocouple to a preset level. When this level is reached, a signal is sent to the alarm or shutdown circuitry alerting the operators to the problem.

A more common approach to abnormal temperature protection is a filled capillary temperature switch. Here a gas-filled probe is used to sense the temperature being monitored. The probe is connected via a protected stainless steel capillary to the switch assembly. As the temperature changes, the gas in the probe and capillary expands or contracts sending a proportional signal back to the switch. When the signal causes the switch to exceed a preset level, a contact is activated. With a filled capillary type temperature switch, the switch housing itself is normally mounted away from the monitoring point, making it easier to wire and also protecting it from vibration or abuse. Capillary lengths are typically 8 to 10 ft. They can be made longer, but 25 ft is probably the upper limit to ensure accuracy and responsiveness.

Vibration.   Because of its design, a reciprocating compressor is subject to vibration. Reciprocating masses, reversing loads, and pulsating gas streams all contribute to a normal vibration level on a compressor. However, if this normal level is exceeded, it indicates that something abnormal is happening and should be investigated. Typical sources of abnormal vibration are pistons hitting a cylinder head from misadjustment or debris in the cylinder, a failed component in the drive train, or even an acoustical vibration being transmitted through the gas pipe into the compressor. Any of these sources can have a detrimental and even catastrophic effect on the compressor. To protect against damaging the compressor, many reciprocating compressors have a vibration switch mounted on their frame.

These vibration switches have typically been the mechanical (spring or magnet) type where increased vibration causes a switch element to be released from the magnetic holding it, thus activating the alarm. The setting of these switches is often the subject of considerable debate. Although certain guidelines can be set up to predict how many *Gs* are acceptable and what level is unacceptable, the best way to protect a compressor is to set the switch sensitivity in the field. The highest normally anticipated operating vibration is typically the jolt of the main drive motor starting, or changes in flow rate or flow direction of process gas streams. In recent years, the considerably greater level of sophistication employed in centrifugal compressor technology has touched the field of reciprocating compressor vibration monitoring. There are now accelerometers, seismic instruments, and noncontacting shaft vibration probes available that have enhanced sensitivity and are well suited for the diagnosis of reciprocating machines.

Flow.   It is often advantageous to monitor gas flow in compressor installations. Small gas flows can be monitored using a simple rotameter. Liquid flows are typically indicated by pinwheels or flapper-type sight flow indicators. Major flows such as process gas flow are typically monitored by means of a calibrated flow orifice and its associated instrumentation.

These flow orifices are normally located in the downstream piping. Protection is generally for loss of flow of a critical fluid. The protective devices are almost always based on loss of pressure against a calibrated orifice, which then triggers a pressure switch.

Liquid level. Monitoring of liquid levels is done with liquid level gauges. For small reservoirs vented to atmosphere, such as the compressor crankcase or cylinder lubricator reservoir, a simple protected transparent plate or tube attached to the side of the reservoir is normally acceptable.

For pressurized applications, such as a separator or knockout drum in the gas stream, a more rugged type of gauge is required. Armored reflex or transparent liquid gauge glasses are designed to take the high pressures, mechanical vibrations, and physical abuse seen in a typical plant environment. These glasses should be isolated from their reservoirs by block valves so that the gauge glass can be removed for maintenance or replacement without depressurizing the reservoir. If required, these gauge glasses can be fitted with illuminators to allow viewing in low-light conditions. Two basic switch types are available for abnormal liquid level protection. The first is referred to as a displacement-type level switch. It uses a float that is raised or lowered by the liquid level in the switch standpipe. When this float goes above or below its set limit, it trips a switch, normally by using a series of magnets. Because this system uses a rising or falling column of liquid, there must be two connections to the switch: one for liquid flow, the other for pressure equalization. As with liquid level gauges, this liquid level switch should have isolating valves.

The other type of switch uses a variable capacitance principle. This solid-state instrument has a single probe, normally made of stainless steel or coated with an inert material, that is inserted through a single connection into the reservoir. As the liquid level moves along the probe, the electronic circuitry senses a change in the capacitance of the probe. From this changing capacitance, it determines how much of the probe is being contacted. By mounting the probe vertically in the tank, the switch can monitor the changing level of the liquid and compare it to a preset level. When the liquid reaches this level, a signal can be sent to activate an alarm or shutdown. The advantages of this type of switch are that it requires only a single connection into the reservoir and that it can continuously monitor the liquid level. Additionally, multiple level alarm points can be set. Another advantage is that solid-state electronics are less sensitive to vibration than positive displacement type switches. If necessary, the probe can even be remotely mounted in a tank and the electronics portion housed in another area.

Another necessity for level protection is to *actively control* the level of liquid in a separator sump or knockout drum, rather than to rely on operator monitoring and manual intervention. There are automatic traps available that monitor liquid level with a float that is linked to a valve mechanism. When the level in this trap gets too high, the float will open the drain valve and keep it open until the float drops to a lower limit. Because these traps are large and heavy, they must be remotely mounted below the sump or knockout drum so that the liquid flows down into them. There also must be a pressure equalization line between the trap and the drum.

While these automatic traps are functional and self-contained (no external power required), they do have limitations. The valve linkage mechanisms are subject to fouling by lubricating oils or other sludge that may form in the liquids. Location of the traps and routing of the associated piping can be cumbersome, particularly for connecting to suction dampener or separators. Additionally, because these automatic traps are typically made of castings, they are limited in their pressure containment capability and many users will not allow them in process plants.

An alternative to the mechanical traps are electromechanical systems. These systems use a liquid level switch as previously described to energize a solenoid-actuated drain valve. When the liquid level reaches a pre-determined high point, the switch makes contact, opening the solenoid drain valve. As the liquid is drained (typically through an orifice to control the flow rate), the liquid level is lowered until it reaches a low-level set point in the switch. Now, the switch signals the solenoid valve to close, and the cycle is ready to repeat. Regardless of which automated system is used, it is wise to retain manual draining capability.

Having reviewed what instrumentation is available and what kind of parameters to instrument, we need to determine which critical systems merit this instrumentation. Again, since the main purpose of the compressor is to compress gas, the gas system is an obvious choice to instrument. Gas pressure and temperature are important to monitor at both suction and discharge for each stage of compression. Most process applications include a high-temperature alarm switch in the discharge gas stream for each stage. If there is more than one cylinder per stage, discharge temperature will be monitored and alarmed for each cylinder discharge. Since the discharge pressure of each stage is normally protected by a pressure relief valve, high-pressure discharge switches are seldom seen. However, low first-stage suction pressure switches are not uncommon and can help to keep from overloading a compressor due to low suction pressure or, in essence, excessive differential pressure.

Mechanically speaking, probably the most important system to the compressor itself is the frame lube oil system. Here the most critical parameter is, of course, oil pressure fed to the main bearings. Standard instrumentation would include a pressure indicator for monitoring the pressure, as well as low-pressure alarm and shutdown switches for protection of the frame and running gear. If an auxiliary lube oil pump is supplied, the low lube oil pressure alarm switch can be wired to simultaneously sound the alarm and start the auxiliary pump. The shutdown switch is normally set a nominal 5 psi below the alarm switch. Should the pressure continue to degrade after alarm activation, the compressor will be shut down before damage is done to the bearings.

Other lube oil systems instrumentation will normally include pressure indicators at the discharge of all oil pumps, temperature indicators monitoring oil temperatures in and out of the oil cooler, a differential pressure gauge and perhaps even a differential pressure switch around the oil filters giving indications as to how dirty the filter is and whether it is starting to restrict the flow. In some cases, an oil temperature alarm switch is furnished downstream of the oil cooler. Indications of high oil temperature might point to the oil viscosity becoming too low for long-term dependable operation. A liquid level gauge glass and sometimes a level switch are installed on the frame that acts as the oil reservoir for the compressor. Temperature monitoring of the bearing surfaces is very useful on reciprocating compressors. Thermocouples, or RTDs, in the bearing caps to monitor main bearing temperatures are not uncommon on large reciprocating process compressors. Thermocouples can also be used in crosshead guides and motor bearings.

Protecting the crankpin or crosshead pin bearings is more difficult. Here, a eutectic device is occasionally installed at the back of the connecting rod bearing cap or in the crosshead pin. The eutectic device contains a fusible element designed to melt at a predetermined temperature and a spring-loaded pin that pops out when the predetermined temperature is reached. When these pins pop out, their motion trips a strategically placed *flapper valve,* venting an auxiliary manifold that, in turn, trips a pressure switch.

Cylinder lubricant is another critical fluid for the compressor that should be instrumented. For a normal pump-to-point lubricator, lubricator drive failure and low reservoir level can be readily monitored. To do this, the manufacturer often adds an extra lubricating pump to the box and pipes it to a pressure switch. This additional pumping unit has a shorter suction straw than the other pumps in the box. The theory here is that if the drive system for the lubricator fails, this extra pump along with all the other pumps will cease to function; thus, the pressure switch will lose pressure and activate an alarm. Additionally, as the level in the lubricator reservoir drops, this extra pump will be the first to starve. This would also activate the alarm while the other pumps would continue to supply lubricant. With the divider-block style lubricator, it is practical to include a pressure indicator in the discharge line from the main pump. A no-flow switch indicating lack of flow from the entire block system can be included on one point. Cycle monitors are available that can monitor the rate of cycling of the divider blocks. To help diagnose failures, each individual feed point from the various divider blocks can be equipped with a pin indicator so that if an individual line is blocked, the pin will indicate which line caused the shutdown.

Other systems and types of instrumentation are sometimes selected. Rod packing thermocouples and RTDs are not uncommon. Sensing of temperature excursions here can indicate that the packing rings are worn or on the verge of failure. Also, rod drop indicators are becoming more popular in the industry. Their purpose is to monitor the position of the piston rod relative to the packing case to give an indication of how the wear or rider bands in the cylinder are degrading. As the wear bands become thinner, the piston drops in the cylinder; thus, the rod drops relative to the packing case.

At least two styles of rod drop indicators are available. The contacting type requires that the rod drop down, contacting a soft metal cap over a pneumatic line mounted at the bottom of the packing case flange. As the rod rubs off the soft metal cap, the air escapes from the pneumatic line thus venting pressure from a switch, which in turn activates an alarm.

There is also a noncontacting style or eddy-current device. In this system, a small probe is mounted on the packing case flange over the piston rod. The probe emits an electronic signal and, by evaluating the change in interference with this signal created by changing proximity to the rod, an electronic circuit determines probe-to-rod distance. By knowing the initial clearance between probe and rod and the allowable wear of the rider band, calculating and presetting alarm points is possible. The advantages of this system include elimination of wear-prone contact between the sensing element and rod in the packing travel area. Also, eddy-current devices facilitate continuous monitoring of rider band wear rates.

3.5.1 Electrics vs. pneumatics

A point made earlier was that any electrical switch must be certified for operation in a particular atmosphere. This normally does not present a problem since most switches on the market carry the requisite approval for typical refinery atmospheres. There are some cases, however, where switches are not available for the proper atmosphere, or where there is a question of suitability between the electrical device and the medium being instrumented. This is generally where pneumatic switches find application.

Pneumatic instruments do not encounter problems with area classifications or intermittent power availability. Pneumatic systems can also be used for remote indication of parameters by use of a pneumatic transmitter. It is not uncommon to use pneumatic transmitters for remote indication of pressure in hydrogen gas or lubricating oil systems. Obviously, the routing of tubing carrying a flammable gas or fluid into a control panel or control room would be considered hazardous.

3.5.2 Switch set points

In determining set points for various protective switches, it is necessary to examine the safety and reliability philosophy of a given plant. In some cases, operating personnel want to be alerted to the fact that there is an upset condition, as with the typical alarm switch. In an oil system, for example, one may set the low oil pressure switch at 25 or 30 psig, low enough to indicate there is a problem that merits operator attention but still within acceptable operating limits. A shutdown switch, on the other hand, needs to be set specifically to protect the equipment from damage. Again, in the case of a typical lube oil system, the low-pressure shutdown switch would be set for 12 to 15 psig. Below this pressure, continued operation could cause damage to the bearings or crankshaft.

It is customary for an equipment manufacturer to suggest set points for switches supplied on a compressor. However, many times these will need to be modified in the field to reflect actual compressor operation.

3.5.3 Control panels

Most process compressors have associated with them some sort of control panel. This can be a master panel in the control room that monitors several of the critical parameters of the compressor, or it can be a dedicated panel standing adjacent to the compressor.

A dedicated panel will normally include everything required to control that compressor. It will have stop-start buttons for the main drive motor as well as switches to control electric lubricators, electric heaters, and auxiliary pumps for prelubrication or main pump backup. The panel may include various pressure or temperature gauges so that the operator can monitor the compressor and its processes from the panel rather than walking around the compressor. Main motor ampere meters may also be on this panel. One of the major features of a dedicated control panel is an annunciator that will be connected to the various switches on the compressor. As the switches send their signals, the annunciators will sound alarm horns and display what malfunction is occurring. As a shutdown signal is received, the annunciator will shut

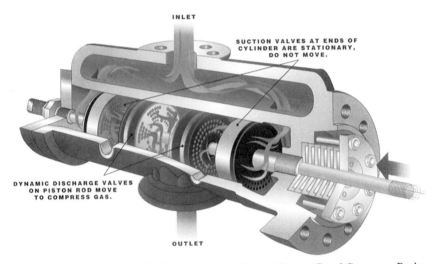

INLET

SUCTION VALVES AT ENDS OF
CYLINDER ARE STATIONARY,
DO NOT MOVE.

DYNAMIC DISCHARGE VALVES
ON PISTON ROD MOVE
TO COMPRESS GAS.

OUTLET

Figure 3.10 Valve-in-piston (VIP) compressor cylinder. *(Dresser-Rand Company, Broken Arrow, Okla.)*

the compressor down and then indicate what caused the shutdown. Compressor capacity can usually be selected from the panel. This can be done manually by turning a multiposition switch or pressing a series of buttons, or even a pair of raise or lower buttons. Operation at a particular compressor capacity step would be displayed on the panel. Capacity control could also be automated in the panel by means of suction or discharge pressure monitors and a logic system set up to load or

Figure 3.11 Conventional compressor frame with unconventional valve-in-piston (VIP) cylinder. *(Dresser-Rand Company, Broken Arrow, Okla.)*

unload the compressor as required to maintain a given suction or discharge pressure.

Panels often include programmable controllers and minicomputers. These can be used to control the loading of a compressor or handle many other decision-making tasks. One of the newer developments for control panels is using minicomputers for compressor diagnostics. Here, the computer monitors various parameters on the compressor and forms a database, or operating history. The computer can then look for combinations of values or rates of change in values that help identify an impending failure before it occurs. A typical example of this is the monitoring of valve temperatures.

Monitoring the operating temperature of valves through the use of thermocouples in the valve chambers makes it possible to record the temperature history of each individual valve throughout the operation of the compressor. If the computer notices an individual valve getting hotter than the average of the other valves or observes the rate of change in temperature increasing, it can alert the operator. This indication of impending valve failure allows the operator to schedule a maintenance shutdown instead of waiting for the valve to disintegrate and potentially damage the compressor.

3.5.4 Valve-in-piston reciprocating compressors

A rather interesting variation of the conventional reciprocating compressor is marketed by Dresser-Rand. Using frames in the 500- to 2000-kW size range, with speeds of 1800 r/min and strokes ranging from 3½ to 5 in (89 to 127 mm), *valve-in-piston* (VIP) cylinders, Figs. 3.10 and 3.11, are available for a number of services.

The gas compression principles in the VIP cylinder are very straightforward. Two inlet or suction valves are stationary and mounted directly in opposite ends of the cylinder bore. The discharge valves are dynamic and mounted on the piston rod. (Fig. 3.10).

As the discharge valves move toward the outer end, the frame end suction valve opens, allowing incoming gas to flow into the void created by the movement of the discharge valve. At the other end, the discharge valve opens as the gas is compressed against the outer end suction valve. Gas flow is direct and simple.

The valves used in the VIP compressor are the same mass-dampened, ported-plate PF-style valves used in Dresser-Rand gas field compressors. The VIP cylinder (Fig. 3.11) is a one-piece, cast, high-strength double-acting cylinder. Basically, this design eliminates the conventional piston because it is both a valve and a piston.

Labyrinth Piston Compressors

Labyrinth piston compressors represent a very important subset of non-lubricated reciprocating machines. Generally vertically oriented, they are typically configured as shown in Figs. 4.1 through 4.3. Virtually every one of the thousands of machines in service worldwide since 1935 has been manufactured by Sulzer-Burckhardt of Winterthur, Switzerland.

4.1 Main Design Features

The main design features are highlighted in Figs. 4.1 and 4.2. Labyrinth piston compressors do not use piston rings or rider bands. Unlike the oil-free reciprocating compressors of traditional design with dry-running piston rings, no friction occurs in the cylinder (1). The same applies, as a rule, for the piston rod stuffing box (4). Instead of piston rings, the labyrinth piston (2) is provided with a large number of grooves that generate a labyrinth-sealing action with regard to the cylinder wall. The cylinder wall is likewise provided with grooves that are, however, finer than those of the piston.

The piston moves within the cylinder with a clearance so that, even in the warm state, contact-free running is assured. Thermostats at the gas outlet detect any overheating that could lead to the piston scraping the cylinder wall. Thanks to this construction, lubrication of the cylinder and of the piston rod stuffing box is not necessary. Moreover, the suction and discharge valves are designed so that no lubrication is required.

The driving mechanisms of the compressors are usually lubricated by a gear pump (8) driven by the crankshaft. Depending on the requirement, compressors can be equipped additionally with an oil pump for preliminary lubrication (driven by an electric motor), an oil cooler, and

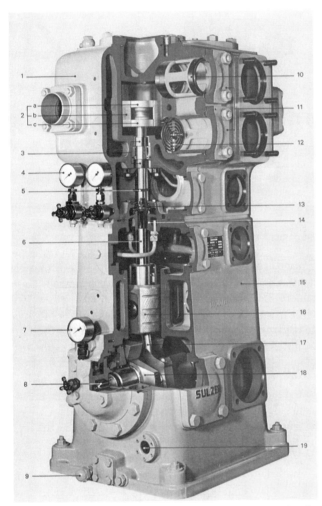

1   Cylinder block
2   Labyrinth piston
2a Upper piston crown
2b Piston skirt
2c Lower piston crown
3   Piston-rod gland
4   Pressure gauge with
    throttle valve
5   Piston-rod shield
6   Piston-rod
7   Oil pressure gauge
8   Geared oil pump
9   Oil discharge
10 Lantern
11 Valve cover
12 Plate valve
13 Oil scrapers
14 Piston-rod guide bearing
15 Frame
16 Crosshead
17 Connecting rod
18 Crankshaft
19 Oil sightglass

Figure 4.1 Labyrinth piston compressor. *(Sulzer-Burckhardt, Winterthur and Basel, Switzerland)*

a high-efficiency oil filter. The lubricating oil system supplies oil under pressure to the main bearings, the lower and upper connecting rod bearings, and the crosshead guide mechanisms. Splash oil lubrication is provided for the piston rod guide bearings (6).

When operating in the normal temperature range, both cylinder and crosshead guide mechanisms are water cooled. This applies also for the piston rod guide bearings. Where low suction temperatures apply, such as in refrigeration applications, cooling of these bearings can be dispensed with.

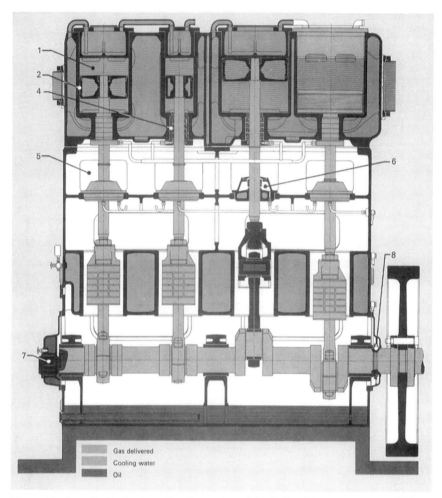

Figure 4.2 Longitudinal view of a three-stage labyrinth piston compressor. *(Sulzer-Burckhardt, Winterthur and Basel, Switzerland)*

The piston is guided from outside the compression space by the piston rod, which is located with relatively small clearance in the guide bearing (6), and by a precise guidance of the crosshead (7). The separation between the oil-free portion and the oil-lubricated crank drive is ensured by oil scrapers (6). To prevent the thin film of oil remaining on the piston rod from creeping upward along it, the rod is provided with an oil slinger. The distance between the drive mechanism and the stuffing box is greater than the length of the piston stroke, so that the oil-wetted portion of the piston rod cannot penetrate into the oil-free stuffing box.

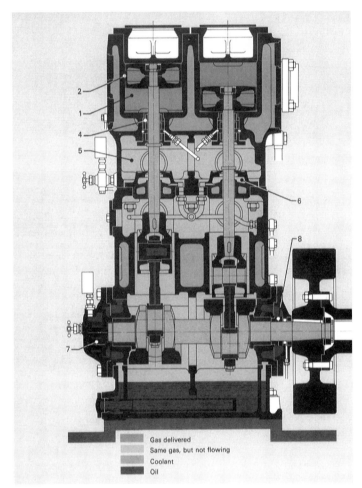

Figure 4.3 Longitudinal view of a single-stage labyrinth piston compressor with completely encapsulated and pressure-tight crankcase. *(Sulzer-Burckhardt, Winterthur and Basel, Switzerland)*

4.2 Energy Consumption

From time to time, the opinion is voiced that labyrinth piston compressors require more energy than dry runners with piston rings. Hence the question is raised as to whether the energy losses occurring in the labyrinths between piston and cylinder and between piston rod and stuffing box are, in fact, greater than those that take place at these points in dry runners because of friction and (as a matter of fact smaller) gas losses. Comprehensive tests have shown that an immediate loss of drive power occurs in the case of the ring-sealed piston because of the unavoidable mechanical friction of the piston rings.

Moreover, this results in unfavorable wall temperature influences on energy requirement and suction capacity. In the case of labyrinth pistons, where no mechanical friction occurs within the cylinder, any power losses at the piston are to be explained principally by leakage losses.

The experience gained from labyrinth piston machines indicates that for average to large volumes swept by the piston and for gases that are not extremely light, the energy losses due to leakage along the labyrinth are approximately equal, in some cases even smaller than those that occur because of friction, sealing leaks, and wall temperature influences in piston ring-type machines. However, the situation of the labyrinth piston, when light gases and small volumes swept by piston are concerned, is less favorable. As a consequence of the slight quantitative difference for air and other similar density gases, the comparison measurements have to be carried out very carefully and while accurately maintaining the same external conditions.

Two cases are described later where such comparisons were established by precise experiment. Labyrinth and plastic piston ring structures were made for two single-stage double-acting cylinder and piston sets. These were incorporated in a vertical single-throw standard crankshaft motion gear with a nominal speed of 750 r/min. Both cylinders had precisely the same dimensions. The only difference was in the surface configuration of the bores. Whereas one piston was of the standard labyrinth type, the other carried three plastic rings. In both cases the same compressor valves, pipelines, measuring instruments, and drive elements were used for the measurements. Differences in operating characteristics could thus be reliably established without disturbing side effects.

As is usual in the compressor sector, the comparison between the two test series was made using the efficiencies:

$$\eta_{adiabatic} = \frac{P_{adiabatic}}{P_{effective}} \tag{4.1}$$

$$\eta_{isothermic} = \frac{P_{isothermic}}{P_{effective}} \tag{4.2}$$

These values are shown in Fig. 4.4 as a function of the pressure ratio, determined for three different speeds of rotation.

4.3 Sealing Problems

Sealing questions are of interest as well. Logically, where a compressor is not fully encapsulated, the piston rod stuffing boxes have to seal the cylinder to the outside. Where these are oil lubricated, the lubricating

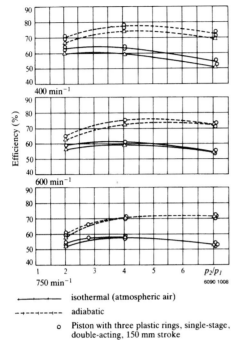

Efficiency (%)

400 min⁻¹

600 min⁻¹

750 min⁻¹

p_2/p_1

6090 1008

——————   isothermal (atmospheric air)

–·––·––·––   adiabatic

○   Piston with three plastic rings, single-stage, double-acting, 150 mm stroke

△   Labyrinth piston

p_1 Suction pressure 1 bar

p_2 Discharge pressure

Figure 4.4 Efficiency comparisons of conventional nonlubricated compressor and labyrinth piston machine. *(Sulzer-Burckhardt, Winterthur and Basel, Switzerland)*

oil acts as a sealing agent. Dry-running packing rings that rub on the piston rod are not as effective as seals, and friction-free labyrinth packings are even less so.

As far as most gases are concerned, sealing problems play a big role in the choice of the compressor design. The labyrinth stuffing box of the labyrinth piston compressor consists principally of a number of graphite rings; these are fitted to the piston rod with longitudinal and transverse clearance in the annular chamber. The rings are self-centering with regard to the piston rod. Labyrinth grooves on the inner surface provide the required sealing effect. Graphite is an ideal material for packing rings, since it has good dry-running characteristics, high chemical stability, low thermal expansion, and is not hygroscopic. Moreover, packing rings made from graphite cannot run hot.

One- or three-part designs of graphite rings are used. Three-part rings, whose constituent parts are held together radially by two garter springs, have the advantage that they can be replaced without the piston and the piston rod having to be pulled. In addition, they can be reworked if they become somewhat worn on the inner surface. The purpose of these springs is not, however, to press the three parts of the ring onto the piston rod, but to facilitate fitting and dismantling. These rings also have clearance with respect to the piston rod.

The advantages of the pure labyrinth stuffing boxes during operation are so convincing that the designer accepts the unavoidable losses via the labyrinth and deviates from the friction-free principle only in special cases. These labyrinth losses are taken up in the lower part of the stuffing box and returned to the suction side of the compressor, so that as far as the environment is concerned, no or only negligible gas losses occur. However, the higher the value of suction pressure above atmospheric, the greater the gas losses through the lowest sealing elements. To keep these losses as low as possible, the lowest ring can be designed as a sliding contact-sealing element. Such a sliding ring is in three parts, smooth on the inner surface, and pressed lightly against the piston rod by garter springs. Special arrangements have been developed for higher suction pressures.

A machine design as illustrated in Fig. 4.5 can be used for the compression of gases that are neither poisonous nor flammable, such as air, carbon dioxide, oxygen, and nitrogen. Slight leakage of such gases to the outside can be tolerated. However, this does not apply for helium

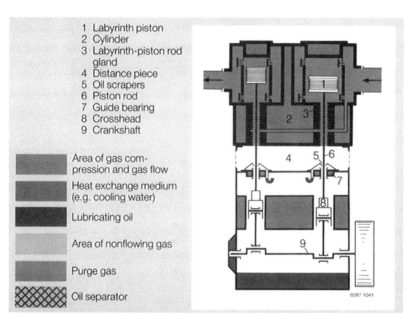

Figure 4.5 Labyrinth piston compressor with open distance piece and nonpressurized crankcase. Typically used for compression of gases, where a strict separation between cylinder and crankcase is essential and where process gas is permitted in the open distance piece (e.g., for O_2, N_2, CO_2, process air; generally in the industrial gas industry). *(Sulzer-Burckhardt, Winterthur and Basel, Switzerland)*

and argon. Although inert and nonpoisonous, their loss is not acceptable because of the high price.

Machines as illustrated in Fig. 4.6 are used for gases that are poisonous, flammable, and incompatible with lubricating oil. These have special stuffing boxes with gas sealing (mainly nitrogen), since a strict separation between crankcase and distance piece (4) is absolutely necessary. The adaptor is flushed out with scavenging gas, and the crank mechanism is filled either with air or with scavenging gas.

Completely encapsulated compressors (Figs. 4.7 and 4.8) are used for gases that are compatible with lubricating oil but have characteristics that do not allow even the smallest amount of loss to the outside, e.g., all hydrocarbons, carbon monoxide, hydrogen, helium, and argon. The machine shown in Fig. 4.8 is of special pressure-resisting design— the K type series. Originally, these machines were developed as refrigeration compressors (K is the abbreviation in German for refrigeration), so that they are also suitable for the compression of all refrigerants.

Attention has to be paid to the standstill pressure where closed refrigeration circuits are concerned. The crank mechanism of these compressors is thus designed for at least 15 bar internal pressure. For certain gases, however, the solubility in lubricating oil imposes a lower limit on the permissible internal pressure. The machines in Figs. 4.7 and 4.8 each feature mechanical seals. Unlike the pressure-resisting crankcase shown in Fig. 4.8, the crankcase in Fig. 4.7 can only accept a low internal pressure. In closed machines, the gaseous medium usually fills the crankcase, where it can mix with the oil mist.

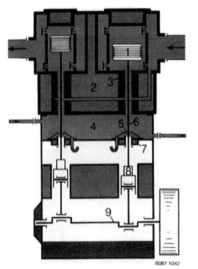

Figure 4.6 Labyrinth piston compressor with closed and purged distance piece. Used for compression of gases, where a strict separation between cylinder and crankcase is essential and where no process gas may leak to the surroundings or no ambient air may enter the distance piece (e.g., for weather protection). *(Sulzer-Burckhardt, Winterthur and Basel, Switzerland)*

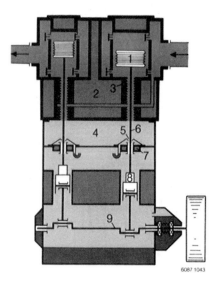

Figure 4.7 Labyrinth piston compressor with gastight crankcase and mechanical crankshaft seal. This design is used for compression of gases that are compatible with the lubricating oil (e.g., for hydrocarbon gases, CO, He, H_2, Ar) and where no process gas may leak to the surroundings. The suction pressure is limited by the design pressure of the crankcase. *(Sulzer-Burckhardt, Winterthur and Basel, Switzerland)*

In the non-gastight machines, there is no danger that oil penetrates into the oil-free zone of the compressor since the pressure in the crankcase is very low. However, where pressure-resisting machines are concerned, gas flow from the crankcase into the cylinder part has to be expected if the suction pressure decreases. Such a flow, nevertheless, passes through the oil separator shown in the center of the crankcase, Fig. 4.8, where the oil mist is retained. An external pressure balancing line (with molecular sieve) can be used, if required, to improve the sep-

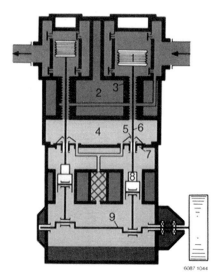

Figure 4.8 Labyrinth piston compressor with gastight and pressure-tight crankcase and mechanical crankshaft seal. Used to compress gases that are compatible with the lubricating oil and where no process gas may leak to the surroundings. Suction pressure may range between subatmospheric and crankcase design pressure. This machine finds its applications in closed cycles, for hydrocarbon gases, refrigerants, VCM, CO, N_2, CO_2, He, H_2, Ar, etc. *(Sulzer-Burckhardt, Winterthur and Basel, Switzerland)*

aration effect still further. Hence, this also can be regarded as an oil-free functioning configuration. Very low leakage rates can be achieved with the pressure-resisting machine design by virtue of a purpose-oriented construction, e.g., baseplate and frame as a one-piece casting, round frame openings with O-ring seals, in conjunction with an especially carefully accomplished casting process. At machine standstill, these leakage rates for helium are in the range of 10^{-3} to 10^{-4} cm^3/s. In many cases, even a leakage rate of 10^{-1} cm^3/s will meet the requirements. Special procedures have been developed to confirm such low leakage rates.

Labyrinth piston compressors are made in at least 40 frame sizes with one, two, three, four, and six cranks for piston strokes of 65 to 375 mm and matching cylinders for one-, two-, three-, and four-stage compression. Available suction capacities range from 20 to 11,000 m^3/h and discharge pressures of up to 300 bar (refer to Fig. 4.9). The permissible

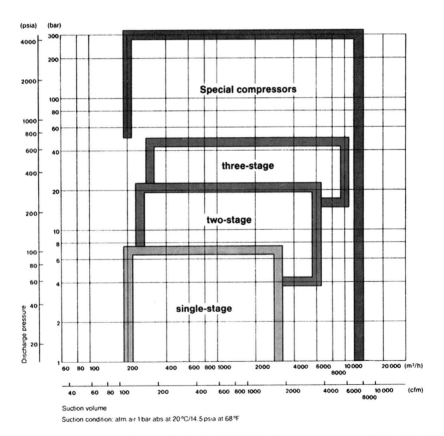

Figure 4.9 Typical application limits for labyrinth piston compressors. *(Sulzer-Burckhardt, Winterthur and Basel, Switzerland)*

power input for the driving mechanism ranges from 20 to somewhat more than 2000 kW. The loadability of the crank mechanism is, however, limited by the permissible loading of the piston rod.

Since no oil can penetrate into the cylinder and no temperature-sensitive materials are used, comparatively high-compression ratios and final temperatures up to more than 200°C are possible. This means that, in many cases, one compression stage less is required than would be necessary for compressor designs with plastic piston rings. It should be noted that the stage pressure ratios are often limited because of safety aspects, e.g., when compressing oxygen, and energy consumption considerations.

Hyper Compressors*

Intimately bound to the chemical industry and based on a long evolution, the main stages of which were the liquefaction of air and the synthesis of ammonia, the technique of using very high pressures was eventually perfected from developments in the manufacture of low-density polyethylene. This is now the only industry requiring large reciprocating compressors for very high pressures, because the pressures necessary for other main chemical processes have been steadily reduced since 1945. For this reason, the present chapter will be restricted to the ethylene compressors.

Considering that the classical designs of reciprocating high-pressure compressors cover an uninterrupted range up to about 1000 atm, very high pressures will imply those greater than 1000 atm.

5.1 Introduction

A characteristic feature of the high-pressure ethylene polymerization process is that a very large difference in pressure is necessary between the inlet gas entering the reactor and the outlet of the recycle gas. The recirculators, generally called secondary compressors (Fig. 5.1), work between two limits, that is, 100 to 300 atm on the suction side and 1500 to 3500 atm on the delivery side, for most of the existing processes. Because the coefficient of reaction lies between 16 and 30 percent, the secondary compressors have to handle 3 to 6 times the fresh gas quantity, being thus by far the most powerful machines in the production stream. Their unit capacities when the industrial expansion first began were of 4 to 5 ton/h; now they exceed 50 ton/h,

---

* Contributed by Sulzer-Burckhardt, Winterthur and Basel, Switzerland. Based on C. Matile's paper "Industrial Reciprocating Compressors for Very High Pressures."

and their power requirement per unit has been increased from 600 hp up to some 20,000 hp.

Because the whole operating range of these secondary compressors takes place well above the critical point of ethylene, the thermodynamic behavior of the fluid lies somewhere between that of a gas and that of a liquid. This peculiar condition has two main effects. The first is a very small reduction of the specific volume with increasing pressure; for instance, at a temperature of 25°C the specific volume is 3 dm³/kg at 100 atm, 2 dm³/kg at 700 atm, and 1.5 dm³/kg at 4500 atm. The second effect is a very moderate rise of adiabatic temperature with increasing pressure; for instance, with suction conditions of 200 atm and 20°C the delivery temperature will reach only 100°C at 2000 atm.

These particular thermodynamic conditions greatly influence the design of high-pressure ethylene compressors. Compared with the conventional reciprocating compressor, the compression ratio is of little practical significance; the important factor is the final compression temperature, which should not exceed 80 to 120°C, depending on process, gas purity, catalyst, etc. to avoid premature polymerization. The influence of the cylinder clearance on the volumetric efficiency is slight because of the small reduction of specific volume, and very high compression ratios are therefore possible with quite admissible efficiency. In addition, the stability of intermediate pressures depends chiefly on the accuracy of temperatures. For instance, in the case of two-stage compression from a suction pressure of 200 atm to a delivery pressure of 2500 atm, a drop in the first-stage suction temperature

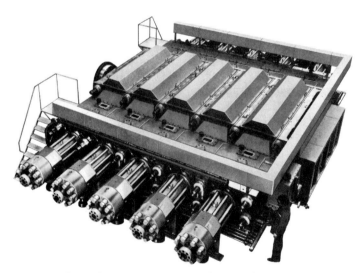

Figure 5.1 Large hyper compressor used in ethylene service. (*Sulzer-Burckhardt, Winterthur and Basel, Switzerland*)

from 40 to 20°C will cause the intermediate pressure to rise from 1000 to almost 1600 atm.

For these reasons, a secondary compressor that has only one or two stages is required, despite the very large pressure differences involved. However, this again compels the designer to face extremely high mechanical strains due to the high amplitude of pressure fluctuation in the cylinders. Finally, an additional and sometimes disturbing feature of ethylene must be mentioned. If the gas reaches a very high pressure and a high temperature simultaneously (which can easily occur in a blocked delivery port because of very low compressibility), it will decompose into carbon black and hydrogen in an exothermic reaction of explosive character.

5.2   Cylinders and Piston Seals

Sealing of the high-pressure compression chamber is a major problem that could be solved by avoiding friction between moving and stationary parts. This has been realized for laboratory equipment and small-scale pilot plants by the use of either metallic diaphragms or mercury sealants in U-tubes, and such arrangements are still in use for research purposes. In addition, they have the advantage of avoiding any contamination of the compressed gas by any lubricant. Unfortunately, chiefly for economic reasons, they proved to be impractical for industrial compressors, at least for the present state of techniques. Thus, because labyrinth seals are out of the question for very high pressures, friction seals have to be accepted; in fact, two solutions are currently used—moving and stationary seals.

Metallic piston rings are the only sort of moving seals used in the large high-pressure reciprocating type of compressor. They are generally made in three pieces: two sealing rings, each covering the slots of the other, and an expander ring behind both of them, which also seals the gaps in the radial direction. The materials used are special grade cast iron, bronze, or a combination of both, with cast iron or steel for the expander. The piston, of built-up design, comprises a series of supporting and intermediate rings with a guide ring on top of them and a throughbolt (two different designs are shown in Fig. 5.2). All parts of the piston are made of high tensile steel, and particular care must be given to the design and to the stress calculation of the central bolt, which is subjected to severe strain fluctuations.

The use of piston rings allows for a simple cylinder design, the main part of which is a liner that has been thermally shrunk to withstand the high variations of the internal pressure (see Figs. 5.3 and 5.4). The inner sleeve, which was previously made of nitrided steel, is now generally of massive sintered material like tungsten carbide. The use of

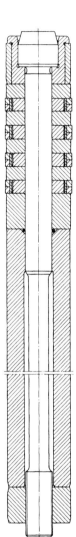

Figure 5.2 High-pressure pistons with piston rings. *(Sulzer-Burckhardt, Winterthur and Basel, Switzerland)*

this expensive material is justified by two beneficial qualities: it possesses an extremely hard surface and has a high modulus of elasticity. The first considerably improves the conditions of friction and greatly reduces the danger of seizure. The high modulus of elasticity of sintered tungsten carbide allows the amplitude of the breathing movement under the internal pressure fluctuation to be much smaller than with steel. The stress variations in the expanded outer sleeves are therefore appreciably reduced. However, because these sintered materials have a very poor tensile strength, care must be taken to ensure

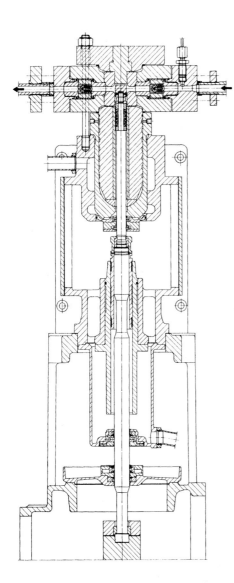

Figure 5.3 High-pressure cylinder for moderate end pressures. *(Sulzer-Burckhardt, Winterthur and Basel, Switzerland)*

that the inner sleeve is always under compression, even if the temperature increases. This is the main purpose of external cooling of the liner and not, as is usual, to dissipate the heat of compression.

Packed plungers are the other answer to piston sealing. Although some manufacturers still use packings of hard plastic materials (nylon or similar), the most widely used packings are the metallic self-adjusting type. They are usually assembled in pairs, the actual sealing ring tangentially split into three or six pieces being covered by a three-piece radially cut section. Both are usually made of bronze, kept closed by

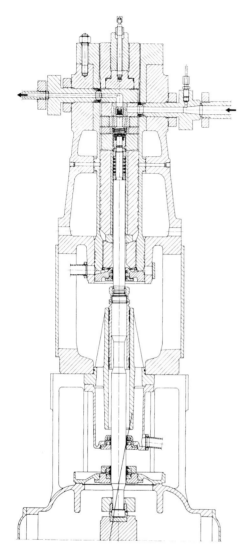

Figure 5.4 High-pressure cylinder for medium end pressures. *(Sulzer-Burckhardt, Winterthur and Basel, Switzerland)*

surrounding garter springs, and held in place by locating and supporting steel plates. These plates must also be thermally shrunk to resist the high variations in internal pressure. Unfortunately, the use of sintered hard materials is restricted by the fact that the supporting plates are subjected, in the axial direction, to heavy bending and shearing forces that these materials generally cannot stand. To improve the friction conditions of the packing rings, the high tensile steel supporting disks are frequently surface hardened or plated with carbide. The plungers are made of nitrided steel for use in moderate pressures, and for higher pressures are of steel, plated with hard materials. For very high pressures the use of solid bars of hard metal is the best wear-resistant solution for both plungers and packings. The disadvantage of the packed plunger design lies in the much larger joint diameters of the static cylinder parts, which require 2 to 3 times higher closing forces than the piston ring design. Large cylinders, such as the one shown in Fig. 5.5, require pretensioning of the cylinder bolts to about 10 times the maximum plunger load. This ratio is higher for smaller cylinders.

For piston rings and packed plungers the optimum number of sealing elements appears to be four or five. In both solutions, it is essential that the piston be accurately centered if the seals are to be effective; this is the reason for the guiding ring within the cylinder and for the additional guide at the connection between the piston and driving rod. At the base of the cylinders an additional low-pressure gland allows gas leaks to be collected and the plunger to be flushed and cooled. Other separate glands positioned on the rod connecting the piston to the drive (see Figs. 5.3 and 5.4) prevent the cylinder lubricant from mixing with the crankcase oil, and because the intermediate space is open to the atmosphere, it is impossible for gas to enter the working parts.

From the point of view of design and maintenance, piston rings would appear to be the most adequate solution, and they are currently used for pressures up to 2000 atm, or in some circumstances up to 3000 atm. The choice between them and the packed plungers depends largely on the process and type of lubricant used. One difficulty is that normal mineral oils are dissolved by ethylene under high pressure to such an extent that they have no longer any lubricating effect. The glycerine used in earlier machines has been widely replaced by paraffin oil, either pure or with wax additives, which is much less diluted by the gas than other mineral oils. However, it is a rather poor lubricant and is inferior to the various types of new synthetic lubricants, which are generally based on hydrocarbons. The basic difference between piston rings and plunger packings is that the latter may be lubricated by direct injection, while piston rings are lubricated indirectly. This may be an advantage since the low polymers carried by the return gas back from the reactor are reasonably good lubricants. However, too large an amount of low polymers causes the rings to stick in their grooves, and some kinds of catalyst carriers also brought back by the gas are excellent solvents for lubricants. Thus, the most convenient solution has to be selected for each specific case. In general, for higher delivery pressures (greater than 2000 to 2500 atm) better results are obtained with the use of packed plungers.

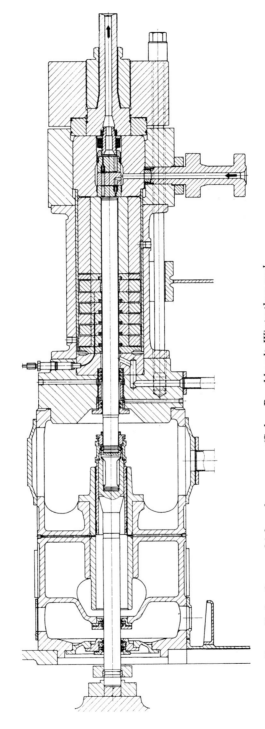

Figure 5.5 Gas cylinder for very high end pressures. *(Sulzer-Burckhardt, Winterthur and Basel, Switzerland)*

5.3 Cylinder Heads and Valves

It is relatively simple to construct a vessel that will resist 2000 atm, but the problem becomes more intricate when the vessel must withstand, for years, a pressure that fluctuates between 300 and 2000 atm at a frequency of 3 to 4 Hz. The leading idea of the designer must be to divide a complicated problem into a series of simple ones, each of which is then accessible to accurate methods of investigation. If this is done properly, it is possible to divide a large piece at the very places where inadmissible changes of stresses would occur and to keep the combined strains in each item within tolerable limits. The examples of cylinders shown in Figs. 5.3 through 5.5 illustrate the result of this method. The striking feature is the very simple shape of all pieces subjected to high pressure.

A first obvious result is that suction and delivery valves have to be located in a separate cylinder head. Figure 5.3 shows one type of cylinder head that can be used for moderate pressure fluctuations (up to amplitudes of about 1200 atm) and moderate cylinder dimensions. The intersection of the gas passages with the main bore is located in a small forged core, shrunk in a heavy outer flange, and pressed by the upper cover in the axial direction. This piece has a symmetrical shape with carefully rounded internal edges. By dismantling only the upper cover, it is possible with this design to pull out the complete piston with its rings through the central hole without disconnecting the gas pipes and without removing the valves. A typical valve for this kind of cylinder head is shown in Fig. 5.6a. The same valve is used on the suction and delivery side, the two end pieces being shaped differently to avoid incorrect assembly. The valve is held against the central head piece by the connection flange of the gas piping as a kind of composite lens.

For higher amplitudes of pressure and larger cylinders, cross bores and duct derivations must be taken away from the area of large pressure fluctuations. This is effected by the use of central valves, combining suction and delivery valves into one concentric set. For cylinders of moderate size this can be done as shown in Figs. 5.4 and 5.6b: the different valve elements are located in a succession of simply shaped disks, with the same diameter as the cylinder liner and piled up on top of it. The lower two disks, which have been produced by the shrinking technique, receive the pressure fluctuation in their central hole, while the upper two, which contain the radial bores for the gas connections, are subjected only to static pressure.

For still larger cylinders, the combined valve is assembled as a separate unit to keep compact dimensions and weights and is inserted into the central hole of the cylinder head core, as shown in Fig. 5.5. The two valves illustrated in Figs. 5.6c and d are designed on the same basic

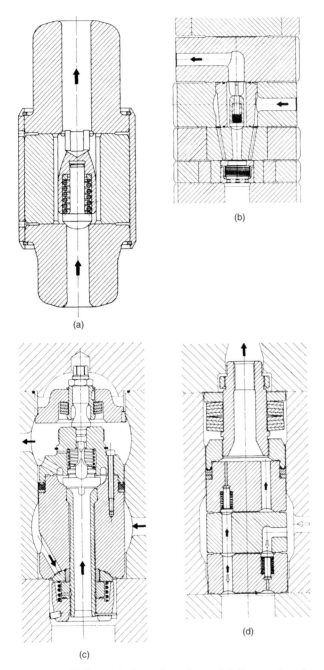

(a)

(b)

(c)

(d)

Figure 5.6   Different designs of suction and delivery valves for hyper compressors. *(Sulzer-Burckhardt, Winterthur and Basel, Switzerland)*

principle; the last one, used in very large cylinders, is fitted with multiple suction and delivery poppets to reduce the moving masses. The entire valve body is subjected to suction pressure on the outside and only to the pressure fluctuations in the longitudinal hole. The suction pipe is connected to the radial bore of the cylinder head core as shown in Fig. 5.5. Separation of suction and delivery pressures is ensured by the circumferential self-sealing ring of hard plastic material, as shown in Figs. 5.6c and d. The entire valve is pressed on the end of the cylinder liner by the difference of pressures: the set of plate springs visible in the figure has only to maintain the valve against pressure drop during periods of operation on bypass. The gas delivery pipe is connected radially to the core piece (like the suction one) for moderate delivery pressures and axially for higher pressures (as shown in Fig. 5.5).

All components subjected to high stresses, particularly the internal cylinder elements under high tridimensional fatigue strains, are generally investigated at the design stage by three different methods. The first is a conventional calculation of combined stresses, based on the classical hypotheses, using computer programs as far as convenient. The second approach is that of the frozen stress technique of photoelasticity applied on resin models cast either on full scale or on slightly reduced scale: it supplies accurate information about the course of the two main stresses in every plane section within the material. The third method is a direct measurement of the superficial stresses by means of strain gauges on the actual component subjected to the full prestressing and internal pressure. A variation of this last method consists of stress-measuring on an enlarged model made of a low-modulus material like aluminum: it provides better information through strain gauges on small rounded edges and allows progressive modification of such places in an attempt to reach an optimum. Comparison of results of these different methods gives a very useful reciprocal check on their exactness and accuracy.

5.4 Drive Mechanism

Different types of driving mechanism are diagrammatically illustrated in Fig. 5.7. The first two (Figs. 5.7a and b) have been extensively used during the initial period of development and are still applied to smaller units. They are characterized by the fact that the high-pressure cylinders have been fitted to frames of conventional design, without substantial modification of the existing equipment. Some manufacturers did not take into account the purely unilateral loading of the crosshead pins—and they generally got into trouble. Others tried to balance the forces by getting additional pistons set under constant or variable gas pressure in the reverse direction; this may work but is a rather unsatisfactory solution because it is expensive and introduces supplementary wearing elements. The best means of application is to use special high-pressure lubrication pumps, fastened to the crossheads and driven by the rocking movement of the connecting rods, which inject the oil directly into the crosshead bearings, thus lifting the pins against the load. This arrangement is well known from the design of large diesel engines, but because the requirements called for higher delivery pressures and larger capacities, the solutions shown in Figs. 5.7a and b appeared to be increasingly unsatisfactory. Since it is impos-

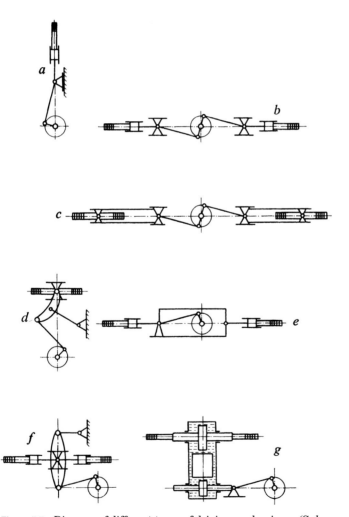

Figure 5.7   Diagram of different types of driving mechanisms. *(Sulzer-Burckhardt, Winterthur and Basel, Switzerland)*

sible to use double-acting pistons on very high pressures, these designs load the driving mechanism with the full gas pressure (instead of the difference between delivery and suction pressures) and are working only on each second stroke. While this was still admissible for small units, it proved to be uneconomical for larger ones, and there was obviously a need for more specialized constructions.

The widespread design represented in Fig. 5.7c is still based on a conventional application of the horizontally opposed reciprocating compressor, but it avoids the above difficulty by having, on each side of the frame, an external yoke that is rigidly connected to the main crosshead by means of solid connecting bars. A pair of opposed pistons (or plungers) is then coupled to each yoke, which is shaped as an outboard crosshead. Because of the long flexible connecting bars, the movement of the yoke is not disturbed by any transverse force, and it allows a full loading of the drive; thus, this is not a bad solution. However, the compressor is becoming extremely wide, and the accessibility of some of these high-pressure cylinders is rather poor.

All the other systems illustrated in the figure are specially designed solutions: Figs. 5.7d and f use a rocking beam bound to a fulcrum by a lever, which gives a linear translation of the rotary movement. Figure 5.7e uses a moving frame surrounding the crankshaft to connect the crosshead to the piston on the opposite side—a solution already applied for more than half a century to high-pressure pumps; and Fig. 5.7g is based on the idea of hydraulic transmission of the driving power. It should be noted that Fig. 5.7d may perform a modification of the stroke of the crankshaft in a fixed predetermined ratio, that Fig. 5.7f reduces the stroke in a fixed ratio, and that Fig. 5.7g can perform a variable reduction of the stroke. Although Fig. 5.7e appears to be the best specific design for a large production compressor, the very special solution of Fig. 5.7g is worth a further explanation.

Fig. 5.8 shows the basic, greatly simplified diagram of operation. By means of two reciprocating columns of fluid, a double-acting primary piston operates a secondary piston located above it. A pair of opposed high-pressure gas pistons are coupled to the latter. Although the hydraulic transmission of power could theoretically work as a closed system, it is actually necessary to renew the fluid continuously through forced-feed recirculation, both for the purpose of cooling and to compensate for possible seal leaks. Figure 5.8 shows a low-pressure feeding system; it has also been made as a high-pressure feed. Since this transmission may be built as a hydraulic intensifier, it is possible to use a comparatively light primary mechanism at rather high speed and to reduce the linear speed and increase the forces on the secondary part. Furthermore, by opening a bypass valve between the two fluid columns, the secondary stroke may be reduced. In this manner, stepless output control can be achieved down to zero. Because the fluid pressures on both sides of the pistons vary according to two opposed indicator diagrams, there are two points on each stroke where they will balance. If such a bypass is opened wide on the first of these points, the fluid will transfer theoretically without losses, and the secondary piston will stand still until the valve is closed. In fact, this is one of the very few ways of realizing power-saving capacity control of reciprocating compressors for very high pressures. This output control can be governed automatically, and applying it separately to each compression stage makes it possible to exactly control the intermediate pressure.

5.5 Miscellaneous

The number of problems posed by industrial reciprocating compressors for very high pressures is almost unlimited. Nearly every question of installation or maintenance needs a special study and an original answer, and all elements and accessories require special design and calculation. Only a few of them will be mentioned here.

It is common practice, when designing large reciprocating compressors, to take into account the three different kinds of strains for selecting the most favorable crank-angle arrangement: (1) the resulting forces and moments of inertia acting on the foundations, (2) the resulting torque diagrams under different conditions of operation (important for the cyclic variations of current consumption of the driving motor),

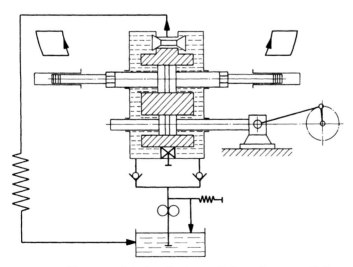

Figure 5.8 Diagram of hydraulic transmission of power. *(Sulzer-Burckhardt, Winterthur and Basel, Switzerland)*

and (3) the forces due to pressure pulsations in the gas piping. For most compressors working at lower pressures, this last consideration may be deleted or answered in a summary way at the initial stage of design, because it may be solved by the use of surge drums. In the case of very high pressures, the gas pulsations, which are capable of destroying the piping system, have to be given first priority in the basic investigations, even if it sometimes leads to acceptance of higher inertia forces.

The most practical ways of studying gas pulsations are to use either an analog computer, which is, in fact, an electroacoustic analogical system where every part is individually adjustable or replaceable or, alternatively, a digital computer study. The first purpose of the analysis is to avoid any resonance between the active systems (the compression cylinders) and the passive systems (the whole piping network); the second purpose is to reduce the amplitudes of the remaining pressure pulsations as far as possible. Theoretically, the means available are: (1) change of diameter and of length of the gas piping, (2) removal of pipe connections or adjunction of additional piping, and (3) use of pulsation snubbers and of orifices at well-selected places. In reality, the possibilities are restricted because of the high speed of sound in the gas (1000 to 2000 m/s), because of the very low compressibility of the gas, and because of the very high price of vessels and piping. However, in many cases it has proved easily possible to reduce a dangerous pulsation (for instance, from 25 percent down to 5 percent) by inexpensive and simple means.

Designers dealing with compressors for very high pressures need to keep in mind at least three basic ideas: (1) safety, (2) large forces (how to apply them), and (3) accessibility. The last two are, of course, chiefly economic, but they are often combined with the aim of safety. For instance, in the design of large compression cylinders, as shown in Fig. 5.5, the long throughbolts connecting the base with the cylinder head are an important safety factor. If, by chance, the gas decomposed in the cylinder, these long bolts, acting as springs, would be elastically lengthened by an appreciable amount without significantly increasing the stresses and would allow the gas to escape between the liner and the head. They must all be equally pretensioned with a very high force. If done by hand, this would be an extremely tiring and time-consuming exercise, and for this reason a hydraulic piston has been incorporated within the cylinder head, which allows, when set under oil pressure, a very quick, easy, and regular tightening and loosening of the bolts. After removal of the outer flange, the whole inside of the cylinder can be removed with the help of a lifting device. It resembles a closed cartridge, as shown in Fig. 5.9. The same figure also shows how major parts of the driving mechanism may be dismantled without removal of the cylinders.

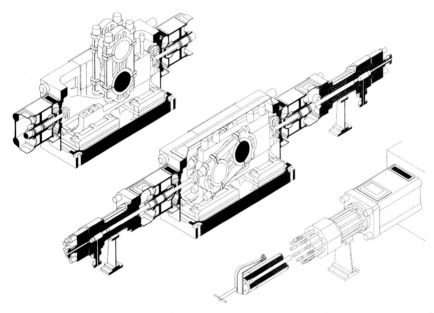

Figure 5.9 Large mechanically driven compressor for very high pressures (sectional views). *(Sulzer-Burckhardt, Winterthur and Basel, Switzerland)*

5.6 Conclusion

Although closely related to other reciprocating compressors, industrial compressors for very high pressures require the construction of a separate group of machines, different in many ways, and call for much greater research, development, and calculation than the others. Being compelled to employ all materials very near to their limits of resistance, designers are bound to keep in close contact with the latest developments in material science and many related technologies.

Metal Diaphragm Compressors*

6.1 Introduction

Gas compressors are mechanical devices that convert energy from one form to another. Energy conversion can be accomplished by the use of different types of machines, but the net result is the same. The pressure of the gas is increased, and therefore the energy level of the gas is increased. All compressors have an element that increases the energy level of the gas. It can take the form of a volume reduction element as in the case of positive displacement compressors or a velocity element as in the case of dynamic compressors.

Metal diaphragm compressors are positive displacement machines in which the compressing element is a metal diaphragm or diaphragm group. The displacing element is a piston having a reciprocating motion within a cylinder. The metal diaphragm reduces (compresses) the volume of the gas causing the gas pressure to be increased. Thermodynamically, this type of compression is considered flow-type work, and it is an adiabatic or polytropic process of a nonideal gas.

6.2 Terminology

Metal diaphragm compressors share many basic elements with positive displacement piston compressors. This section defines some terms unique to metal diaphragm compressors.

cavity A contour of either single or multiple radii machined in a flat plate or disk. Contours are machined in both the process and hydraulic cavity plates and are usually mirror images. The sum of the volume of these two cavities is the displacement of the diaphragm group.

---

* This segment developed and contributed by Pressure Products Industries, Inc., Warminster, Pa.

cavity plate   The plate in which the cavity is machined. Cavity plates are either *process* or *hydraulic*. Process cavity plates also contain cooling passages for removing the heat of compression.

clearance volume (dead volume)   The volume present in one compressor cylinder or head in excess of the net volume displaced by the piston or diaphragm during the stroke.

diaphragm   A thin, metal membrane that isolates the gas from the hydraulic fluid.

diaphragm group   A group of three metal diaphragms that isolates the gas from the hydraulic fluid. This group consists of a process, middle, and hydraulic diaphragm. The middle diaphragm contains either grooves or slots radiating from the center of the diaphragm to the circumference of the diaphragm to channel gas or hydraulic fluid to the leak detection groove in the event of a diaphragm failure.

displacement   The net volume displaced by the piston or diaphragm at the rated machine speed, generally expressed as a volumetric flow rate. For a single-stage compressor, it is only the displacement of the compressing end. For multistage compressors, it is the displacement of the first stage.

head assembly   The metal diaphragm compressor subassembly containing the cavity plates, support heads, diaphragms, process check valves, O-ring seal sets, and hydraulic piston. Different types of head assemblies are used, depending on the pressure and size of the cavity.

hydraulic injection pump   A pump attached to an extension of the power frame crankshaft that injects a measured amount of hydraulic fluid into the hydraulic pulsing system at a predetermined time during the suction portion of the compressor cycle.

lock ring   A threaded member designed to retain the forces (pressure and seal loads) in the metal diaphragm compressor inserted head assembly. The lock ring contains thrust bolts that provide the preload on the head assembly seals.

lower head   A structural element in which the hydraulic cavity plate sits and the hydraulic piston reciprocates. The part is used on high-pressure diaphragm compressors with intensifier assemblies.

main nut   A threaded member designed to retain the forces (pressure and seal loads) in the metal diaphragm compressor intensifier head assembly.

pressure limiter   The device that is used to control hydraulic system volume and pressure so that volumetric efficiency of the metal diaphragm compressor is maximized. *It is not a safety device to protect the compressor or the process system.*

stuffing box   The main structural member used in inserted and intensifier head assemblies.

support head   A structural element similar to a flange that supports either the process cavity plate or the hydraulic cavity plate. The upper support head supports the process cavity plate; the lower support head supports the hydraulic cavity plate.

6.3 Description

A metal diaphragm compressor, Fig. 6.1, is a positive displacement compressor. Gases are isolated from the reciprocating and hydraulic parts of the compressor by three flexible, thin, metal disks called *diaphragms*. The motion of the reciprocating piston is transmitted to the diaphragms by the hydraulic fluid. This motion causes the diaphragms to move into the process cavity, thereby reducing the volume and increasing the gas pressure.

The compression cycle of the metal diaphragm compressor is not unlike the positive displacement piston compressor. Both use a reciprocating piston to convert mechanical energy to flow work in the gas. Both use spring-loaded check valves that open only when the proper differential pressure exits across the valve. In each design, the clearance volume (dead volume) influences the volumetric efficiency of the compressor. However, diaphragm compressors differ in the way the compression cycle is managed, although a pV diagram for the two types of compressors would be virtually identical.

The pV diagram for positive displacement piston compressors was shown earlier in Figs. 1.1 through 1.5. Point 1 is the *start of compression,* and the cylinder is filled with gas at the suction pressure. The check valves are closed and the piston is at bottom dead center of its stroke.

Figure 6.1 Metal diaphragm compressor. *(Pressure Products Industries, Inc., Warminster, Pa.)*

On the *compression* portion of the stroke, the piston has moved, reducing the volume in the cylinder with an accompanying rise in pressure (point 1 to point 2). The valves remain closed and the cylinder pressure has reached the upstream piping pressure.

Point 2 to point 3 is the *discharge* or *delivery* portion of the stroke. Compressed gas is flowing out of the discharge check valve and into the discharge piping. When the piston reaches point 3, the discharge valve will close. The piston is at top dead center of its stroke. Gas at pressure P_2 is still in the cylinder.

From point 3 to point 4, the piston is in reversal, the suction and discharge valves remain closed, and the gas trapped in the clearance volume begins to expand, resulting in a pressure reduction. This is the *expansion* portion of the cycle.

The cylinder pressure eventually drops below the suction pressure, P_1 at point 4. The suction valve will then open and gas will flow into the cylinder until the piston reaches the reversal point of its stroke, point 1.

The pV diagram of a metal diaphragm compressor is identical to a piston compressor for the *gas compression cycle*. Differences occur during the compression cycle of the hydraulic fluid. The hydraulic fluid compression cycle, often referred to as the *mechanical compression cycle,* accounts for all pressure changes in a metal diaphragm compressor. A graph of the mechanical compression cycle for a metal diaphragm compressor is shown in Fig. 6.2.

The mechanical compression cycle shown in Fig. 6.2 traces the hydraulic system pressure from the process suction pressure to the process discharge pressure and then to the hydraulic pressure limiter setting and back to the process suction pressure.

Starting at 0° of compressor crankshaft rotation (the reciprocating piston is at bottom dead center), the diaphragm group is fully deflected into the hydraulic cavity plate (Fig. 6.3). The metal diaphragm compressor head assembly is filled with gas at the suction pressure. The check valves are closed. This compares to point 1 on the pV diagram (Fig. 1.5).

On the *compression* portion of the stroke, the hydraulic piston moves from bottom dead center, compressing the hydraulic fluid and forcing the diaphragm group into the cavity in the process cavity plate (Fig. 6.4). Gas volume in the process cavity plate is reduced with an accompanying rise in pressure. The valves remain closed until the process cavity pressure reaches the upstream piping pressure. This compares to points 1 and 2 on the pV diagram.

Compressed gas is flowing out of the discharge check valve and into the discharge piping during the *discharge* portion of the stroke. When the diaphragm group is fully deflected or displaced into the process cav-

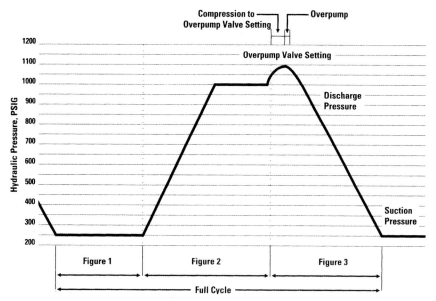

Figure 6.2 Mechanical compression cycle for a metal diaphragm compressor. *(Pressure Products Industries, Inc., Warminster, Pa.)*

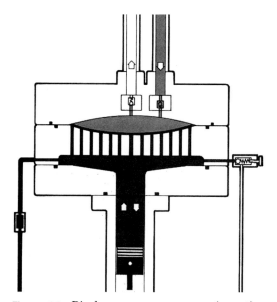

Figure 6.3 Diaphragm compressor reciprocating fluid piston at 0° position. The hydraulic piston is at bottom dead center. The hydraulic system has just been filled with fluid by a single stroke of the automatic injection pump. The process gas, entering through the inlet check valve at suction pressure, has moved the diaphragm group to the bottom of the cavity. The cavity is now filled with the process gas. *(Pressure Products Industries, Inc., Warminster, Pa.)*

ity plate, the discharge check valve will close. This compares to points 2 and 3 on the *pV* diagram. Gas at pressure P_2 is still in the cylinder.

Differences occur between the metal diaphragm compressor and the positive displacement piston compressor at this point in the cycle. The positive displacement piston compressor would now start its reversal and go into the expansion portion of the cycle. On the other hand, the metal diaphragm compressor hydraulic piston still has a distance to travel since the volume of the hydraulic system is slightly greater than the volume of the process system (Fig. 6.5). The hydraulic system has received extra volume during the suction portion of its stroke from the *hydraulic system injection pump*. This extra volume is required to ensure that the diaphragm group is fully deflected or displaced into the process cavity plate. Without this extra volume, the diaphragm group

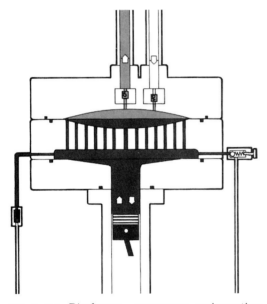

Figure 6.4 Diaphragm compressor reciprocating fluid piston advancing. As the crankshaft rotates, the piston moves from bottom to top dead center, and the hydraulic pressure increases. As the hydraulic pressure reaches the pressure level of the process gas in the cavity, the diaphragm group moves toward the top of the cavity, compressing the gas. When the pressure of the process gas within the cavity reaches the pressure level downstream of the discharge check valve, the check valve opens and the gas is discharged. Pressure in the hydraulic system continues to increase, which moves the diaphragm group completely through the cavity, thereby ensuring maximum gas displacement and efficiency. *(Pressure Products Industries, Inc., Warminster, Pa.)*

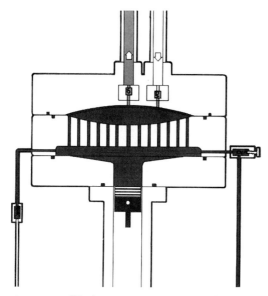

Figure 6.5 Diaphragm compressor reciprocating fluid piston at limit of stroke. When the diaphragm group has completely moved through the cavity, the piston must travel still further to reach its top dead center position. As this takes place, hydraulic fluid is forced through the hydraulic overpump valve, which is set at a pressure level higher than the desired discharge process pressure. The compression portion of the cycle is now completed, and the piston begins to move toward bottom dead center. As the hydraulic piston moves toward bottom dead center, the expansion of residual gas combined with gas entering the cavity at suction pressure deflects the diaphragm group toward the bottom of the cavity, and the cycle is complete. *(Pressure Products Industries, Inc., Warminster, Pa.)*

would never attain full deflection or displacement and therefore would not reach maximum discharge pressure. The volumetric efficiency of the compressor would be reduced because of an increase in the clearance volume. The extra volume is discharged through the hydraulic pressure limiter once the hydraulic system reaches the pressure limiter setting. The extra volume discharged through the hydraulic pressure limiter is called *overpump*.

The expansion cycle of the metal diaphragm compressor begins once the hydraulic pressure limiter has closed and the hydraulic piston has started its reversal. The suction and discharge valves remain closed, and the gas trapped in the clearance volume begins to expand, resulting in a pressure reduction. This compares to points 3 and 4 on the pV diagram.

The cavity pressure eventually drops below the suction pressure. The suction valve will then open, and gas will flow into the process cavity until the diaphragm group reaches its maximum deflection in the hydraulic cavity plate. It is during this phase of the cycle that the hydraulic injection pump will add the extra volume that will eventually become the *overpump* at the end of the discharge portion of the cycle.

The power requirements of a metal diaphragm, it is important to note, are not based solely on the work imparted to the gas. The mechan-

Flanged and Bolted Head Design

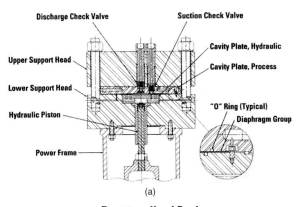

(a)

Bootstrap Head Design

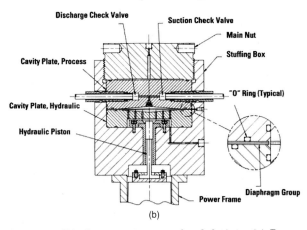

(b)

Figure 6.6 Diaphragm compressor head designs. (*a*) Represents the simplest and most commonly used design; this is limited to pressures of 5000 psi (345 bar) and below. (*b*) Provides a positive seal with a simple, low-torque closure. This design is used when the combination of pressure and diameter make it the most efficient closure. (*Pressure Products Industries, Inc., Warminster, Pa.*)

ical energy required during the mechanical compression cycle and the thermodynamic work of the gas compression cycle must both be considered to determine the power requirements of the metal diaphragm compressor.

Figure 6.6 illustrates the head components of a typical metal diaphragm compressor. The head assembly consists of the upper head (process cavity plate), diaphragm group, lower head (hydraulic cavity plate) and lower support head. Not shown is the power frame, which would include hydraulic piston, hydraulic pressure limiter, hydraulic injection pump, and suction and discharge check valves. Triple diaphragm construction and the leak detection port are shown enlarged in Fig. 6.7. Note the static O-ring seals for the process side, hydraulic side, and secondary containment seal.

Diaphragm compressors compress gas with no contamination and virtually no leakage. More specifically, leakage is less than 1×10^{-5} standard mL/s helium with O-ring seals. Rates of less than 1×10^{-7} can be obtained with metal-to-metal seals.

Under normal operating conditions, the gas is completely isolated from the hydraulic fluid by the diaphragm group. This permits toxic, flammable, pure, or expensive gas to be compressed safely, without contamination or leakage. Triple diaphragm construction ensures product purity since the three diaphragms will continue to isolate the gas from the hydraulic fluid, even under abnormal conditions such as a diaphragm or seal failure. The leak detection system retains any effluent during abnormal conditions. The system senses any diaphragm or

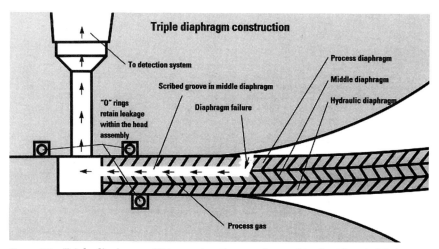

Figure 6.7 Triple diaphragm. This construction can be combined with suitable instrumentation to rapidly detect diaphragm failure. *(Pressure Products Industries, Inc., Warminster, Pa.)*

seal abnormality and gives the operator a visual indication and shut-down or alarm capabilities by use of the system pressure switch.

Metal diaphragm compressors are usually available in single-stage and two-stage models. Most compressor displacements range from 0.032 to 110.8 cfm (0.054 to 188.27 m^3/h) based on an operating speed of 400 r/min. Standard discharge pressures range from 25 to 30,000 psi (172 kPa to 207 MPa).

Lobe-Type and Sliding Vane Compressors

7.1 Descriptions

Lobe-type* or rotary positive blowers, also called *rotary piston machines* or *gas pumps* are intended for use with steam and noncorrosive gases. Basic models, Figs. 7.1 and 7.2, are usually designed with integral shaft ductile iron impellers (Fig. 7.3) having an involute profile. The alloy steel timing gears are taper mounted on the shafts, and cylindrical roller bearings are generally used. Both ends of the unit are splash oil lubricated. The casing, headplates, gear cover, and end cover are typically made of gray cast iron. Piston ring seals form a labyrinth between the compression chamber and cored vent cavities. The vent cavities are valved for purge or drain.

On many modern lobe-type machines, high-performance mechanical seals are installed at each bearing to control gas and oil leakage and are suitable for vacuum or pressure service (Fig. 7.4). Some models of lobe-type blowers or gas pumps incorporate a proprietary design that reduces noise and power loss by using an exclusive wraparound flange and jet to control pressure equalization—eliminating rapid backflow of gas into the pump from the discharge area.

The operating principle of lobe-type machines is illustrated in Fig. 7.5. Incoming gas (right) is trapped by impellers. Simultaneously, pressurized gas (left) is being discharged (*a*).

As the lower impeller passes the wraparound flange, a portion of the gas (white arrow) equalizes pressure between trapped gas and discharge area, thus aiding impeller movement and reducing power (*b*).

---

* Based on information provided by Aerzen USA, Coatesville, Pa. and Dresser Industries, Inc. Roots Division, Connersville, Ind.

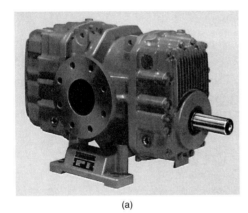

(a)

(b)

(c)

Figure 7.1   Typical small-to-moderate size lobe-type blowers. *(Aerzen USA Company, Coatesville, Pa.)*

(a)

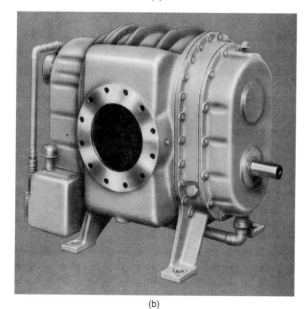

(b)

Figure 7.2 Basic lobe-type blowers. *(Dresser Industries, Inc., Roots Division, Connersville, Ind.)*

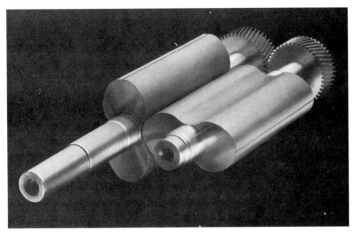

Figure 7.3 Integral shaft ductile iron rotors and impellers for lobe-type blowers. *(Aerzen USA Company, Coatesville, Pa.)*

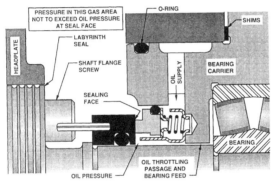

Figure 7.4 Mechanical seal installed on modern lobe-type blower. *(Dresser Industries, Inc., Roots Division, Connersville, Ind.)*

Incoming gas (right) is trapped by impellers. Simultaneously, pressurized gas (left) is being discharged.

As lower impeller passes wraparound flange, Whispair jet (white arrow) equalizes pressure between trapped gas and discharge air, aiding impeller movement and reducing power.

Impellers move gas into discharge area (left). Backflow is controlled, resulting in reduction of noise relative to conventional gas pumps.

(a)   (b)   (c)

Figure 7.5 Operating principle of lobe-type blowers. *(Dresser Industries, Inc., Roots Division, Connersville, Ind.)*

The impellers now move gas into the discharge area (left). Backflow is controlled, resulting in reduction of noise relative to conventional gas pumps (c).

Main fields of application for rotary piston blowers, or lobe-type machines, are pneumatic conveying plants for bulk materials in vacuum and pressure operating systems. The smallest blowers are mounted on bulk-carrying vehicles; the largest machines (Fig. 7.6) are used in pneumatic elevators for unloading of vessels. The hourly output of these plants is up to 1000 tons. Another frequent application is in aeration ponds of sewage treatment plants. Other lobe-type machines are found in power plants or facilities requiring high-pressure gas circulation with pressure-tight machines up to a maximum of 25 bar internal pressure. Typical pressure rise capabilities are 12 psi (0.8 atm).

A large variety of sizes and models cover the capacity range from 30 to 85,000 m³/h (approximately 18 to 50,000 cfm). Drivers include electric motors, internal combustion engines, and hydraulic motors.

Figure 7.6 Large lobe-type rotary piston blower. *(Aerzen USA Company, Coatesville, Pa.)*

Figure 7.7   Sliding vane compressor and principal components: rotor and shaft (1), bearings (2), blades (3), mechanical seals (4), cylinder and housing (5), heads and covers (6), gaskets (7), lube supply line (8), coupling (9). *(A-C Compressor Corporation, Appleton, Wis.)*

Sliding vane compressors* (Fig. 7.7) are typically found in such applications as air blast hole drilling, pneumatic conveying, chemical and petroleum vapor recovery, gas transmission, and small plant air systems. Each unit has a rotor eccentrically mounted inside a water-jacketed cylinder. The rotor is fitted with blades that are free to move radially in and out of longitudinal slots. These blades are forced out against the cylinder wall by centrifugal force. Figure 7.8 illustrates how individual cells are thus formed by the blades, and the air or gas inside these cells is compressed as the rotor turns.

---

* Based on information provided by A-C Compressor Corporation, Appleton, Wis.

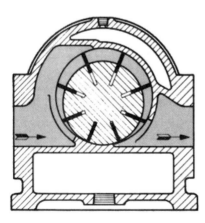

Figure 7.8   Operating principle of sliding vane compressor. *(A-C Compressor Corporation, Appleton, Wis.)*

Sliding vane compressors are available in single- and multistage geometries. Typical single-stage capacities are ranging through 3200 cfm and 50 psig; two-stage compressors deliver pressures from 60 to 150 psig and flows up to approximately 1800 cfm.

Liquid Ring Compressors

Liquid ring compressors (Fig. 8.1) represent a subgroup of the two major compressor categories, dynamic and displacement. Since these machines use a liquid to displace gases, they are often classified as volumetric compressors with liquid displacers. Although considerably larger machines have been produced, the overwhelming majority fit in the size range where 15 to 150 kW drivers are needed to compress gases to about 100 psig, or approximately 7 bar discharge pressure (Fig. 8.2).

In general, liquid ring compressors are the functional equivalent of liquid ring pumps. The principal difference is double-lobe construction in a compressor, which balances radial forces imposed on the rotor (Figs. 8.3 and 8.4). Liquid ring compressors tolerate carryover, as incoming soft solids and liquids are cushioned by the seal liquid and washed through to discharge. Liquid ring air compressors sealed with water actually scrub out particles as small as airborne bacteria with high efficiencies.

That portion of the seal liquid that passes on through the pump is removed from the discharge stream by a separator, which the compressor manufacturer furnishes as part of the system. A continuous supply of makeup seal liquid maintains the rotating ring. Liquid from the separator is usually cooled and recirculated to provide this makeup supply.

Heat of compression raises seal liquid temperature only about 10 to 15°F during its passage from makeup to discharge, and temperatures inside a liquid ring compressor remain far below the peaks that adiabatic compression would produce. If the gas mixture is explosive, the liquid ring compressor can serve as a flame arrester when it is sealed with a nonflammable liquid, such as water.

If the inlet gas mixture contains vapors that will condense at seal liquid temperature, a capacity bonus is obtained. Vapor condensation

Figure 8.1 Liquid ring compressor. *(Nash Engineering Company, Norwalk, Conn.)*

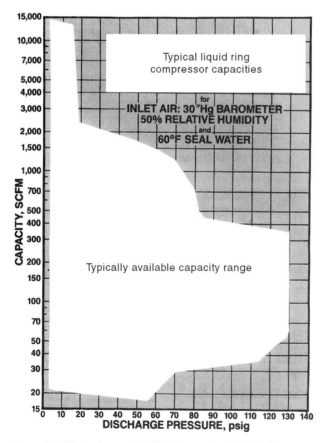

Figure 8.2 Typical capacity field for modern liquid ring compressors. *(Nash Engineering Company, Norwalk, Conn.)*

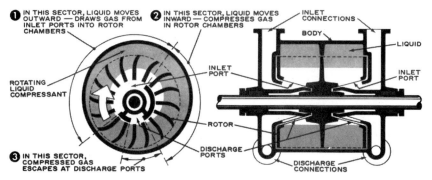

Figure 8.3 Functional schematic of liquid ring compressor with circular casing. *(Nash Engineering Company, Norwalk, Conn.)*

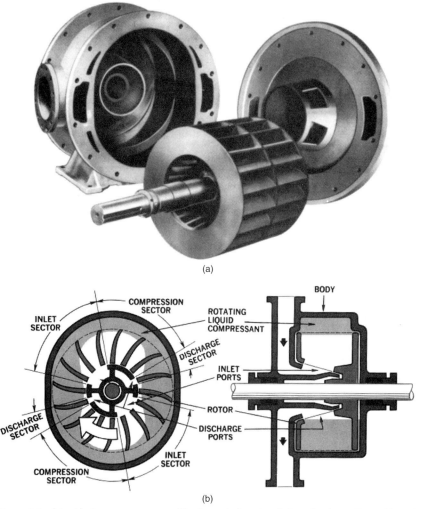

Figure 8.4 Liquid ring compressor with elongated casing (*a*), and schematic section at inlet and discharge sectors (*b*). *(Nash Engineering Company, Norwalk, Conn.)*

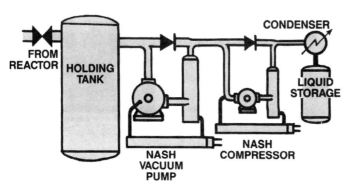

Figure 8.5 Vinyl chloride recovery system using liquid ring equipment. *(Nash Engineering Company, Norwalk, Conn.)*

reduces the volume that the pump or compressor must handle. Condensate flows out along with discharged seal liquid, and accumulated excess liquid is then drawn out of the separator system.

Water is an excellent seal liquid and is most often used for its convenience. There are many situations, though, in which some other liquid yields important advantages. Some products cannot tolerate even trace amounts of water. If condensate recovery is desired, that same liquid, or one compatible with it, may be used as the seal liquid. Sometimes, a liquid that inhibits the attack of a corrosive gas mixture will improve equipment service life or make costly construction unnecessary. If low-temperature cooling is not available, the seal liquid's vapor pressure may limit attainable vacuum. In that case, an oil seal or an engineered synthetic fluid can eliminate the need for cooling water and will extend the vacuum range.

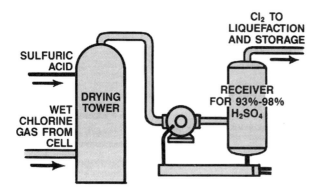

Figure 8.6 Corrosive gas compression system using liquid ring compressor. *(Nash Engineering Company, Norwalk, Conn.)*

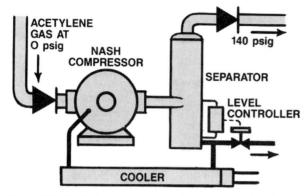

Figure 8.7 Explosive gas compression system using liquid ring compressor accommodates 10:1 compression ratio in a single stage. *(Nash Engineering Company, Norwalk, Conn.)*

Common, but greatly simplified arrangement drawings are shown in Figs. 8.5 through 8.7. Figure 8.5 represents a monomer recovery system. In one of several batch monomer recovery systems, unreacted vinyl chloride (PVC) is first transferred into the evacuated holding tank. A liquid ring vacuum pump scavenges gas out of the PVC and delivers it to the compressor inlet at or near atmospheric pressure. The single-stage compressor then compresses the gas for condensation and storage as a pressurized liquid.

A corrosive gas compression system is depicted in Fig. 8.6. This layout and equipment choice minimizes the attack of corrosive gases on compressors and vacuum pumps, often without resorting to the use of expensive construction materials. Selecting appropriate seal liquids plays a part in this success. As one of many examples, concentrated sulfuric acid is used as the seal liquid in many liquid ring compressors handling chlorine gas. Another approach is used with dry hydrogen chloride gas, which is compressed with an oil seal. Water is used to seal stainless steel compressors handling carbon dioxide, sulfur dioxide, or hydrogen sulfide.

Typical of arrangements for handling dangerously explosive gases, the compressor system shown in Fig. 8.7 keeps acetylene cool and saturated with water, which is used as the seal liquid. Tests with intentionally propagated explosions both upstream and downstream of liquid ring compressors confirm that they can function effectively as flame arresters. A single-stage liquid ring compressor can handle a discharge pressure as high as 140 psig (9.5 bar). It should be noted that multistage compression is available for higher discharge pressures.

Rotary Screw Compressors

9.1 Twin-Screw Machines

Rotary screw compressors are typically configured as shown in Figs. 9.1 through 9.4. Two counterrotating helical screws are arranged in a compressor casing; gas inlet and discharge nozzles are at opposite ends. Three-, four-, and five-lobe rotors are produced (Fig. 9.5).

9.1.1 Working phases

The screw compressor is a positive displacement machine and as such has distinct working phases: suction, compression, and discharge. We will limit our description of the working phases to just one lobe of the male rotor and one interlobe space in the female rotor. Once the operation is understood, it is not particularly difficult to envision the relative interaction of all of the lobes and interlobe spaces with resulting uniform, basically nonpulsating, continuous gas flow through the compressor.

The suction phase is depicted in Fig. 9.6a. As the lobe of the male rotor begins to unmesh from an interlobe space in the female rotor, a void is created, and gas is drawn in through the inlet port. As the rotors continue to turn, the interlobe space increases in size, and gas flows continuously into the interlobe space. The inlet port is large, and the filling takes place over a large portion of each rotation. Just prior to the point at which the interlobe space leaves the inlet port of the suction end, the entire length of the interlobe space is open from end to end— the lobes and interlobe space being completely unmeshed. The interlobe space is thus completely filled with drawn-in gas.

The transfer phase is a transitional phase between suction and compression where the trapped pocket of gas within the interlobe space is

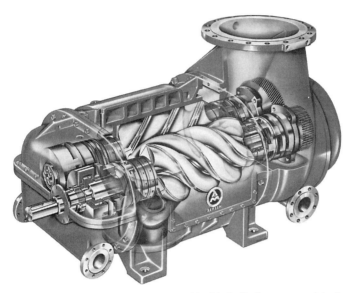

Figure 9.1 Rotary screw compressor (double-helical screw machine). *(Aerzen USA Company, Coatesville, Pa.)*

Figure 9.2 Small, packaged rotary screw compressor. *(Aerzen USA Company, Coatesville, Pa.)*

Figure 9.3 Medium size rotary screw compressor package. *(Aerzen USA Company, Coatesville, Pa.)*

isolated from inlet and outlet ports and is merely transported radially through a fixed number of degrees of angular rotation at constant suction pressure.

Figure 9.6*b* shows the compression phase. As can be seen, further rotation meshes a male lobe (not the same lobe as previously disen-

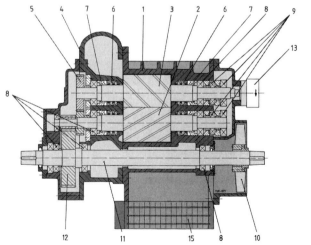

1	housing
2	male rotor
3	female rotor
4	intake side plate
5	timing gears
6	carbon ring-shaft sealing
7	oil sealing
8	radial bearing
9	axial bearing
10	ventilation fan
11	driving shaft
12	step-up gear
13	oil pump
15	oil cooler

Figure 9.4 Oil-free rotary screw compressor with integral step-up gears. *(Aerzen USA Company, Coatesville, Pa.)*

gaged because of the four to six relationship) with the gas-filled interlobe space on the suction end and compresses the gas in the direction
of the discharge port. The meshing point moves axially from the inlet to
the discharge end; thus, the occupied volume of the trapped gas within
the interlobe space is decreased, and the gas pressure consequently
increased.

The discharge phase is illustrated in Fig. 9.6c. At a point determined
by the designed built-in compression ratio, the outlet port is uncovered,
and the compressed gas is discharged by further meshing of the lobe
and interlobe space. While the meshing point of a pair of lobes is moving axially, the next charge is being drawn into the unmeshed portion,
and thus the working phases of the compressor cycle are repeated.

9.1.2   Areas of application

Rotary screw compressors have been around for many decades and are
very likely the equipment of choice for either oil-free or oil-wetted compression of air in mining, construction, industrial refrigeration, and a
host of other applications where their relative simplicity, general reliability, and high availability are appreciated.

What is less well known is that rotary screw machines are equally
suited to compress such process gases as ammonia, argon, ethylene,
acetylene, butadiene, chlorine, hydrochloric gas, natural and synthetic
pipeline gases, flare gas mixtures, blast furnace gas, swamp and
biomass gases, coke oven or coal gas, carbon monoxide, town gas,
methane, propane, propylene, flue gas, crude or raw gas, sulphur dioxide, nitrous oxide, vinyl chloride, styrene, and hydrogen.

Modern sealing and liquid injection technology has been partly
responsible for making rotary screw units capable of competing in
applications previously reserved for other compressor types. And as a
result of sophisticated contour machining and enhanced metallurgy,
single- or multistage rotary screw compressors today cover a range of
suction volumes from 300 to 60,000 std m³/h (176 to 35,310 scfm), with
discharge pressures up to 40 bars (580 psi). For vacuum applications,
an absolute pressure of 0.09 bar (1.3 psia) is achievable.

9.1.3   Dry vs. liquid-injected machines

Two slightly different types of rotary screw compressors can be
employed in process plants. They are the dry machines, and the wet,
liquid-injected units. Liquid-injected rotary screw compressors are
further divided into oil-injected machines and machines using other
liquids.

Dry compressors typically use shaft-mounted gears to keep the two
rotors in proper mesh. Prevalent in the pharmaceutical and high-purity
chemical industries, these machines may also be used in aeration ser-

vices in the brewing industry and other applications where complete absence of entrained air and other contaminants is mandatory.

Oil-injected rotary screw compressors are generally supplied without timing gears. Other liquid-injected compressors usually require gearing to keep the two counterrotating screws in the proper mesh. The injected liquid could be water, a heat-removing fluid, or some other liquid. In oil-injected machines, the lubricant provides a layer separating the two screw profiles even as one screw drives the other. All liquid-injected machines offer the following advantages:

- The injected liquid provides internal cooling. Certain gases are thus kept from polymerizing, or from operating in an explosion-prone temperature range

- Compared to their dry counterparts, these units achieve considerably higher compression ratios. This capability is, in part, attributable to the fact that liquid-injected machines do not use seals between the rotor chamber and the bearings. This reduces the bearing span, and therefore, the rotor deflection. For example, a single liquid-injected compressor stage can do the job of two or more stages of dry compression.

9.1.4   Operating principles

High-performance screw compressors use a twin-shaft rotary piston to combine positive displacement with internal compression (Fig. 9.5 and 9.6). Gas entering at the suction flange is conveyed to the discharge port and entrapped in continuously diminishing spaces between the convolutions of the two helical rotors. The result is compression of the gas to the final pressure before it is expelled via the discharge nozzle.

The position of the edge of the outlet port determines the inherent or built-in volume ratio, v_i, which is the ratio of the volume of a given mass of gas at the discharge and suction ports. The corresponding built-in compression ratio, π_i (i.e., gas pressure at discharge over the pressure at the suction port) is calculated using the following equation:

$$\pi_i = v_i^{\,k} \tag{9.1}$$

where k is the ratio of specific heats of the gas at constant pressure and volume, respectively.

The compression process is shown in the theoretical pressure-volume diagram (Fig. 9.7). A rotary screw compressor is designed for an anticipated compression ratio, π_i. If the machine discharges into a receiver with a compression ratio in excess of π_i, the compressor end wall will be exposed to that pressure.

When operated at compression ratios higher than the designed value, a centrifugal compressor is likely to undergo *surging* or periodic

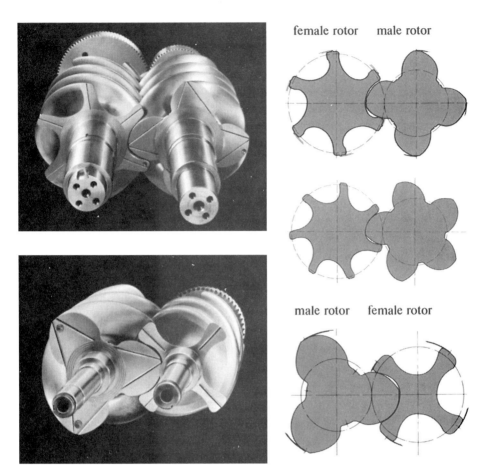

female rotor   male rotor

male rotor   female rotor

Figure 9.5 Typical screw compressor rotor combinations. *(Aerzen USA Company, Coatesville, Pa.)*

reverse flow, causing significant decline in machine performance. The screw compressor, on the other hand, is subject only to the constraints of machine component strength and input power. Thus it can easily produce the increased compression ratio or discharge pressure. Rotary screw compressors can also accommodate less than built-in compression ratios. In this case, however, some efficiency will be sacrificed. These efficiency losses are identified as shaded areas in Fig. 9.7.

9.1.5 Flow, power, and temperature calculations

The induced volume flow of the gas may be calculated from any compression ratio, if the data applicable to the particular compressor being considered are known. One revolution of the main helical rotor conveys

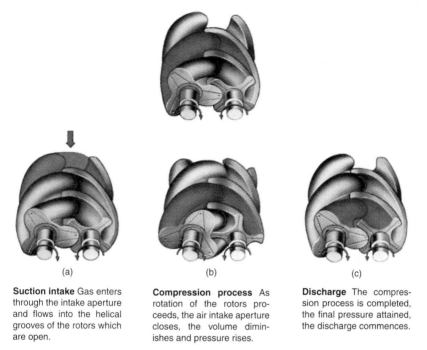

(a)   (b)   (c)

Suction intake Gas enters through the intake aperture and flows into the helical grooves of the rotors which are open.

Compression process As rotation of the rotors proceeds, the air intake aperture closes, the volume diminishes and pressure rises.

Discharge The compression process is completed, the final pressure attained, the discharge commences.

Figure 9.6 Working phases of rotary screw compressors. *(Aerzen USA Company, Coatesville, Pa.)*

the unit volume q_0, L/r. From this, one calculates the theoretical volume flow, Q_0, in std m³/min for the compressor running at n r/min as follows:

$$Q_0 = \frac{n \times q_0}{1000} \tag{9.2}$$

The actual volume flow Q_a is lowered by the amount of gas Q_v flowing back through the very small clearances between machine components. Thus

$$Q_a = Q_0 - Q_v \tag{9.3}$$

Q_v (also known as the volume flow lost via component slippages) is mainly dependent on the following factors:

- Total cross section of clearances
- Density of the medium handled
- Compression ratio
- Peripheral speed of rotor
- Built-in volume ratio

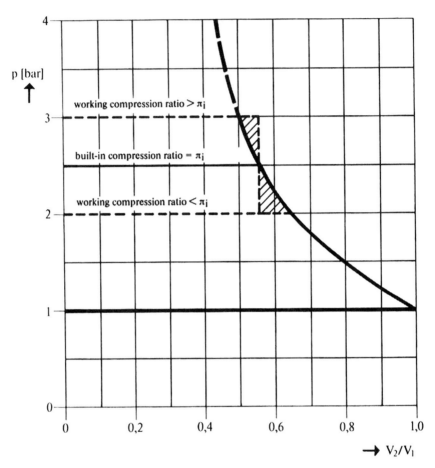

Figure 9.7 The pV diagram of a modern helical screw (rotary screw) compressor. *(Aerzen USA Company, Coatesville, Pa.)*

9.1.6 Power calculation

The volumetric efficiency, η_v, is expressed as

$$\eta_v = \frac{Q_a}{Q_0} = 1 - \frac{Q_v}{Q_0} \tag{9.4}$$

The theoretical power input, W_0, kW required to compress the induced flow volume Q_a is given by

$$W_0 = \frac{10^{-3}}{60} \times \rho_a Q_0 H_a \tag{9.5}$$

where ρ_a, expressed in kg/std m³ is the gas density at inlet conditions; and H_a represents the amount of energy required for the adiabatic compression of one kg of gas from pressure P_1 to P_2.

Alternatively, the theoretical power input could be obtained from

$$W_0 = (10^4 \times Q_a\, P_1/6000) \times (k/k - 1) \times [(P_2/P_1)^{(k-1)/k} - 1] \qquad (9.6)$$

where Q_a is expressed in std m^3/min, and P_1 is in bar.

In practice, the theoretical power input is just a part of the actual power, W_a, transmitted through the compressor coupling. W_a should include the dynamic flow loss, W_d, and the mechanical losses, W_v. The mechanical losses—typically amounting to 8 to 12 percent of the actual power—refer to viscous or frictional losses due to the bearings, the timing, and step-up gears.

The dynamic flow losses typically amount to 10 to 15 percent of the actual power. A crucial factor in determining these losses is designated N_{id}, which is a function of the built-in compression ratio and the Mach number (ratio of gas velocity over the velocity of sound) at compressor inlet conditions. One can use the following formula to estimate dynamic flow power loss:

$$W_d = C_f \times (L/D) \times (k/1.4) \times (P_1/1.013) \times (Q_0/60 \times N_{id}) \qquad (9.7)$$

where C_f is an empirical factor (obtained from Fig. 9.8), L is the rotor length, D is rotor diameter, N_{id} is another empirical factor (obtained from Fig. 9.9). The reference conditions assumed in the above formula and the associated charts are $Q_0 = 60$ std m^3/min, $L/D = 1.0$, wrap angle described by a point on the thread of a screw as the point travels from the bottom to the top of the rotor (inset in Fig. 9.9) = 300°, $P_1 = 1.013$ bar, and $k = 1.4$ (corresponding to air).

Thus the actual power requirement for the compressor is given by

$$W_a = W_0 + W_d + W_v \qquad (9.8)$$

In North America, screw compressors for important process applications are typically built in compliance with the American Petroleum Institute (API) Standard 619. No negative tolerance is permitted on capacity, and the power requirement may not exceed the quoted horsepower by more than 4 percent.

Screw compressors made in Europe can easily comply with this requirement. However, their customary Verein Deutscher Ingenieure (VDI; Soc. of German Engineers) Specification 2045 would allow a different margin of deviation to accommodate tolerances resulting from the usual operational limits of the manufacturing process.

9.1.7 Temperature rise

For a dry compressor, the temperature (in °C) of the compressed gas at final compression is calculated as follows:

$$\Delta T_0 = (T_1/\eta_v) \times [(P_2/P_1)^{(k-1)/k} - 1] \quad T_2 = T_1 + \Delta T_0 \qquad (9.9)$$

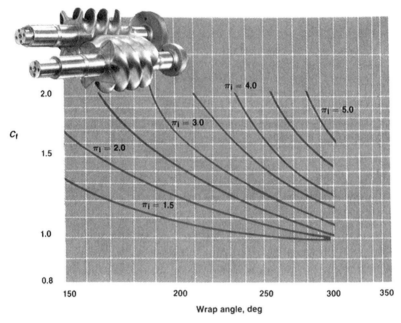

Figure 9.8 Empirical loss factor C_f as a function of compression ratio π_i and wrap angle of screw compressor rotors. *(Aerzen USA Company, Coatesville, Pa.)*

When operating in the oil-free, dry-running condition, a screw compressor may come up to a maximum final compression temperature of 250°C. When air is the compressed medium, this temperature (with adiabatic exponent $k = 1.4$) corresponds to a compression ratio $(P_2/P_1) \approx$ 4.5. On the other hand, gases with $k = 1.2$ will permit, within the same temperature limits, a compression ratio as high as 7.0.

In an oil-injected screw compressor (Figs. 9.10 and 9.11), most of the heat of compression is carried away by the oil. The amount of oil injected is adjusted to ensure that final discharge temperatures do not exceed 90°C (194°F). If air is taken in under atmospheric pressure, compression ratios as high as 21 are obtainable.

9.1.8 Capacity control

Because screw compressors are positive displacement machines, the most advantageous method of achieving capacity or volume flow control is that obtained by variable speed. This may be done by variable-speed electric motors, a torque converter, or a steam turbine drive.

Speed may be reduced to about 50 percent of the maximum permissible value. Induced volume flow and power transmitted through the coupling are thus reduced in about the same proportion.

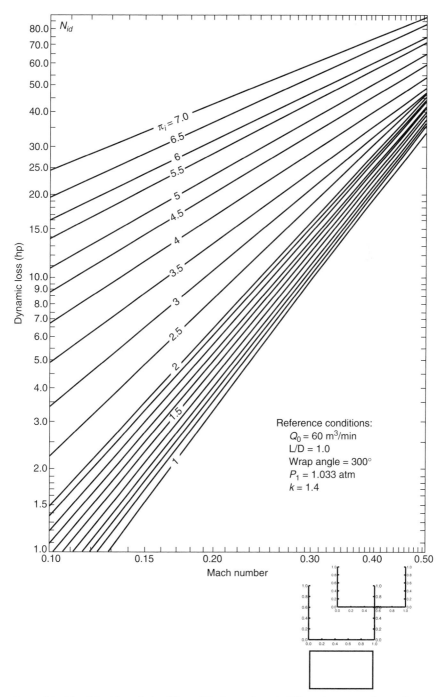

Figure 9.9 Empirical loss factor N_{id} vs. Mach number at different compression ratios π_i. The effect of compressor inlet conditions on the dynamic-flow power losses is shown as a function of the Mach number of the gas. *(Aerzen USA Company, Coatesville, Pa.)*

Figure 9.10 Oil-injected rotary screw compressor. *(Aerzen USA Company, Coatesville, Pa.)*

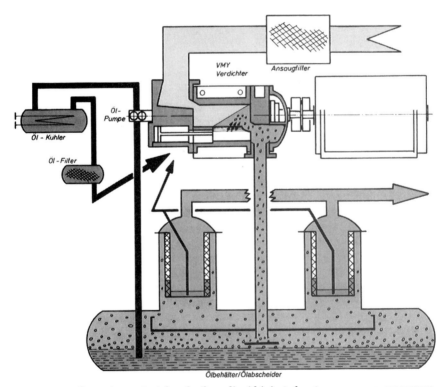

Figure 9.11 Operating principle of oil or liquid-injected rotary screw compressors. *(Aerzen USA Company, Coatesville, Pa.)*

Another method of capacity control is by using a bypass, in which the surplus gas is allowed to flow back to the intake side by way of a device that is controlled by the allowable final pressure. An intermediate cooler reduces the temperature of the surplus gas down to the level of the inlet temperature.

Use of a full-load and idling-speed governor is yet another means of capacity control. In this mode, as soon as a predetermined final pressure is attained, a suitable transducer operates a diaphragm valve (Fig. 9.12) that opens up a bypass between discharge and suction sides of the compressor.

When this occurs, the compressor idles until pressure in the system drops to a predetermined minimum value. This will cause the transducer to initiate closure of the diaphragm valve, and the compressor will again be fully loaded.

Control by suction throttle and discharge unloading is particularly suitable for air compression in the manufacture of industrial gases. As in the case of the full-load and idling-speed control method, a predetermined maximum pressure in the system (in a compressed air receiver, for example) causes pressure on the discharge side to be relieved down to atmospheric pressure. At the same time, the suction side of the system is throttled down to about 0.15 bar (2.2 psia). When pressure in the whole system drops to a predetermined minimum value, full load is restored once again.

Because the temperature at the final compression stage is governed by the injected oil, it is possible to operate an oil-flooded screw compressor over a wider range of compression ratios than would be feasible with dry machines. In addition, suction throttling in a screw compressor brings about a drop in the inlet pressure, thereby increasing the compression ratio.

Consequently oil-injected machines can achieve smooth adjustment of volume flow with relative ease. Some machines, however, may not be designed for this increased compression ratio. It is thus necessary to ensure that achievable pressures do not exceed the mechanical limitations of a given compressor.

Larger compressors can be readily equipped with an internal volume-regulating device (Fig. 9.13). It consists of a slide that is shaped to match the contours of the housing. By moving the slide in a direction parallel to the rotors, the effective length of the rotors can be shortened. The range of this smooth, infinitely variable control extends from 100 percent down to 10 percent of full compressor capacity. Also, slide controls offer stepless flow adjustment combined with power savings. These controls are primarily applied on screw compressors where the injected liquid has lubricating properties.

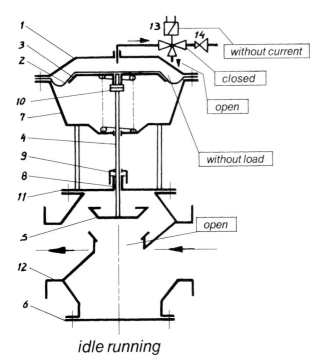

1. Diaphragm cover
2. Diaphragm
3. Diaphragm disk
4. Spindle
5. Valve cone
6. Blind flange
7. Diaphragm housing
8. Guide bush
9. Stuffing box nut
10. Counter nut
11. Guide cover
12. Valve housing
13. Diaphragm valve
14. Pressure reducing
     valve

idle running

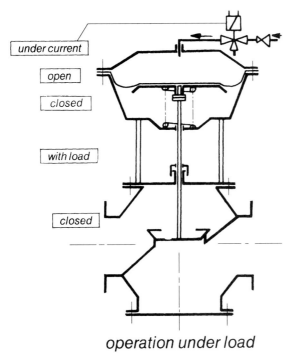

operation under load

Figure 9.12 Diaphragm valve associated with constant-speed unloading devices in rotary screw compressors. *(Aerzen USA Company, Coatesville, Pa.)*

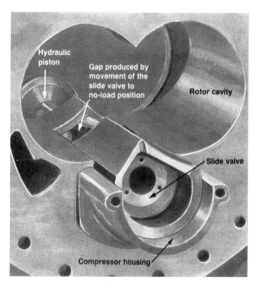

Figure 9.13 Internal volume regulating device for oil-injected rotary screw compressors. The position of a slide valve can be shifted in a direction parallel to the axes of the rotors. This provides control of the volume flow of the compressed gas. *(Aerzen USA Company, Coatesville, Pa.)*

9.1.9 Mechanical construction

Rotary screw compressors designed for high speeds and pressures incorporate sleeve bearings and self-adjusting, multisegment thrust bearings (Fig. 9.14). These machines can also be equipped with the type of sealing system best suited for a particular process gas service. For example, carbon ring seals are used in conjunction with buffer gas injection and leakoff ports that are connected back to compressor suction (Fig. 9.15).

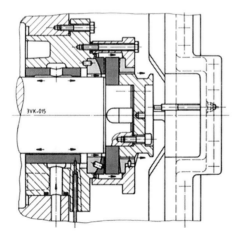

Figure 9.14 Radial and thrust bearings furnished with large rotary screw compressors. *(Aerzen USA Company, Coatesville, Pa.)*

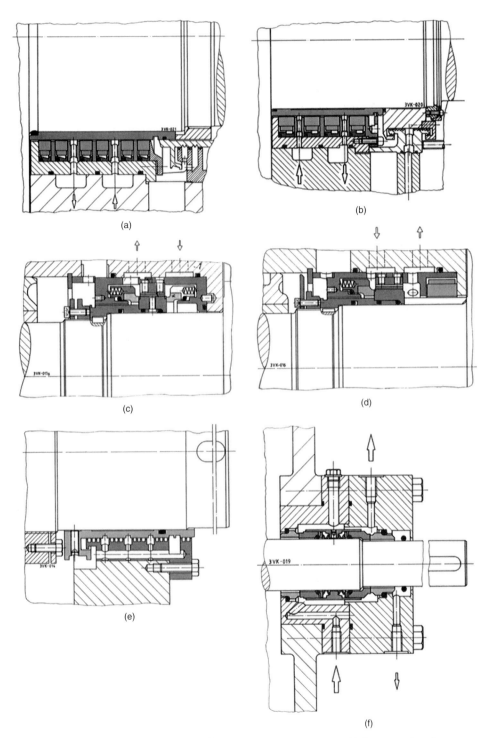

Figure 9.15 Sealing arrangements for rotary screw compressors. At the conveying chamber: (*a*) carbon labyrinth seal, (*b*) water-sealed floating rings, (*c*) double-acting slide ring seal, (*d*) combined floating ring and slide ring seal. At the drive shaft: (*e*) labyrinth seal, (*f*) double-acting slide ring seal. *(Aerzen USA, Coatesville, Pa.)*

Floating ring seals containing barrier water (Fig. 9.15*b*) allow a certain amount of water to reach the compression space. This water functions as a sealing, cooling, flushing, or gas scrubbing medium. Typically, most of the barrier water is returned to its supply system for reuse.

Stationary double-mechanical seals, lubricated with pressurized water or a suitable oil (Fig. 9.15*c*) are used in many applications where emissions must be minimized. Alternatively, a stationary single seal combined with a floating sleeve element (Fig. 9.15*d*) works very well in machines that feature high-differential pressures. The more traditional labyrinth or mechanical seals can be used on the transmission side of the casings for geared rotary screw units (Figs. 9.15*e* and *f*).

9.1.10   Industry experience

The capabilities of modern two-stage screw compressors that use water injection are being exploited in several European coke gas-producing plants. Conventional centrifugal compressors are highly vulnerable to performance degradation due to rapid polymerization of this relatively dirty, hydrogen-rich gas. In fact, centrifugal units require frequent cleaning of the internals, causing costly downtime every 6 to 8 weeks.

For example, in one coal gasification plant in Europe centrifugal compressors have been replaced with the three two-stage, water-injected screw units shown in Fig. 9.16. The track record (in terms of availability and reliability) of these multistage screw compressors, driven by a 5.5-MW (7400-hp) electric motor, has proven to be remarkably good. Each of the three machines compresses about 33,000 std m³/h (19,420 scfm) of coke-oven gas varying in pressure from 1 to 12 bars (14.5 to 174 psia). They also operate at considerably lower cost than the centrifugal compressors they replaced. Both power and overall maintenance costs have been reduced.

Figure 9.16   Large (approximately 7000 hp) rotary screw compressor installation at a German coal gasification plant. *(Aerzener Maschinenfabrik, Aerzen, Germany)*

The decision to use screw compressors with water seals must be based on sound technical and thermodynamic considerations. In dry compression, for example, the discharge temperature may be well in excess of 100°C (212°F). As a result, the higher hydrocarbons may evaporate, leaving asphaltlike residues in the gas to form a coating on the rotors and the housings. These deposits can adversely affect throughput and efficiency. In particular, the sticky residue may fasten the two rotors together in a dry screw compressor whenever the machine is brought down and cooled for any reason.

Water injection limits the final gas temperature to 100°C. In addition, feeding the right amount of water into the compression space of the rotary screw compressor prevents polymerization by removing the heat of compression with the evaporating water.

Much of the water is supplied through the four shaft seals of each stage. The remainder is sprayed into the inlet nozzle of each stage, with the injection rate controlled by a temperature transducer at the compressor discharge. Refer to Fig. 9.17 for a flow schematic of a two-stage unit with water barrier seal.

The gas temperature is thus regulated to just below the dew point. This ensures that the deposits are flushed away by the excess water present. The water is finally drained off at the discharge silencer and in the water separators of the intercoolers and aftercoolers of each stage.

Because the gas contains corrosive components, such as hydrogen sulfide, ammonia, hydrogen cyanide, and carbon dioxide, the materials of construction of choice for the compressor have been chromium-nickel alloy steels. While ensuring erosion resistance, these steels are also resistant to chemical attack by the gas. However, precautions must be taken to avoid corrosion-related wear that can occur over a period of time as a result of the injection of water.

An inspection carried out after 1 year of continuous service in the coking plant has revealed that the rotors and housings are free of dirt deposits. In addition, they have shown no signs of physical or chemical damage due to erosion or corrosion.

Employing what is known as *intermediate pressure regulation,* the screw compressor matches the volume flow to the variable, momentary requirement of the process gas. Gas not required by the final receiver or downstream process is returned after the first stage via a bypass to the intake side. This results in significant power savings while providing a continuous regulation of volume flow ranging from about 20,000 to 33,000 std m³/h (11,770 to 19,420 scfm). For example, at 20,000 std m³/h, the power consumption is about 3200 kW (4290 hp) in comparison with 3950 kW (5295 hp) at 30,000 std m³/h.

The compressor inlet system has been fitted with separate superchargers that offer the additional capability of boosting the inlet pres-

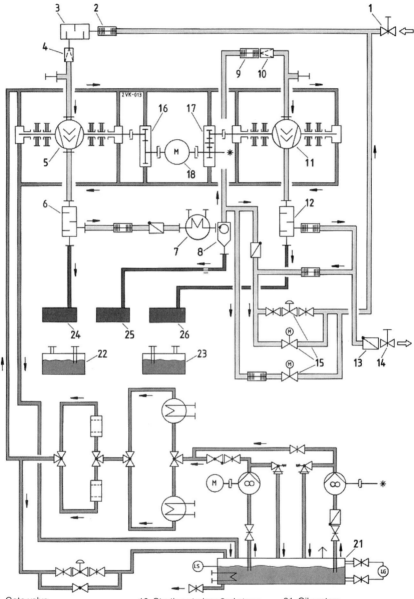

1. Gate valve	10. Starting strainer 2nd stage	21. Oil system
2. Lateral compensator	11. Screw compressor 2nd stage	22. Barrier water system
3. Intake silencer 1st stage	12. Discharge silencer 2nd stage	23. Water injection system
4. Starting strainer 1st stage	13. Non-return valve	24. Condensate tank 1
5. Screw compressor 1st stage	14. Gate valve	25. Condensate tank 2
6. Discharge silencer 1st stage	15. Control devices	26. Condensate tank 3
7. Intercooler	16. Gear box 1st stage	27. Drive motor
8. Separator	17. Gear box 2nd stage	
9. Safety relief valve 1st stage	18. Safety relief valve 2nd stage	

Figure 9.17 Flow diagram of a two-stage rotary screw compressor unit with barrier water seal. *(Aerzener Maschinenfabrik, Aerzen, Germany)*

sure to about 1.6 bar (23 psia). This results in an intake volume flow of about 46,000 std m³/h (27,070 scfm). Thus, combining intermediate pressure regulation with supercharging provides a continuous range of regulation from 20,000 to 46,000 std m³/h. With this wide range of operability, the coking plant can adapt at any time to changing gas requirements.

9.1.11   Maintenance history

The maintenance history of the three two-stage, 5.5-MW rotary screw compressors indicates that they went through partial dismantling every 5 years. And at that time, only the carbon seal rings needed replacement because they suffered some physical damage. All other parts, including bearings, have been reinstalled in the compressors without modification or repair.

In 1988, maintenance costs are estimated at $80,000 for the 5-year period. This estimate, which includes labor and materials, is orders of magnitude below the expenditures incurred with centrifugal compressors in this type of service.

A second installation that operates two three-stage rotary screw compressors has had its first turnaround inspection after 35,000 h of uninterrupted service. Several carbon seal rings showed minor wear, and all other parts were in excellent condition. However, because the carbon rings were still quite serviceable, this installation has increased its projected turnaround intervals to 45,000 to 50,000 operating hours.

A third installation (also in Europe) has been less than satisfactory, however. The screw compressors in this plant have experienced accelerated erosive wear of carbon seal rings. This is attributed to unacceptable water quality. The lesson to be learned is that water-injected rotary screw compressors require demineralized water. In this respect, they resemble the common steam turbine.

Of course, applications of screw compressors are in no way limited to coking plants. They can be used for delivery and compression of contaminated gases as well. Moreover, these units are ideally suited for compression of gases that tend to polymerize at relatively low temperatures.

9.1.12   Performance summary

The performance of a screw compressor is influenced by factors such as gas properties, internal clearances, length-to-diameter ratio of the rotors, built-in compression ratio, and operating speed. Although compressor manufacturers have not yet found practical ways to present compressor performance for various machine sizes, inlet and outlet conditions in a single graphical layout or diagram, one can still make rule-of-thumb estimates using the graphs and charts shown later for conventional centrifugal compressors.

Typical adiabatic efficiency is almost always between 70 and 80 percent. The maximum allowable compression ratio for one stage of a screw compressor corresponds to the value that will not cause the final compression temperature to rise above the permitted value of 250°C (482°F). To a large extent, this will depend on the k-value (the ratio of specific heats) of the gas to be compressed.

Compressor speeds can vary from about 2,000 to 20,000 r/min, depending on unit size. The peripheral speed of the rotor determines the magnitude of the rotational speed. This peripheral speed ranges from 40 to 120 m/s (131 to 394 ft/s) up to a maximum of 150 m/s (492 ft/s) for gases with low molecular weight.

As would be the case with virtually any other specific type or category of fluid machinery, rotary screw compressors embody advantages as well as disadvantages over other equipment competing for market share. Listing first the advantages, the application engineer would wish to consider the following:

- Considerably reduced sensitivity to molecular weight changes compared to centrifugal machines.

- Much greater tolerance for polymerizing service than other compressors, except perhaps liquid ring machines.

- Capable of accepting more liquid and fine solids entrainment than other compressors, except liquid ring compressors.

- Higher efficiency and less maintenance than liquid ring machines.

- Estimated availability in excess of 99.5 percent. This may approach or, in certain services, exceed that of centrifugal and axial compressors.

- Smaller size and lower cost than reciprocating compressors in same capacity range.

- Lower cost than centrifugal compressors in the small and moderate size ranges (below approximately 3000 kW, or about 4000 hp).

- Higher pressure capability than other types of rotary positive displacement machines.

Among the disadvantages found are some that are perceived and others that are real. They, thus, merit more detailed examination.

- Sensitivity to discharge temperature that could affect close clearances and, hence, operability and availability: Proper temperature control instrumentation and generous sizing of cooling water or liquid injection facilities make this a nonissue for modern liquid injected screw compressors.

- Performance affected by rotor and casing corrosion or erosion. Increased clearances promote internal recycle or gas slip effects— not a serious concern with water and oil-injected rotary screw compressors.

- Noise level is high enough to require silencing—a factor that must be taken into account. Capable rotary screw compressor manufacturers are fully equipped to provide well-engineered means of reducing environmental noise to meet even the most stringent requirements (see Fig. 9.18).

- Rotary screw compressor systems require pulsation suppression. While not as severe as piping pulsations encountered with equivalent reciprocating compressors, a properly engineered screw compressor system would incorporate appropriate pulsation bottles and, for high discharge temperatures, pipe expansion loops.

- Choice of rotor and casing materials more limited than for centrifugal compressors. This observation is related to the intricacies and close tolerance requirements of the machining process. Also, a knowl-

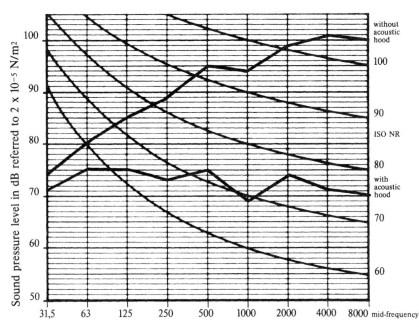

Figure 9.18   Typical sound levels obtained from rotary screw compressors with and without acoustic enclosures. *(Aerzen USA, Coatesville, Pa.)*

edgeable manufacturer is cognizant of certain nonlinearities in the coefficients of expansion of different stainless steels. This might impose experience-based temperature limitations on certain metallurgies and service conditions.

- Maintenance cost and duration of downtime higher than for centrifugals—highly service dependent and not always so; merits closer investigation.

- Flow control flexibility inferior to that of centrifugals and reciprocating compressors—a serious misconception that neglects to take into account the full spectrum of available options given earlier.

9.2 Oil-Flooded Single-Screw Compressors

Oil-flooded, single-screw compressors are available for gas flows in the vicinity of 1000 cfm (1700 m^3/h) and discharge pressures approaching 800 psi (55 bar). As can be seen from Fig. 9.19, these machines incorporate features that make them a hybrid between other, sometimes competing, compressor types. Cooling oil, which circulates through the compressor to absorb the heat of compression also provides sealing of the gas in the compression spaces and lubrication of the rolling ele-

Figure 9.19 Single-screw, oil-injected compressor. *(Dresser-Rand Company, Broken Arrow, Okla.)*

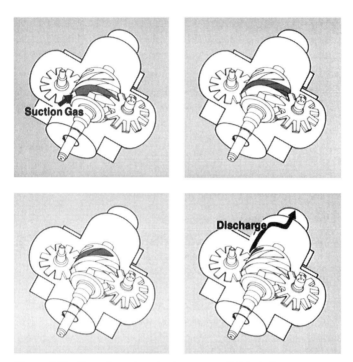

Figure 9.20 Gas flow in single-screw, oil-injected compressor. *(Dresser-Rand Company, Broken Arrow, Okla.)*

ment bearings. Synthetic or special lubricants are used for applications with corrosive gases or when high condensation rates are encountered.

The intermeshing of three principal rotating parts accomplishes the continuous compression process in oil-flooded single-screw machines. The gas flow can be visualized from Fig. 9.20.

Suction gas flows into the inlet passage and fills a screw groove. The inlet gas is trapped in the groove when the gate rotor tooth meshes with the screw groove and seals the groove. As the screw continues to rotate, the trapped gas is compressed as the length and volume of the groove is reduced. Injected oil seals the running clearances to prevent leakage of gas.

When the screw rotates far enough, the groove passes the discharge port, delivering gas to the discharge manifold. The location of this port determines the internal volume reduction and thus the internal compression ratio.

Since there are two gate rotors, compression occurs simultaneously on both sides of the screw rotor. Thus compressive forces are radially balanced. Thrust forces are minimal since suction pressure is ported to both ends of the screw. Because the gate rotor axes are perpendicular

Process Flow Schematic

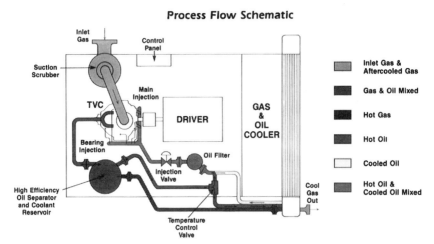

Figure 9.21 Process flow schematic showing single-screw, oil-injected compressor. *(Dresser-Rand Company, Broken Arrow, Okla.)*

to the screw axis, virtually no torque is transmitted to the gate rotors, so wearing forces are kept low.

The schematic of Fig. 9.21 shows the basic relationship between the gas to be compressed and the cooling-lubricating oil. Inside the compressor, oil and gas are mixed, and then delivered to a high-efficiency gas-oil separator. Clean gas is then delivered to the skid edge, usually aftercooled. The oil collected in the separator is cooled, filtered, and reinjected into the compressor. The compressor discharge temperature is held constant during the compression cycle, but the operating discharge temperature selected will vary according to the application.

Capacity control is accomplished through slide pistons, item 7, contained in the compressor casing of Fig. 9.19.

By means of rack and pinion gears, the two slides are moved axially and controlled via a stub shaft (B) which protrudes from the side of the casing near the discharge flange.

Figure 9.22 shows one capacity slide at 100 percent capacity (top) and 0 percent capacity (bottom). The screw rotor has been removed for clarity. In the 100 percent position, the slide is positioned so that no leakage of gas can occur during compression; thus all the gas that enters the screw groove is delivered to the triangular discharge port. At less than 100 percent, the slide valve is moved so that some of the gas that had entered the groove returns to suction prior to compression. In the same motion, the discharge port moves away from the return port. This delays the groove from passing the discharge port, preserving the internal volume reduction in the compressor.

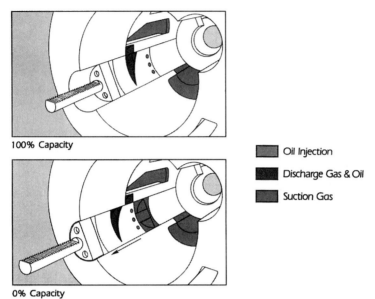

100% Capacity

0% Capacity

■ Oil Injection

■ Discharge Gas & Oil

■ Suction Gas

Figure 9.22   Capacity control slides in single-screw, oil-injected compressor. *(Dresser-Rand Company, Broken Arrow, Okla.)*

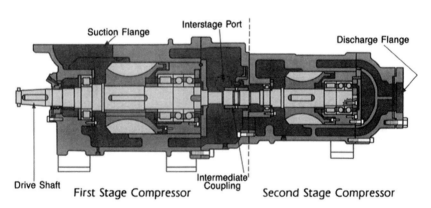

Figure 9.23   Two-stage version of an oil-flooded single-screw compressor. *(Dresser-Rand Company, Broken Arrow, Okla.)*

Oil-flooded, single-screw compressors are also available in single-shaft, two-stage designs for high-compression ratio service. The second stage of compression is connected to the first through a gastight transition piece that delivers gas and oil between stages, and a through shaft with a splined coupling that transmits torque to the second-stage screw (Fig. 9.23). Two-stage compressors are equipped with sidestream capability (e.g., refrigeration economizing) and 40 percent to 100 percent infinitely variable capacity control.

Unlike oil-flooded twin-screw compressors which inevitably incorporate separate circuits for *lube oil* and oil injected into the *compression space*, oil-flooded single-screw compressors use the same oil for bearing lubrication and gas space cooling. This fact must be considered in the selection process since certain gases may contaminate the oil and adversely affect the life of rolling element bearings and other compressor components.

Chapter 1

Reciprocating Compressor Performance and Sizing Fundamentals

From the preceding chapters we recall that most reciprocating compressors encountered in process, gas or oil production, and gas transmission applications use double-acting cylinders. This simply means that compression occurs on both the outward and inward strokes of the piston. This is accomplished using a packed piston rod firmly attached to a crosshead. The crosshead, in turn, is attached to the connecting rod via a wrist pin.

Reciprocating compressors offer the following advantages to the user:

1. Flexibility in design configuration

2. Good efficiency even in small sizes, at high pressures, and at part loads

3. Operating flexibility over a wide range of conditions for a given configuration

Our objective is now to introduce the reader to the basics of how to calculate reciprocating compressor performance and to present a methodology of estimating compressor size and power requirements. The methods given here are approximate by necessity, and the reader is encouraged to communicate with compressor vendors if more accurate results are required.

We will begin by reviewing basic capacity and horsepower calculations and demonstrating the effect of design variables on the results of the calculations. Basic equations will be presented that will enable readers to estimate capacity and horsepower along with frameload. Further information is given that will allow estimation of compressor size given some general information on compressor cylinders, strokes, rotative speeds, horsepower capacity, and rod load capability. The standard nomenclature used by the majority of U.S. compressor manufacturers has been selected for our calculations.

10.1   Capacity and Leakage Considerations

10.1.1   Theoretical maximum capacity

The theoretical maximum capacity of a reciprocating compressor cylinder is given by Eq. (10.1).

$$Q = 0.0509 \times \frac{P_s}{T_s} \times \frac{Z_{std}}{Z_s} \times \text{DISP} \times [1 - \text{CL} (R^{1/N} - 1)] \qquad (10.1)$$

where
Q = capacity, million standard ft³/day (Ref. 14.7 psia, 520°R)
P_s = suction pressure, psia (flange)
T_s = suction temperature, Rankine
Z_{std} = compressibility factor at standard conditions
Z_s = compressibility factor at suction conditions
DISP = cylinder displacement, ft³/min
CL = cylinder clearance volume as decimal fraction of displaced volume
R = pressure ratio across cylinder (flange to flange)
N = isentropic volume exponent at operating conditions (specific heat ratio for ideal gas)

The critical portion of Eq. (10.1) is the theoretical volumetric efficiency defined, in Eq. (10.2), as:

$$\text{VE} = 1 - \text{CL} (R^{1/N} - 1) \qquad (10.2)$$

This equation describes the variation of a given compressor capacity as a function of residual clearance volume, pressure ratio, and gas. The trends are:

1. Decreases with increasing clearance

2. Decreases with increasing pressure ratio

3. Increases with increasing volumetric exponent

The other variables in the capacity equation are related to fluid density at the compressor cylinder inlet. Capacity increases with increasing inlet density.

Figures 10.1 and 10.2 demonstrate the variation of compressor capacity with clearance and pressure ratio. The compressor chosen for these and most of the other figures that follow has the following specifications:

Bore = 20 in (double acting)

Stroke = 15 in

Clearance = 15%

Rod diameter = 3 in

Rotative speed = 327 r/min

Gas = methane

P_s = 15 psia

T_s = 560°R (100°F)

10.2   Capacity Losses

Real compressors with valves, piston rings, packing, heat transfer, and attached piping do not pump ideal capacity. There are a number of loss factors that generally reduce the capacity. This section covers these factors and how they vary with design and operating parameters. The pV diagrams reproduced earlier in Fig. 2.43 graphically demonstrate the effect of most of the loss factors on capacity.

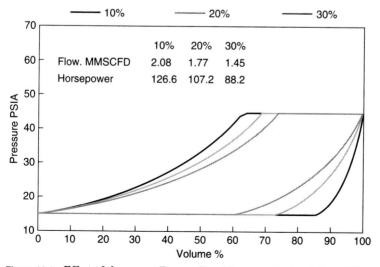

	10%	20%	30%
Flow. MMSCFD	2.08	1.77	1.45
Horsepower	126.6	107.2	88.2

Figure 10.1   Effect of clearance. *(Dresser-Rand Company, Painted Post, N.Y.)*

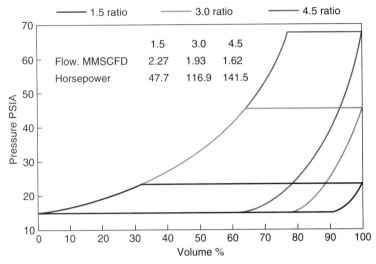

Figure 10.2 Effect of compression ratio. *(Dresser-Rand Company, Painted Post, N.Y.)*

10.3   Valve Preload

As mentioned earlier, reciprocating compressor valves are essentially spring-loaded check valves. Manufacturing tolerance and reliability considerations cause the designer to introduce a positive preload. This means that the compressor must develop a small pressure drop across the valve in the direction of flow before the sealing element will begin to move. The effect of this is to increase the pressure ratio across the compressor cylinder. The pressure trapped in the cylinder will be higher than discharge pressure at minimum volume and lower than suction pressure at maximum volume. The net effect is twofold: (1) The pressure ratio across the cylinder is higher than expected which decreases capacity by decreasing volumetric efficiency. (2) The gas density in the cylinder at maximum volume is lower than expected while the density at minimum volume is higher than expected, resulting in reduced capacity.

Figure 10.3 demonstrates the effect. This effect is most pronounced at low suction pressures and decreases to the point where it is negligible at higher pressures. The designer, however, must know the details of the valve design to be used to accurately predict the preload effect on capacity.

10.4   Valve and Gas Passage Throttling

Compressor valves and cylinder gas passages must see the flow that goes through the bore. There are pressure losses associated with this flow. The suction losses have an effect on capacity. As each increment of

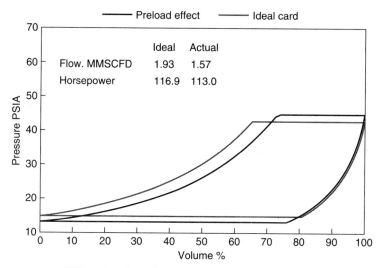

Figure 10.3 Effect of preload. *(Dresser-Rand Company, Painted Post, N.Y.)*

fluid flows into the cylinder bore, it experiences a pressure drop. The process may be closely approximated as isenthalpic. Once each increment of fluid is in the cylinder, it must be recompressed to the pressure that exists at maximum volume. The work required to do this increases the temperature of the fluid so that it is higher at maximum volume than the suction temperature. This reduces the density at maximum volume and, therefore, the capacity.

The effect of discharge valve and passage flow losses on capacity is determined by whether all of the fluid that should flow out of the cylinder does so by the end of the stroke (minimum volume). There is a similar effect on the suction side. The impact of this on capacity is similar to the effect of valve preload.

Figures 10.4 and 10.5 demonstrate the capacity loss due to throttling. Figure 10.4 shows the same compressor configuration as previously used. Figure 10.5 reflects the effect of increased rotating speed by shortening the stroke to 5.5 in and increasing the speed to 892 r/min. The bump at the beginning of the valve event is an inertia effect. It is more pronounced at the higher rotative speed.

The throttling capacity loss has an inverse relationship to flow efficiencies built into the valves and cylinder flow passages. Large and open valves and passages mean low pressure losses and low capacity loss. They generally mean higher clearances that also reduce capacity. Thus, low valve losses do not always improve capacity.

To accurately predict the effect of valves on capacity, we must know the details of the design being used. While we generally know the average velocity through the lift area, we must know how efficiently the valve design uses that area. The true loss is determined by what can be termed the effective flow area. This is defined as the product of geometric lift area and flow coefficient. The flow coefficient can vary widely from one valve design to another.

10.5   Piston Ring Leakage

The compressor piston uses seals to minimize leakage. However, ring leakage does occur and has a detrimental effect on capacity. In a double-acting compressor, the effect is twofold: (1) As fluid leaks from the higher pressure side of the ring, the capacity is reduced by loss of mass in the high pressure end. This also increases the mass in the low pressure end, thus decreasing the mass that will flow in through the suction valves. (2) The leakage process is closely approximated as isenthalpic. This has a tendency to increase the temperature in the low pressure end of the cylinder. The result is lower density at maximum volume, further reducing capacity.

When the piston reverses and the high-pressure end becomes the low-pressure end, the process reverses. Thus, a small amount of gas is essentially trapped in the cylinder.

The effect of piston ring leakage is shown in Fig. 10.6 for a double-acting cylinder.

A single-acting cylinder will generally show higher leakage than a double-acting cylinder because the time average pressure drop in one

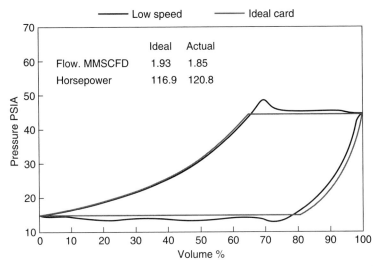

Figure 10.4   Effect of valve and passage flow losses, low speed. *(Dresser-Rand Company, Painted Post, N.Y.)*

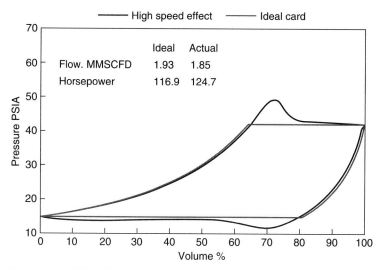

Figure 10.5 Effect of valve and passage flow losses, high speed. *(Dresser-Rand Company, Painted Post, N.Y.)*

direction is higher. The temperature effect is also there if the inactive end of the cylinder is vented to the suction of the active end. This is normally the case.

The effects of design and operating parameters on ring leakage are such that leakage:

1. Decreases with increasing rotative speed

2. Decreases with increasing bore

3. Increases with decreasing molecular weight

4. Increases with pressure ratio

5. Decreases with the number of rings

6. Is higher with nonlube construction

10.6 Packing Leakage

Compressors that use piston rods have packing. As in the case of piston rings, the designer attempts to minimize the leakage, but a small amount will occur unless the packing has a buffer pressure introduced that can negate leakage from inside the cylinder.

When the packing leaks, the effect on capacity is limited to the loss of fluid from the packed end of the cylinder. The effect is shown in Fig. 10.7.

The effects of operating and design parameters cause packing leakage to:

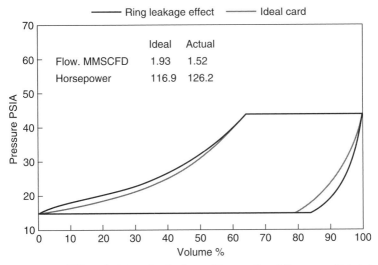

Figure 10.6   Effect of piston ring leakage. *(Dresser-Rand Company, Painted Post, N.Y.)*

1. Increase with increasing pressure level
2. Increase with rod diameter
3. Increase with decreasing molecular weight
4. Decrease with increasing rotative speed

10.7   Discharge Valve Leakage

Figure 10.8 demonstrates the effect of discharge valve leakage. Fluid leaks back into the cylinder bore from the discharge passage. This not only distorts the *pV* diagram but also lets hot gas back into the cylinder. Both effects decrease capacity.

As in the case of other leakage effects, the designer attempts to minimize valve leakage. The effect is so dramatic that zero leakage is a worthwhile goal.

Operating and design parameters have the following effect on discharge valve leakage:

1. Increases with increasing pressure ratio
2. Increases with decreasing molecular weight
3. Decreases with rotative speed
4. Decreases when plastic sealing elements are used

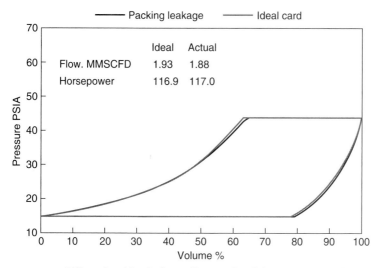

Figure 10.7   Effect of packing leakage. *(Dresser-Rand Company, Painted Post, N.Y.)*

10.8   Suction Valve Leakage

Figure 10.9 shows the effect of suction valve leakage. Fluid leaks from the cylinder bore to the cylinder suction passage. The effect on capacity is similar to discharge valve leakage, and zero leakage is a worthwhile goal.

The effects of operating and design parameters are identical to discharge valve leakage.

10.9   Heating Effects

Heating effects on capacity can be divided into two categories: (1) External heat transfer between the surroundings and the fluid prior to entering the cylinder and (2) Internal heat transfer between the cylinder walls, cooling medium, and the fluid while it is in the cylinder.

Heat transfer to or from the fluid prior to entering the cylinder bore has a direct effect on capacity by raising or lowering the temperature of the fluid trapped in the cylinder at maximum volume. As explained earlier, raising the temperature lowers the capacity, while lowering the temperature has the opposite effect. In fact, many operators intentionally lower inlet temperature to increase compressor throughput. This should be done with caution because there may be effects such as increased condensation that have undesirable side effects.

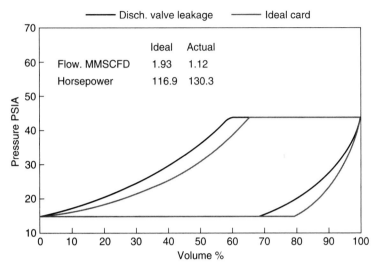

Figure 10.8 Effect of discharge valve leakage. *(Dresser-Rand Company, Painted Post, N.Y.)*

Heat transfer internal to the cylinder bore can affect capacity both by affecting the temperature of the trapped fluid and the shape of the pV diagrams. Figure 10.10 demonstrates the effect of net heat transfer from the gas while in the bore.

While heat transfer and its resulting effects on capacity are complicated and extremely dependent on details of design, the following generalizations may be made and debated:

1. For a given cylinder design, increasing rotative speed decreases heat transfer effects.

2. For a given cylinder design at constant speed, increased fluid mass flow decreases heat transfer effects.

3. For a given fluid at a given density, increasing cylinder bore decreases heat transfer effects.

4. Increasing the differential temperature between the fluid and surroundings and/or cylinder coolant increases heat transfer effects.

10.10 Pulsation Effects

Reciprocating compressors are unsteady flow machines. This time-varying flow is repeatable from one crankshaft rotation to the next. The resulting pressure variations in the connecting pipe work are called pulsations, and they affect capacity. The basic effect is determined by the pressure the pulsations impose on the cylinder bore at

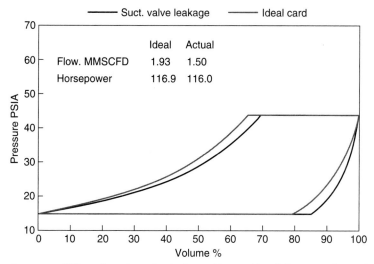

Figure 10.9 Effect of suction valve leakage. *(Dresser-Rand Company, Painted Post, N.Y.)*

maximum and minimum volumes. Pulsations, like heat transfer, can either increase or decrease capacity. The characteristics are:

1. Higher pressure in the suction passage at maximum cylinder volume increases capacity.

2. Higher pressure in the discharge passage at minimum cylinder volume decreases capacity.

3. Lower pressure in the suction passage at maximum cylinder volume decreases capacity.

4. Lower pressure in the discharge passage at minimum cylinder volume increases capacity.

Figure 10.11 shows the effect of pulsations for a particular set of circumstances.

Predicting pulsations at the time of compressor sizing is virtually impossible. Therefore, the effects on capacity are limited by controlling pulsations to acceptable levels at a later stage of the design cycle.

With all of the potential effects of operating conditions and design features on compressor capacity, how are we to estimate compressor sizes without detailed knowledge? Fortunately, the effect of some parameters increases with speed, while the effect of other parameters decreases with speed. Generally, slow-speed compressor capacity is

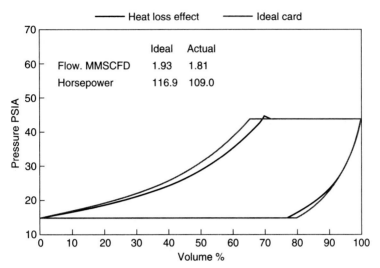

Figure 10.10 Effect of internal heat transfer. *(Dresser-Rand Company, Painted Post, N.Y.)*

governed more by leakage and heat transfer effects, while high-speed compressor capacity is governed more by valve effects. We can, therefore, write a general equation to use for all reciprocating compressors as a first estimate. Expressed in MMSCFD, this equation is:

$$Q = 0.0509 \times \frac{P_s}{T_s} \times \frac{Z_{std}}{Z_s} \times \text{DISP} \, (0.95 - \text{CL} \, (R^{1/N} - 1)) \quad (10.3)$$

This essentially states that any reciprocating compressor has a built-in 5 percent capacity loss. In some cases, this is a conservative estimate and in others, it is liberal. This is a starting point only. For nonlube and/or single-acting service, an additional loss of 4 percent should be used in each case.

10.11   Horsepower

We base compression efficiency on theoretical isentropic horsepower. Equation (10.4) calculates isentropic horsepower.

$$\text{HP} = 43.67 \times Q \times \frac{N}{N-1} \, (R^{(N-1)/N} - 1) \quad (10.4)$$

Inspection of this equation indicates that the major effects on isentropic power are the capacity Q and the pressure ratio R: (1) horsepower increases with capacity and (2) horsepower increases with pressure ratio until the decrease in capacity with increasing ratio becomes the overriding factor.

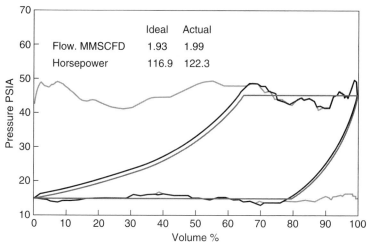

Figure 10.11 Effect of pulsation. *(Dresser-Rand Company, Painted Post, N.Y.)*

Figures 10.1 and 10.2 demonstrated that theoretical horsepower depends on volumetric clearance and pressure ratio.

10.12 Horsepower Adders

The same factors that affect capacity have an adverse effect on horsepower. It is appropriate to call these adders, as opposed to the more popular term *losses*. The reason is that the horsepower actually consumed by a compressor is almost always higher than isentropic.

Horsepower adders have three effects on the horsepower required for compression. The first effect is to add power to the suction and discharge portions of the pressure-volume diagram. This is caused by valve losses. The second effect is to distort the compression and reexpansion lines. This is caused by leakage and internal heat transfer. The third effect is to simply add the power required to overcome mechanical friction.

Summing up the horsepower adders, we find major effects from the valving, while other effects from leakage and heating are relatively minor in all but a few applications. These are confined to low-density and low-suction pressure applications such as vacuum pumps. Friction can be treated as a direct adder. With these considerations, we can use the following equation to estimate horsepower requirements of a compressor cylinder and its associated running gear.

$$\text{BHP} = 43.67 \times Q \times \frac{N}{N-1}\,(R^{(N-1)/N} - 1) \times \frac{1}{N_c} \times \frac{1}{N_m} \qquad (10.5)$$

where N_c = compression efficiency
N_m = mechanical efficiency

Compression efficiency varies with many factors, and a unique relationship is difficult to define. However, companies such as Dresser-Rand recommend using 0.85 as a first guess for lubricated service. Deduct an additional 0.05 for nonlube and/or single-acting service.

Mechanical efficiency is generally accepted to be approximately 0.95, and this value is recommended by most manufacturers.

Readers requiring a more accurate analysis of horsepower may wish to contact the compressor vendors. There are many design variations between vendors that can have significant influence on actual power.

10.13   Gas Properties

10.13.1   Ideal gas

As described earlier in Sec. 1.15, many compression processes can be described using the ideal gas law.

$$P \times V = m \times T \times \frac{10.73}{MW} \qquad (10.6)$$

where P = pressure, psia
V = volume, ft^3
m = mass, lbm
MW = molecular weight, lbm/mol
T = temperature, Rankine

This, plus a description of an isentropic process enables us to theoretically describe a compression process. This equation is:

$$P \times V^N = \text{constant} \qquad (10.7)$$

where N = specific heat ratio

The discharge temperature may be estimated as the isentropic temperature at the end of the compression process. This is given by

$$T_d = T_s \times R^{(N-1)/N} \qquad (10.8)$$

where T_d = discharge temperature, Rankine
T_s = suction temperature, Rankine
R = pressure ratio

10.13.2 Real gas

No gas is truly ideal. To approximate an ideal gas, we use a concept called *compressibility factor.* This factor is used to represent the difference between a real gas and an ideal gas. Using this factor, we rewrite Eq. (4.1) as:

$$P \times V = Z \times m \times T \times \frac{10.73}{MW} \tag{10.9}$$

where Z = compressibility factor

This factor is obtained using a suitable equation of state that more accurately describes gas properties.

The isentropic process is still described as

$$P \times V^N = \text{constant} \tag{10.7}$$

N is now the isentropic volumetric exponent defined by real gas properties.

Discharge temperature is described by

$$T_d = T_s \times R^{(N_t - 1)/N_t} \tag{10.10}$$

where N_t = isentropic temperature exponent defined by real gas properties

10.14 Alternative Equations of State

Over the years, there have been many equations of state proposed. For most gases, equations of state or compressibility methods based on work by Redlich-Kwong, The American Petroleum Institute, or Peng-Robinson are recommended. Although detailed discussion of these calculation methods is outside the scope of this text, we should note that every one of these methods falls apart for all gases when operating in the dense phase. This is generally above and to the left of the liquid-vapor dome on the temperature-entropy diagram. When this is the case, we usually resort to National Bureau of Standards T-S diagrams for pure gases.

For gas mixtures in the dense phase, it is recommended to use specialized methods such as the National Bureau of Standards DMIX software for defining real properties.

An alternative source of real gas properties for unusual mixtures might be a user of these mixtures. Users have often conducted research to define these properties to accurately design processes.

10.15   Condensation

Many fluids that must be compressed contain saturated water and/or hydrocarbon vapors. When this is the case, we must evaluate the quantity of fluid that will condense in the heat exchangers used to remove the heat of compression downstream of the compressors. Evaluating condensed water vapor is relatively easy, and most computerized performance models handle this.

Hydrocarbon condensation can be evaluated by several software packages. Among them are NGPSA (National Gas Processors Association), PROCESS (Simulation Sciences), and CHEMSHARE. Most compressor vendors have one or more of these programs and use them when condensation is suspected.

10.16   Frame Loads

Most compressors have limitations on the loading imposed by the compression process. This loading results from the differential pressure across the piston. The loading seen by the frame is dependent on pressures internal to the cylinders. However, at the early stage of compressor sizing, the investigator will only know the pressures at the cylinder flanges. We can get a good idea on what class of machine will be required from calculations based on flange pressures.

For a double-acting compressor, the frame loads are calculated by:

$$\text{Tensile load} = P_d \times (A_p - A_r) - P_s \times A_p + P_a \times A_r \text{ lb} \qquad (10.11)$$

$$\text{Compressive load} = P_d \times A_p - P_s \times (A_p - A_r) - P_a \times A_r \text{ lb} \quad (10.12)$$

where P_d = discharge pressure, psia
P_s = suction pressure, psia
P_a = atmospheric pressure, psia
A_p = piston area, in^2
A_r = piston rod area, in^2

Most compressors require the load to reverse so that the load is tensile in one part of the cycle and compressive during the rest of the cycle. Failure to reverse will result in crosshead pin and/or bearing problems. The degree of reversal required for reliable operation depends on details of design. If the ratio of higher load to lower load is 5:1 or less, reversal will be adequate to allow proper lubrication of the pin and bearing. Some designs allow higher ratios, but the higher the ratio, the more sensitive the design will be. Caution is recommended in applying this criterion as crosshead pin reversal is affected by reciprocating inertia as well.

If reversal problems are encountered, special design considerations are in order. As mentioned earlier, single-acting cylinders, tailrods, divided cylinders, or tandem cylinders can be used to overcome reversal problems.

Table 10.1 gives a typical selection of strokes, rod diameters, speeds, BHP per crank, number of cranks, and maximum cylinder bores to use in initial sizing calculations. The frame loads given are less than maximum allowable to leave a margin for internal pressures and relief valve considerations. The compressor speeds are given as recommended maximum 60-Hz synchronous speeds for electric motor drives. Also included are horsepower per throw and maximum number of throws available.

10.17 Compressor Displacement and Clearance

Compressor displacement for a double-acting cylinder is calculated by

$$\text{DISP} = (2 \times A_p - A_r) \times S \times \frac{\text{RPM}}{1728} \text{ CFM} \qquad (10.13)$$

where A_p = piston area, in^2
A_r = piston rod area, in^2
S = stroke, in
RPM = rotative speed

TABLE 10.1 Typical Frame Sizes and Geometries Available from Major Reciprocating Compressor Manufacturers

Frame symbol	Frame load, lb	Stroke, in	Speed, r/min	Maximum number throws	BHP per crank	Rod dia, in	Maximum cylinder bore, in
High-Speed Separable Frames							
A	26,500	5.0	1200	4	480.0	2.00	22.50
B	50,000	6.0	1200	6	1000.0	2.50	26.50
Electric Drive Frames							
C	10,000	6.0	720	2	75.0	1.50	14.00
D	22,000	12.0	400	4	600.0	2.00	27.50
E	44,000	15.0	360	6	800.0	3.00	42.00
F	72,000	15.0	360	8	1900.0	3.50	42.00
G	90,000	15.0	360	10	2400.0	4.00	42.00
H	145,000	15.0	360	10	3300.0	5.00	42.00
I	170,000	15.0	360	10	4900.0	5.25	42.00
Integral Engines							
J	80,000	19.0	300	5	1000.0	4.00	17.50
K	105,000	19.0	330	8	1200.0	4.50	17.50

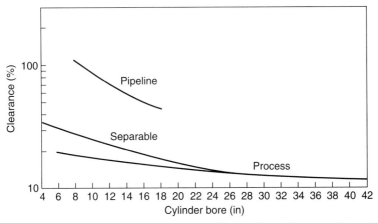

Figure 10.12   Type of cylinder vs. clearance. *(Dresser-Rand Company, Painted Post, N.Y.)*

The volumetric cylinder clearance is an important factor in calculating compressor capacity. For initial sizing purposes, it is recommended that the values given in Fig. 10.12 be used to calculate capacity. This figure gives approximate clearance values for three classes of compressors. This allows the user to select from high-speed machines used mainly in gas fields, or medium- and low-speed machines used in process and large enhanced oil recovery projects, and low-speed natural gas transmission pipeline machines.

10.18   Staging

As was brought out earlier in this text, the first decision to be made is how many stages of compression to use. This is dependent on many factors and the evaluation is based on the following considerations:

1. Discharge temperature

2. Process considerations (sidestreams)

3. Overall efficiency

4. Frame loads

5. Volumetric efficiency

The two factors that have the biggest effect on compressor reliability are discharge temperature and volumetric efficiency.

When discharge temperatures exceed 300°F (149°C) on compressors using mineral oil for cylinder lubrication, we find that reliability decreases. This is primarily due to breakdown of lubricating oil and the resulting deposits causing high wear and valve problems. Diester-base synthetic lubricants would provide increased reliability and should be given serious consideration.

We had seen earlier that whenever the volumetric efficiency gets too low, some of the factors that affect capacity and horsepower become quite large compared to the ideal conditions. This and compressor valve reliability on the discharge side causes us to limit the discharge volumetric efficiency at any operating condition to a minimum of 0.1 or 10 percent. To calculate the discharge volumetric efficiency, evaluate the suction volumetric efficiency from Eq. (10.14). The discharge volumetric efficiency is given by:

$$\text{VED} = \frac{\text{VE}}{R^{1/N}} \tag{10.14}$$

In many applications that could be handled with a single-stage compressor, power savings are available by using a two-stage approach with an intercooler between stages. The intercooler removes the heat of compression after the first stage. This increases the gas density, which reduces the bore requirement for the second stage. The result is lower power required in the second stage. However, the savings are not as good as one would calculate on an isentropic basis. There are interstage pressure drops and a second set of valve effects to take into account.

Remember that the equations presented earlier relate to cylinder flange pressures. It will be necessary to add pressure drops for initial suction, interstages, and final discharge to account for pulsation vessel, cooler, and piping losses. It is customary to use 1 percent of line pressure for each element in the system. For example, for initial suction pressure, assume a 1 percent loss to account for a suction vessel.

For an interstage system close-coupled to the cylinders use 1 percent for the first-stage discharge vessel, 1 percent for the intercooler, and 1 percent for the second-stage suction vessel for a total of 3 percent. If the intercooler is remote mounted, use an additional 1 percent for the piping for a total of 4 percent. For the final discharge use 1 percent for the vessel, 1 percent for the aftercooler (if used), and 1 percent for piping if the aftercooler is remote mounted.

10.19   Fundamentals of Sizing

The following step-by-step method is recommended for compressor sizing. While it may appear that sizing is a matter of following cookbook rules, considerable judgment and experience are required to do it well.

To perform reciprocating compressor screening and sizing calculations, the following information must be given:

1. Capacity requirement (convert to MMSCFD)
2. Initial suction pressure and temperature
3. Cooling medium temperature and desired approach temperature (determines interstage suction temperatures)
4. Final discharge pressure
5. Sidestream capacity, temperature and pressure (if used)
6. Gas analysis
7. Any special application information (relates to particular user requirements)

10.19.1   Determine the number of stages

Calculate the total pressure ratio R_t. If it exceeds 5, two or more stages are probably required. Assume no more than an average of 3 as the ratio per stage unless the discharge temperature allows. Use Eq. (10.15) to determine the number of stages.

$$R_s = (R_t)^{1/N_s} \qquad (10.15)$$

where R_s = stage pressure ratio
R_t = total ratio
N_s = number of stages

Use Eq. (10.7) and the suction temperatures given to check for acceptable discharge temperatures. If the discharge temperatures are not acceptable, increase the number of stages until they are. Remember to add the appropriate pressure losses previously mentioned.

10.19.2   Determine the approximate horsepower

Using the capacity and number of stages, use Eq. (10.5) to calculate the horsepower. This can be done by calculating each stage independently using the appropriate isentropic volumetric exponent. This may change for each stage dependent on pressures, temperatures, and gas analysis. Adding the individual stages together will give the total horsepower.

Using Table 10.1 allows us to determine what frame size is needed based on horsepower per throw and the total number of throws available. If the individual stage horsepowers and/or the total horsepower exceed what is available, we will need to use more than one throw per stage and/or more than one machine.

Another consideration at this point is the number of capacity steps demanded required by process considerations. If more than five are required, we will generally need more than one cylinder per stage. If the process designers want multiple steps, challenge them. Multiple steps will drive the cost of the compressor up and in many cases, will increase compressor size while decreasing efficiency and reliability.

10.19.3 Determine cylinder bore requirements

Using Fig. 10.12 and stroke and speed information from Table 10.1, determine the cylinder bores for each stage. This is done by trial and error using Eq. (10.3) and (10.13) to calculate cylinder bores required for various clearances as given in Fig. 10.12. This may best be accomplished by guessing a bore above what is required and one below, calculating the capacity, and plotting capacity vs. bore. The required bore can then be determined by the required capacity. Remember that the required capacity must be divided by the total number of cylinders used for each stage. A useful hint is that the compressor vendor can usually add clearance to a given cylinder to meet specific operating conditions. This is usually preferred to selecting nonstandard bore diameters.

Figure 10.13 indicates pressure ratings for various cylinder bores and classes of machine. This information is for cast cylinders only. Higher pressure forged steel cylinders are available. Check this against the discharge pressure for each stage. If any pressure ratings are exceeded, either increase the number of cylinders per stage or unbalance the pressure ratios on various stages to meet the requirements. This latter move means backing up to our earlier segment dealing with determination of stages, checking temperatures, and proceeding from there again.

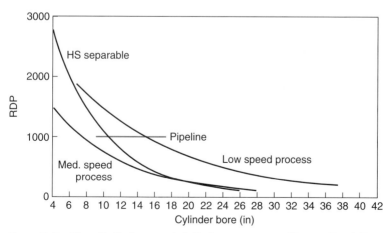

Figure 10.13 Size of cylinder vs. rated discharge pressure. *(Dresser-Rand Company, Painted Post, N.Y.)*

10.19.4   Check frame load

Using the calculated bores and Eq. (10.11), check that the frame loads meet the limits in Table 10.1 and the reversal requirements explained earlier.

If frame loads are exceeded, more cylinders will be required on the stages violating the maximums. The alternative may be to unbalance the pressure ratios if one stage exceeds the limits by a small amount. If the ratios are unbalanced, return to the segment explaining how stage requirements are arrived at, recheck the temperatures, and proceed again.

If the frame load reversal criteria are not met, the remedies suggested under Frame Loads (Sec. 10.16) may be tried. Remember that if a single-acting cylinder is used to correct reversals, the horsepower per throw in Table 10.1 is limited to 60 percent of the indicated value. If a divided cylinder is used, the horsepowers of the two stages must be added together. In any event, the frame load equation, Eq. (10.11), and displacement equation, Eq. (10.13), must be modified to reflect the geometric situation. Again, if the reversal criteria are marginally violated, it may be possible to unbalance ratios to correct it. Return to the segment explaining how stage numbers are determined, check temperatures and proceed.

10.19.5   What next?

If all of the above criteria are met, you have selected a compressor that will fit the application. The next problem is getting confirmation from the compressor vendors that this selection is valid. Remember, we have outlined approximate methods and approximate information. There are many design variations available that may affect this preliminary sizing. What you have is a reasonable estimate of what you will get.

10.20   Sizing Examples

We are now ready to look at two examples of compressor sizing. They are intended to demonstrate the principles and methodology presented in the preceding pages.

The equations used to make the calculations are either given directly or the equation numbers are given in parentheses after the result. It should be noted that some of the equations given earlier may be rearranged to find the unknown value. This is particularly true of the equation for piston displacement, Eq. (10.13).

It is recommended that the reader follow through these examples in detail to gain a feel for the methods used and the influence on performance introduced by varying the important parameters. We have elected to use only three significant figures in the calculations since this exceeds the possible accuracy of these methods. Obviously, the accuracy of more detailed performance calculation methods, including those of PC-based computer programs, would be considerably greater.

Example 10.1

1. *Given:*   Capacity required = 20 MMSCFD
Suction pressure = 75 psia
Suction temperature = 100°F
Air-to-gas coolers
Max. ambient = 110°F
Approach = 30°F
Discharge pressure = 500 psia
Barometer = 14.7 psia
No sidestream
Gas is pure methane
Design requirements:
1% initial suction pressure drop
3% interstage pressure drop
2% final pressure drop

2. Determine number of stages
Total ratios = 500/75 = 6.67 [from Eq. (10.15)]
Stage ratio (2 stages) = 2.58

Correcting for pressure drops (from Staging, Sec. 10.18)
1st suction = 75 × 0.99 = 74.3 psia
1st discharge = 75 × 2.58 × 1.03 = 199 psia
2d suction = 199 × 0.97 = 193 psia
2d discharge = 500 × 1.02 = 510 psia

1st stage ratio = 199/74.3 = 2.68
2d stage ratio = 510/193 = 2.64

Checking discharge temperatures [from Eq. (10.8)]
1st stage = 235°F
2d stage = 282°F

3. Determine approximate horsepower
1st stage (N_c = 0.85, N_m = 0.95) [from Eq. (10.5)]
BHP = 1190

2d stage (N_c = 0.85, N_m = 0.95)
BHP = 1170

Total BHP = 2360

From Table 10.1, we select the higher power, separable frame and plan on using four throws.

4. Determine bore sizes
1st stage (2 cylinders)

Rated discharge pressure = 199 psia (minimum)
Maximum bore (Fig. 10.13) = 21.5 in
Clearance (Fig. 10.12) = 13%
Displacement = 3000 CFM/CYL [from Eq. (10.13)]
Capacity (2 cylinders) = 32.4 MMSCFD [from Eq. (10.3)]

New displacement = 3000 × 20/32.4 = 1850 CFM/CYL
Approximate piston area = 224 in² [from Eq. (10.13)]
Bore = 16.9 in

We will use a 17-in bore,
Clearance (Fig. 10.12) = 19%
Displacement = 1870 CFM/CYL [from Eq. (10.13)]
Capacity (2 cylinders) = 18.4 MMSCFD [from Eq. (10.3)]

New displacement = 1870 × 20/18.4 = 2030 CFM/CYL
Approximate piston area = 246 in² [from Eq. (10.13)]
Bore = 17.7 in

We will use a 17.75-in bore,
Clearance (Fig. 10.12) = 18%
Displacement = 2040 CFM/CYL [from Eq. (10.13)]
Capacity (2 cylinders) = 20.4 MMSCFD [from Eq. (10.3)]

For rough sizing, this is accurate enough.

2d stage (2 cylinders)
Rated discharge pressure = 510 psia (minimum)
Maximum bore (Fig. 10.13) = 15 in
Clearance (Fig. 10.12) = 20%

Since we have the first stage sized, we can quickly estimate the second stage using the suction density ratio.

Estimated displacement = 2040 × 74.3 × 600/(193 × 560) = 841 CFM/CYL
Approximate piston area = 103 in² [from Eq. (10.13)]
Bore = 11.45 in

We will use a 11.5-in bore,
Clearance (Fig. 10.12) = 23%
Displacement = 845 CFM/CYL [from Eq. (10.13)]
Capacity (2 cylinders) = 19.1 MMSCFD [from Eq. (10.3)]

New displacement = 845 × 20/19.1 = 885 CFM
Approximate piston area = 109 in² [from Eq. (10.13)]
Bore = 11.8 in

We will use a 12-in bore,
Clearance (Fig. 10.12) = 22%
Displacement = 922 CFM/CYL [from Eq. (10.13)]
Capacity (2 cylinders) = 21.1 MMSCFD [from Eq. (10.3)]

At this point, we could further iterate on cylinder clearance and find that a 12-in bore with 25 percent clearance would produce 20.1 MMSCFD. However, for our purposes, we have the information we need to complete our job.

5. Check frame loads and reversals
1st stage
Tensile load = 30,000 lb [from Eq. (10.11)]
Compressive load = 31,100 lb [from Eq. (10.12)]

2d stage
Tensile load = 33,400 lb [from Eq. (10.11)]
Compressive load = 36,700 lb [from Eq. (10.12)]

These frame loads are well within the 50,000-lb application limit we set for this frame, and the reversals are excellent. There is the potential for saving money on this application if a lighter frame were available to do this job.

Our final results are:

1st stage (2 cylinders)
17.75-in bore × 6-in stroke

2d stage (2 cylinders)
12-in bore × 6-in stroke

Driver capability
Minimum 2360 hp at 1200 r/min

Example 10.2

1. *Given information:*   Capacity = 45.4 MMSCFD
Suction pressure = 93.3 psia
Suction temperature = 110°F
Cooling water = 100°F
Approach temperature = 10°F
Final discharge pressure = 1940 psia
Sidestream capacity = 20.4 MMSCFD
Sidestream pressure = 208 psia
Sidestream temperature = 110°F
Gas analysis:
Mainstream 83.2% H2, MW = 6.83, T_c = –327 F, P_c = 262 psia
Sidestream 82.4% H2, MW = 7.79, T_c = –315 F, P_c = 266 psia
Design conditions:
Max. piston speed = 850 ft/min
Initial pressure drop = 1%
Interstage pressure drop = 3%
Final pressure drop = 1%
Max. discharge temp. = 250°F
Atmospheric pressure = 14.4 psia

2. Determine the number of stages
The first stage is determined by the sidestream

1st stage suction pressure = $0.99 \times 93.3 = 92.4$ psia
1st stage discharge pressure = $1.03 \times 208 = 214$ psia
Pressure ratio = $214/92.4 = 2.32$
The gas properties:
$N_v = 1.33$ (API)
$N_t = 1.32$
$Z_s = 1.00$
Discharge temperature = 239°F [from Eq. (10.10)]

The rest of the machine is now considered.
Total ratio $R_t = 1940/208 = 9.33$
Stage ratio (2 stages) = 3.05 [from Eq. (10.15)]
2d stage gas properties:
$N_v = 1.32$ (API)
$N_t = 1.30$
$Z_s = 1.00$
2d stage discharge temperature = 277°F [from Eq. (10.10)]

The discharge temperature is too high. We will use three stages.

Stage ratio (3 stages) = 2.11

2d stage suction pressure = 208 psia
2d stage discharge pressure = $1.03 \times 208 \times 2.11 = 452$ psia
2d stage ratio = $452/208 = 2.17$
2d stage discharge temperature = 222°F [from Eq. (10.10)]

3rd stage suction pressure = $0.97 \times 452 = 438$ psia
3rd stage discharge pressure = $1.03 \times 2.11 \times 438 = 952$ psia
3rd stage ratio = $952/438 = 2.17$
3rd stage gas properties:
$N_v = 1.35$
$N_t = 1.31$
$Z_s = 1.01$
3rd stage discharge temperature = 225°F [from Eq. (10.10)]

4th stage suction pressure = $0.97 \times 952 = 923$ psia
4th stage discharge pressure = $1.01 \times 1940 = 1960$ psia
4th stage pressure ratio = $1960/923 = 2.12$
4th stage gas properties:
$N_v = 1.42$
$N_t = 1.32$
$Z_s = 1.02$
4th stage discharge temperature = 224°F [from Eq. (10.10)]

All stages meet the temperature criteria.

3. Determine the approximate horsepower
Using compression efficiency of 0.85 and mechanical efficiency of 0.95, we find, from Eq. (10.5):
1st stage horsepower = 2,570
2d stage horsepower = 3,390

3rd stage horsepower = 3,450
4th stage horsepower = 3,420
Total = 12,800

The piston speed maximum of 850 ft/min excludes using 15-in stroke at 360 r/min. Therefore, we will use 15-in stroke at 327 r/min (818 ft/min). The 145,000-lb frame is just short on horsepower per throw (2930 at 327 r/min), so we will try the 170,000-lb frame with four throws.

4. Determine cylinder bore requirements
1st stage:
Minimum rated discharge pressure (RDP) = 214 psia
Maximum bore (Fig. 10.13) = 37 in
Clearance (Fig. 10.12) = 12%
Displacement = 6040 CFM [from Eq. (10.13)]
Capacity = 42.1 MMSCFD* [from Eq. (10.3)]

This is less than the required capacity. We will need two cylinders and a minimum five-throw frame.

New displacement = 6040 × 45.4/42.1 = 6510 CFM
Piston area (1 cylinder) = 584 in^2 [from Eq. (10.13)]
Bore = 27.3 in

We will use 28 in
Clearance = 13%
Displacement (2 cylinders) = 6870 CFM [from Eq. (10.13)]
Capacity (2 cylinders) = 47.3 MMSCFD [from Eq. (10.3)]

We will use this bore and add clearance to match capacity.

2d stage:
Minimum RDP = 452 psia
Maximum bore (Fig. 10.13) = 26 in
Clearance (Fig. 10.12) = 13%
Displacement = 2950 CFM [from Eq. (10.13)]
Capacity = 46.4 MMSCFD [from Eq. (10.3)]

This is less than the required capacity. We will need two cylinders and a minimum six-throw frame.

New displacement (2 cylinders) = 2950 × 65.8/46.4 = 4180 CFM
Piston area (1 cylinder) = 379 in^2 [from Eq. (10.13)]
Bore = 22.0 in

We will use 22.0 in
Clearance (Fig. 10.12) = 14%
Displacement (2 cylinders) = 4190 CFM [from Eq. (10.13)]
Capacity = 65.2 MMSCFD [from Eq. (10.3)]

---

* Note Z_{std} = 1.0.

This is within 1% of the required capacity, and we will use this bore.

3rd stage:
The approximate displacement is determined by the second-stage displacement times the density ratio.

Approximate displacement = 4190 × 208/438 = 1990 CFM
Piston area = 361 in² [from Eq. (10.13)]
Bore = 21.4 in

A minimum RDP of 952 psia is required. The maximum bore from Fig. 10.13 is 16 in. We will need two cylinders and a minimum seven-throw frame.

Piston area (1 cylinder) = 186 in² [from Eq. (10.13)]
Bore = 15.4 in

We will use 15.5 in
Clearance (Fig. 10.12) = 15%
Displacement (2 cylinders) = 2020 CFM [from Eq. (10.13)]
Capacity = 65.2 MMSCFD [from Eq. (10.3)]

This is within 1 percent of the required capacity, and we will use this bore.

4th stage:
The approximate displacement is determined by the third-stage displacement times the density ratio. We will follow the lead set by the other three stages and use two cylinders requiring an eight-throw frame.

Approximate displacement (2 cylinders) = 2020 × 438/952 = 929 CFM
Piston area (1 cylinder) = 92.6 in² [from Eq. (10.13)]
Bore = 10.9 in

A minimum RDP of 1960 psia is required. This is above the RDP for a 10.9-in bore. We will have to use a forged steel cylinder to get the required RDP.

We will use an 11-in bore
Clearance (Fig. 10.12) = 18%
Displacement (2 cylinders) = 956 CFM [from Eq. (10.13)]
Capacity (2 cylinders) = 65.7 MMSCFD [from Eq. (10.3)]

This is within 1 percent of the required capacity, and we will use this bore.

5. Check frame loads
1st stage
Tensile load = 70,600 lb [from Eq. (10.11)]
Compressive load = 76,600 lb [from Eq. (10.12)]

2d stage
Tensile load = 83,300 lb [from Eq. (10.11)]
Compressive load = 96,900 lb [from Eq. (10.12)]

3rd stage
Tensile load = 76,700 lb [from Eq. (10.11)]
Compressive load = 106,000 lb [from Eq. (10.12)]

4th stage
Tensile load = 56,400 lb [from Eq. (10.11)]
Compressive load = 118,000 lb [from Eq. (10.12)]

These loads are well below the 170,000-lb maximum we selected. We can use the 145,000-lb frame with adequate margin. The reversals are also within acceptable limits.

6. The selection
Our compressor has the following characteristics:

Frame: 8-throw, 15-in stroke, 145,000-lb frame load

1st stage cylinder: 28-in bore, two required
2d stage cylinder: 22-in bore, two required
3rd stage cylinder: 15.5-in bore, two required
4th stage cylinder: 11-in bore, two required

Driver: 327 r/min synchronous motor with a minimum power capability of 12,800 hp. A 13,000-hp motor would be selected to give adequate margin for relief valve setting.

While this basically manual selection approach seems at first tedious, the responsible engineer will quickly become proficient. Most importantly, there is no substitute for going the nonautomated, noncomputerized route when it comes to acquiring a thorough knowledge of the numerous interrelating factors that lead to intelligent equipment selection.

Dynamic Compressor Technology

Dynamic compressors are based on the principle of imparting velocity to a gas stream and then converting this velocity energy into pressure energy. These compressors are frequently called turbocompressors, and centrifugal machines comprise perhaps 80 or more percent of dynamic compressors. The remaining 20 or less percent are axial flow machines intended for higher flow, lower pressure applications as illustrated earlier in the figure in the introduction to Part 1.

Centrifugal Compressor Overview

Centrifugal compressors are relatively troublefree, dependable gas movers. Almost any gas can be compressed by these machines, and their extensive size and pressure ranges made modern process plants and efficient production of bulk chemicals possible in many instances.

Thousands of centrifugal compressors are single-stage machines, either direct-driven or geared (see Fig. A), and thousands are executed in multistage configuration, Fig. B. Both single- and multistage machines are generally made up of standardized components. There are two principal casing types: (1) horizontally split casing, Fig. C and (2) vertically split casing (barrel-type compressors), Fig. D. The nozzle configurations can be selected over a wide range.

To date, many machines have been built for intake volumes between 500 and 200,000 m^3/h (294 to 117,000 cfm) at discharge pressures up to 160 bar (2352 psi). Barrel-type compressors for higher pressures have been designed and are

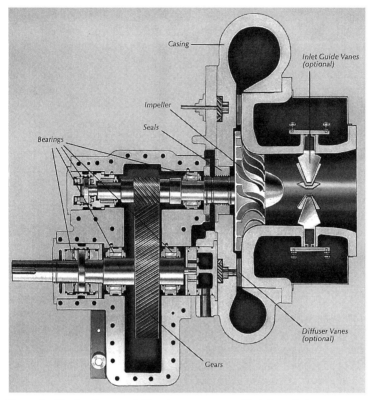

Figure A Single-stage centrifugal compressor with integral step-up gearing. *(Dresser-Rand Company, Olean, N.Y.)*

Figure B Multistage centrifugal compressor in a petrochemical plant. *(Elliott Company, Jeannette, Pa.)*

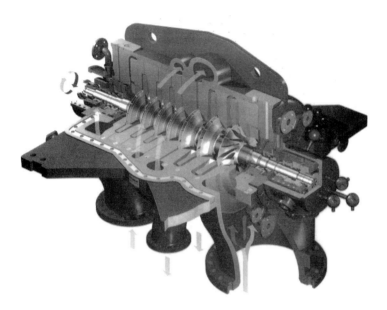

Figure C   Centrifugal compressor with horizontally split casing construction. *(Mannesmann-Demag, Duisburg, Germany)*

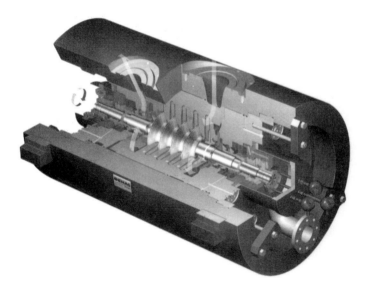

Figure D   Centrifugal compressor with vertically split (also called radially split) design. *(Mannesmann-Demag, Duisburg, Germany)*

Figure E   Axial compressor set for an aircraft test bed in France. These machines can be used to generate compressed air or vacuum. The installation comprises six identical axial compressors. Capacity: 244,000 Nm³/h in compression; 38,000 Nm³/h in vacuum mode. *(Sulzer, Ltd., Winterthur, Switzerland)*

Figure F   Typical axial flow compressors. *(Sulzer, Ltd., Winterthur, Switzerland)*

operating very successfully. Depending on the volume flow and the compression ratio, two, three, or more casings can be arranged in line even with intervening gears. Drive is usually provided directly from a steam turbine, gas turbine, or expander turbine, as well as by an electric motor with gears, or with variable-speed drivers.

With regard to the volume and compression ratio, and in the selection of the materials for the casing, impellers, and other components, the design is extremely flexible. As will be seen

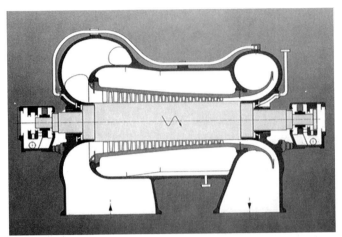

Series A with fixed stator blades (FIXAX).

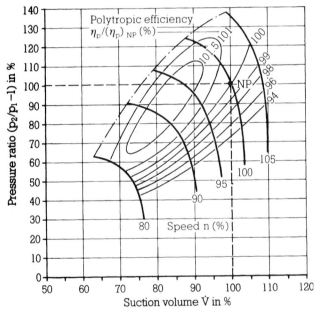

Figure G   Performance maps for axial compressors with speed variation. *(Sulzer, Ltd., Winterthur, Switzerland)*

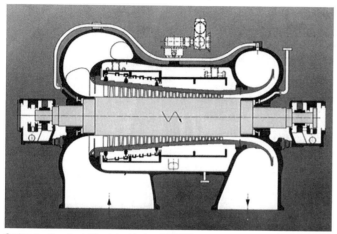

Series A with adjustable stator blades (VARAX).

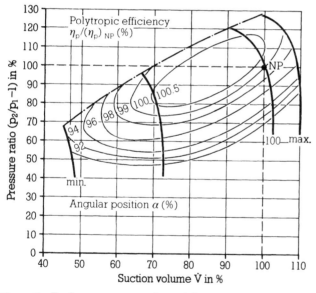

Figure G Performance maps for axial compressors with adjustable stator blades. *(Sulzer, Ltd., Winterthur, Switzerland)*

later, labyrinth seals, mechanical contact seals, floating seals, or dry gas seals can be provided for shaft sealing. Consequently, centrifugal compressors can be used for practically every gas compression requirement.

Axial Compressor Overview

Axial flow compressors can handle large flow volumes in relatively small casings and with favorable power requirement. They are available in sizes producing pressures in excess of 7 bar (about 100 psi) at intake volumes between 40,000 and 1,000,000 m^3/h (23,500 and 588,500 cfm).

Axial flow compressors, Figs. E and F, are most often used for blast furnace air and air separation services but can also be used for nitric acid plants, natural gas liquefaction, etc. Typical performance maps for these compressors are depicted in Fig. G.

Drive is provided by steam turbines or electric motors. In the case of direct electric motor drive, low speeds are unavoidable unless sophisticated variable frequency motors are employed.

Simplified Equations for Determining Performance of Dynamic Compressors*

The following data comprise the fundamental equations that are used in the determination of brake horsepower, operating speeds, and discharge temperature of centrifugal and axial gas compressors.

11.1 Nonoverloading Characteristics of Centrifugal Compressors

Impellers having backward-leaning vanes have a head capacity characteristic that, at constant speed, the discharge pressure decreases gradually with increasing capacity. Thus, at rated suction temperature and pressure, it is not possible to overload a properly selected prime mover since both the head and brake horsepower will decrease appreciably as the capacity increases above 120 percent of the rated capacity.

11.2 Stability

Stability is defined in conjunction with a so-called surge point. Dynamic compressors surge, or undergo a reversal of flow direction, when the gas throughput drops below a certain value that is uniquely defined by compressor geometry, operating conditions, gas properties, and other variables. This flow reversal usually takes place at or near

---

* Narrative and illustrations contributed by Dresser-Rand Company, Olean, N.Y., except as noted

the impeller tip; it can cause process upsets and/or serious mechanical damage to compressor internals. It will be further discussed under "Control," in Chap. 16.

The percent of change in capacity between the rated capacity and the surge point, at rated head, is measured as the stability of centrifugal compressors (Fig. 11.1). This value will vary from approximately 70 percent for compressors developing very low pressure ratios to as low as 30 percent for compressors developing very high ratios. In the initial design of a compressor, provision can be made for high stability at slightly reduced efficiency, if it is likely that partial loads will be of long duration. When the design load is sustained most of the time, the efficiency can be improved at the expense of stability.

Figure 11.1 also shows how the performance of dynamic compressors is influenced by different control methods: (1) operation at speeds ranging from 70 to 105 percent of original design; (2) operation at constant speed, but with different guide vane settings; and (3) operation at constant speed and suction valve throttling. In each case, the vertical axis represents either head or pressure developed, while flow (in acfm or m³/h) is represented on the horizontal axis.

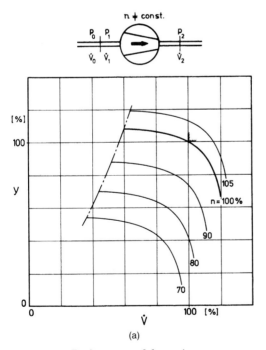

Figure 11.1 Performance of dynamic compressors: (a) variable-speed. *(Sulzer, Ltd., Winterthur, Switzerland)*

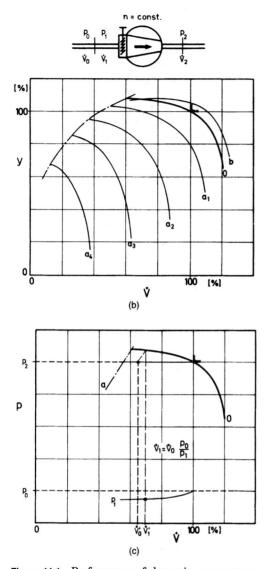

Figure 11.1 Performance of dynamic compressors:
(b) variable inlet guide vanes; (c) suction valve
throttling. *(Sulzer, Ltd., Winterthur, Switzerland)*

11.3 Speed Change

The large difference in head with small changes in speed is illustrated
by the head capacity curve, Fig. 11.1a. This shows that centrifugal compressor prime movers are sometimes designed for operation between
70 and 105 percent of the rated speed. Operation without speed change

results in maintaining a head vs. flow relationship as described by the performance curve of Fig. 11.2. Note that operation to the left of the surge limit and in the choke flow, or "stonewall" region, is not feasible.

11.4 Compressor Drive

The type of prime mover that is used for centrifugal compressors will be determined in most cases by the economics of the application. There are four different classes of prime movers that are considered most suited for centrifugal compressors.

Steam turbine

Electric motor

Expansion turbine (expander)

Combustion gas turbine

Most centrifugal compressors have been built for drive by the four types mentioned, illustrating the flexibility with which they may be applied. Driver selection is determined primarily by the following factors:

1. Water requirements (steam consumption)
2. Operating speeds

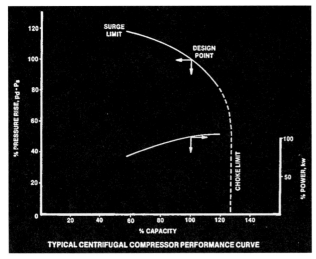

Figure 11.2 Head capacity curve for centrifugal compressor. *(Dresser-Rand Company, Olean, N.Y.)*

3. Process control

4. Process steam supply

5. Fuel or energy costs

6. Reliability

Adjustable inlet guide vanes or suction throttling are available for constant-speed motor drive. The electric motor may also be used for variable-speed operation by the use of hydraulic couplings or other means.

11.5 Calculations

The three items usually to be determined in centrifugal compressor calculations are:

1. Shaft horsepower

2. Operating speed

3. Discharge temperature

Determination of horsepower and speed is predicated on the calculation of the head required for the compression. Head, which actually represents the work being done per pound of fluid being handled, is expressed in terms of feet (i.e., ft · lb/lb or Nm/kg) just as for a liquid pump, and is fundamentally defined by the following:

$$H = k_1 \int V dP \tag{11.1}$$

where H = head, ft (m)
V = specific volume, ft^3/lb (m^3/kg)
P = pressure, psia (bar, absolute)
k_1 = different constants, for English or metric conversions and expressions

For a liquid pump, where the specific volume or density is constant, Eq. (11.1) is readily integrated to:

$$H = k_1 V(P_2 - P_1) = \frac{k_1(P_2 - P_1)}{\rho} \tag{11.2}$$

where ρ = density, lb/ft^3 or kg/m^3.

For a centrifugal compressor, where the specific volume is a variable, a somewhat more complex relation is obtained. If it is assumed that the compression is polytropic and may be represented by the equation:

$$PV^n = \text{constant}$$

Equation (11.1) may be integrated and rearranged to the familiar form:

$$H = \frac{k_1 P_1 V_1}{(n-1)/n} \left[\left(\frac{P_2}{P_1} \right)^{(n-1)/n} - 1 \right] \tag{11.3}$$

where n = polytropic exponent of compression.

Equation (11.3) may alternatively be expressed in the form:

$$H = \frac{ZRT_1}{(n-1)/n} \left[\left(\frac{P_2}{P_1} \right)^{(n-1)/n} - 1 \right] \tag{11.4}$$

where R = gas constant = $\dfrac{1545}{\text{mol. wt.}}$ or $\dfrac{8314}{\text{mol. wt.}}$

T_1 = suction temperature, °R or °K
Z = average compressibility

For ease in calculation, Eqs. (11.3) and (11.4) may be expressed in the form:

$$H = k_1 P_1 V_1 \beta = ZRT_1 \beta \tag{11.5}$$

where $\beta = \dfrac{(P_2/P_1)^M - 1}{M}$

$M = \dfrac{n-1}{n}$

From Eqs. (11.4) and (11.5), it is therefore apparent that the head, and hence the horsepower, for a given compression varies directly with the absolute suction temperature and inversely with the molecular weight of the gas being handled. Since there is a limit to the amount of head that a single impeller will develop, as will be subsequently demonstrated, it follows that gases having a high molecular weight will require fewer impellers (i.e., stages) than gases having a low molecular weight when being compressed through the same ratio. This is indicated in Fig. 11.3 for several common gases.

For perfect gas compression, the compressibility factor is unity. For real gas compression, this factor deviates from unity. In those instances where the amount of this deviation is not large, i.e., where the average compressibility factor varies between 0.95 and 1.02, or where it remains fairly constant over the range of compression, an average value of the compressibility factor may be used in Eq. (11.4) with negligible error. In other instances, where the compressibility factor is subject to larger variation over the range of compression, the head may be approximated from the following relation:

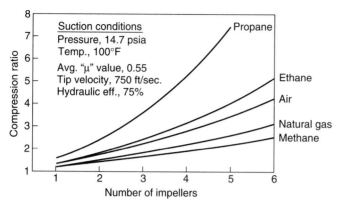

Figure 11.3 Compression ratio vs. number of impellers. *(Dresser-Rand Company, Olean, N.Y.)*

$$H = k_2 \left[(P_1 V_1) + (P_2 V_2) \right] \log_{10} \frac{P_2}{P_1} \qquad (11.6)$$

where k_2 represents different constants in the English and metric systems of measurement.

Equation (11.6) is not strictly correct. It is based on the assumption that the log mean value of PV equals the arithmetic mean. This will result in an error of 1.2 percent for high compression ratios and is based only on the assumption that the compression is polytropic and can be represented by a single exponent (n).

To be correct, the following formula should be used:

$$H = k_1 \log_{10} \left(\frac{P_2}{P_1} \right) \left[\frac{P_2 V_2 - P_1 V_1}{\log_{10} (P_2 V_2 / P_1 V_1)} \right] \qquad (11.7)$$

Equation (11.6) is of particular utility for hydrocarbon gases at moderate or high pressures and/or low temperatures.

It is to be noted that the successful usage of Eqs. (11.3) and (11.4) is dependent on the determination of the polytropic exponent. This may be readily obtained from the following equations, which follow from the definition of hydraulic efficiency:

$$\eta = - \frac{\int V dP}{\Delta h} = \frac{\dfrac{P_1 V_1}{(n-1)/n} \left[\left(\dfrac{P_2}{P_1} \right)^{(n-1)/n} - 1 \right]}{\dfrac{P_1 V_1}{(K-1)/K} \left[\left(\dfrac{P_2}{P_1} \right)^{(n-1)/n} - 1 \right]} \qquad (11.8)$$

$$\eta = \frac{(K-1)/K}{(n-1)/n} \qquad (11.9)$$

where η = hydraulic efficiency
Δh = change in enthalpy, Btu/lb, or kJ/kg
K = isentropic exponent (c_p/c_v)

The hydraulic efficiency is established by tests and is generally a function of the capacity at suction conditions to the compressor.

The head that a centrifugal compressor stage, consisting of an impeller and diffuser, will develop may be related to the peripheral velocity by the following:

$$H = \mu \frac{u^2}{g} \qquad (11.10)$$

where μ = pressure coefficient
u = peripheral velocity, ft/s, or m/s
g = gravitational constant, 32.2 ft/s^2 or 9.81 m/s^2

The value of the pressure coefficient referred to previously is a characteristic of the stage design. An average value for one stage of a multistage centrifugal compressor is 0.55. If a peripheral velocity of 770 ft/s (235 m/s) is assumed, it can be seen that the head per stage is approximately 10,000 ft (3050 m). This permits ready approximation of the number of stages required to develop the head corresponding to the particular compression process.

The power required for the compression of a gas may be calculated from the following:

$$\text{GHP} = \frac{(W)(\Delta h)}{33,000} \qquad (11.11)$$

or

$$\text{KW} = \frac{(m)(H)}{60,000}$$

where   GHP = gas horsepower
W = gas flow, lb/min
KW = gas power, kW
m = mass flow, kg/min
H = differential head, Nm/kg (metric), ft (English)

From Eq. (11.8), however:

$$\Delta h = \frac{\int V dP}{\eta} = \frac{H}{\eta} \qquad (11.12)$$

Therefore:

$$GHP = \frac{(W)(H)}{(33,000)(\eta)}$$

and

$$KW = \frac{(m)(H)}{(60,000)(\eta)}$$

The compressor shaft horsepower is, of course, equal to the gas horsepower divided by the mechanical efficiency. For most centrifugal compressor applications, the mechanical losses are relatively small, and an average mechanical efficiency of 99 percent may be used for estimating purposes.

The rotative speed of a centrifugal compressor is fixed by the peripheral velocity of the impellers and their diameter. As previously indicated, the peripheral velocity is determined by the head to be developed; the impeller diameter is determined by the capacity to be handled, as measured at suction conditions.

From Eq. (11.10):

$$u = \sqrt{\frac{Hg}{\mu}} \qquad \text{also} \qquad u = \frac{\pi DN}{720} \ \text{ft/s}$$

where D = impeller diameter, in
$\quad\quad\ N$ = rotative speed, r/min
$\quad\quad\ H$ = head per stage, ft

Alternatively, $u = \pi DN/60$ m/s, where D and H are expressed in meters. Thus:

$$N = \frac{720u}{\pi D} = \frac{720\sqrt{Hg/\mu}}{\pi D} \tag{11.13}$$

or:

$$N = \frac{1300}{D}\sqrt{\frac{H}{\mu}} \ \text{r/min}$$

where D and H = in and ft, respectively; and:

$$N = 59.82/D\sqrt{\frac{H}{\mu}} \ \text{r/min}$$

where D and H are expressed in meters

As previously indicated, for approximate calculations an average pressure coefficient of 0.55 may be assumed.

The discharge temperature for an uncooled compression process may be calculated from the fundamental relation:

$$T_2 = T_1\left(\frac{P_2}{P_1}\right)^M \tag{11.14}$$

For those applications where the discharge temperatures for an uncooled compression process would be prohibitive, internal diaphragm cooling, or external interstage cooling may be used. If internal diaphragm cooling is used, the average exponent of compression is approximated by the isentropic exponent, and the head may be estimated on the basis of this. With external interstage cooling, each stage, or compressor body, is dealt with separately.

12

Design Considerations and Manufacturing Techniques

12.1 Axially or Radially Split?

Figures C and D in the introduction to Part 2 illustrated these two options that are available for many centrifugal process gas compressors. The decision as to whether an axially or radially split casing should be used depends on a number of factors that will be highlighted next.

12.2 Tightness

The radially split design has circular casing joints or flanges with a perfectly even load distribution (Figs. 12.1 and 12.2). The leakage of gas at the two covers can thus be prevented most effectively. Besides metal-to-metal contact, endless O-rings are inserted in grooves on the two covers. By monitoring the pressure between two adjacent rings, the tightness can be controlled. For toxic, flammable, and explosive gases the barrel design is therefore always of advantage. For this reason, the API 617 Standard specifies the radially split casing construction for gases containing hydrogen if the hydrogen partial pressure exceeds 13.8 bar (200 psig).

12.3 Material Stress

The cylindrical design with the smallest possible inner diameter is obviously the most suitable construction. With axially split casings the available space for bolting is further restricted at the two shaft penetrations. To achieve the required tightness, a high contact pressure at

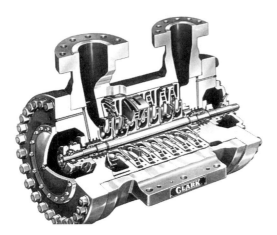

Figure 12.1 Radially split compressor with bolted-on heads, suitable for low-pressure service. *(Dresser-Rand Company, Olean, N.Y.)*

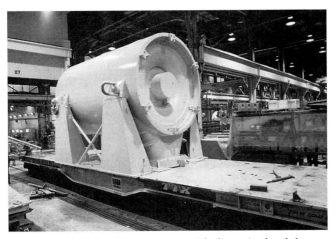

Figure 12.2 Radially split compressor with shear-ring head closure. *(Dresser-Rand Company, Olean, N.Y.)*

the joints is required. The necessary forces in the bolts are often higher than would be required by the static gas forces if the casing flanges were perfectly rigid and flat. For large compressor frame sizes the radially split design can therefore be the only possible solution even at moderate pressures.

12.4 Nozzle Location and Maintenance

For operating pressures where an axially split design would be perfectly adequate, barrel compressors are sometimes preferred since the nozzles can be arranged in any radial direction. If the necessary space in the axial direction at the nondriven shaft end is available for the horizontal pullout of the inner casing cartridge, inspections, rotor changes, or complete cartridge replacements can be accomplished quickly without removing any process piping to and from the compressor. For tandem units, on the other hand, the first or low-pressure casing should be of the axially split design up to the highest possible operating pressure. A barrel compressor coupled at both shaft ends has to be removed for overhaul or replacement of the rotor.

Seals and bearings in state-of-the-art barrel compressors can, however, always be serviced and replaced with the barrel casing remaining in place. Finally, it should be noted that only at moderate pressures, a ring of sturdy bolts can successfully attach the end walls to the outer casing (Fig. 12.1). As pressure increases, however, it becomes mechanically impractical to provide a sufficient number of bolts of sufficiently large diameter to contain the pressure. This is when shear-ring enclosures, Figs. 12.2 and 12.6, are the preferred solution.

Compressor manufacturers are often able to give graphical guidelines or plots that allow the purchaser to zero in on probable casing recommendations. This is done in Fig. 12.3 where the potential client can determine which Elliott compressor frame should be selected to compress 33,000 icfm (56,200 inlet m³/h) of process gas to a gauge pressure of 450 psi (31 bar).

Plots of pressure and flow determine that the horizontally split 46M frame, with a capacity of 22,000 to 34,000 icfm (37,000 to 58,000 m³/h), will serve. If a very low molar mass, hydrogen-rich gas is involved, the vertically split 46MB (barrel-type) frame would be selected.

A vertical line drawn to the speed plot establishes that compressor speed can range approximately from 4600 to 6000 r/min. Since we are approaching the maximum flow capacity of this particular compressor frame, it would be best to operate in the upper half of the speed limit.

12.5 Design Overview*

12.5.1 Casings

Horizontally split centrifugal compressors consist of upper and lower casing halves that are fastened together by stud bolts through mating flanges at the horizontal centerline. Where moderate pressures are encountered, this type of construction may offer advantages in maintainability. Access to the internals of the compressor is gained by a single vertical lift of the upper casing half. Figure 12.4 illustrates this design.

---

* "Design Overview" and remainder of chapter developed and contributed by Harvey Galloway and Arthur Wemmell, Dresser-Rand Company, Olean, N.Y.

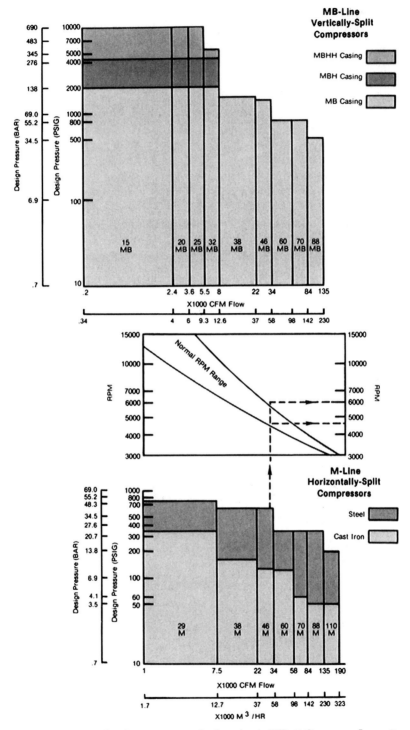

Figure 12.3 Centrifugal compressor selection chart. *(Elliott Company, Jeannette, Pa.)*

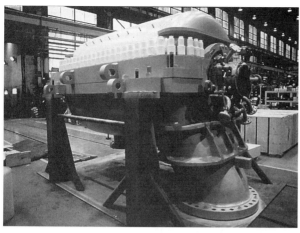

Figure 12.4 Horizontally split centrifugal compressor. *(Dresser-Rand Company, Olean, N.Y.)*

Access to the radial bearings, thrust bearings, and seal for inspection or maintenance does not require removal of the upper casing half.

As depicted in Fig. 12.5, vertically split compressors consist of a case that is formed in the shape of a cylinder and open at either end. End closures or heads are attached at either end. Access to the internals of the compressor is gained by removing the outboard head.

Refer to Fig. 12.6 for a typical cross section of a centrifugal compressor showing the major features that will be discussed, including the thrust and radial bearings, the seals, the rotor and impellers, and the various gas flow path configurations available.

Figure 12.5 Vertically split centrifugal compressor. *(Dresser-Rand Company, Olean, N.Y.)*

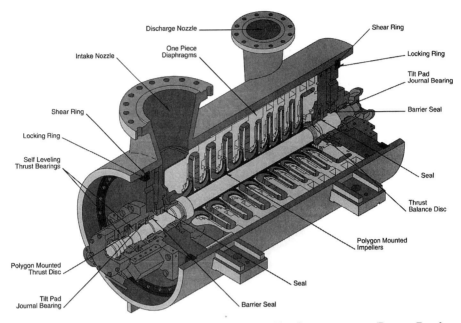

Figure 12.6 Major components of multistage centrifugal compressors. *(Dresser-Rand Company, Olean, N.Y.)*

While our emphasis is on high-pressure multistage centrifugals, we do want to acknowledge the use of axial compressors in the petrochemical, refinery, and chemical industry. Accordingly, Fig. 12.7 depicts the combination of an axial flow compressor at the initial gas intake with a radially bladed compressor section prior to gas discharge.

Figure 12.7 Combination axial-radial flow compressor. *(Dresser-Rand Company, Olean, N.Y.)*

The multitude of compressor applications has resulted in a variety of case and nozzle configurations. The more common case and nozzle configurations are illustrated in Fig. 12.8.

Looking first at the flow path in Fig. 12.9, this compressor has a *straight-through* flow path, meaning that the gas enters through the main inlet of the compressor, passes through the guide vanes into the impeller, is discharged from the impeller into the diffuser through the return bend and into the next impeller, and so on until the total flow is discharged through the nozzle at the other end of the compressor.

Side Streams
Side stream nozzles permit introducing or extracting gas at selected pressure levels. These flows may be process gas streams or flows from economizers in refrigeration compression. Side "loads" may be introduced through the diafram between two stages or, if flow is high, into the area provided by omitting one or two impellers.

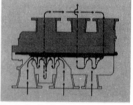

First Section Double-Flow
Capacity of this compressor is significantly increased by arranging double-flow compression in the first section. This is only feasible where head requirements are moderate to low, as in applications with low-pressure rise and/or high molar mass (propane or propylene refrigeration, for example).

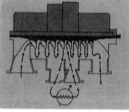

Iso-Cooling (cooling between stages)
When gas temperatures within the compressor reach 200 to 400 °F (90 to 200°C), efficiency is improved and compressor power reduced by cooling the gas. Iso-Cooling nozzles permit the hot gas to be taken from the compressor to an external heat exchanger, then returned to the following stage at reduced temperature for further compression.

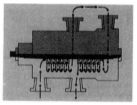

Back-to-Back
Minimizes thrust when a high-pressure rise is to be achieved within one casing. Note how the thrust forces acting across the two "banks" of impellers tend to neutralize one another. A high-pressure, barrel-type casing.

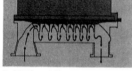

Straight-Through Flow
A convenient arrangement. May employ as many as 12 stages. Most often used for low-pressure process gas compression.

Double-Flow
Doubles the possible maximum flow capability of the compressor. Since the number of impellers handling each inlet flow is only half that of an equivalent straight-through machine, the maximum head capability is necessarily reduced.

Double Iso-Cooling
High molecular mass gases such as chlorine (mass 70.91) heat much more rapidly during compression than do low molar mass gases. In such service, in order to maintain compression efficiency (and to avoid deposits within the compressor due to polymerization), it may be necessary to remove the hot gas from two points within the compressor, passing the gas stream through two intercoolers.

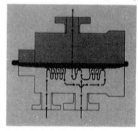

Back-to-Back plus Recirculation Feature
This unit is typical for combined feed gas and circulator service. Last stage flow is combined with recycle flow. Mixing of flows can be either internal or external by means of an added discharge nozzle. Crossover flow is internal through passages in the inner casing.

Figure 12.8 Compressor casing and nozzle configuration. *(Elliott Company, Jeannette, Pa.)*

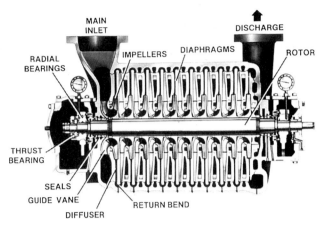

MAIN
INLET

DISCHARGE

RADIAL
BEARINGS

IMPELLERS   DIAPHRAGMS   ROTOR

THRUST
BEARING

SEALS

GUIDE VANE

RETURN BEND

DIFFUSER

TEN STAGE CENTRIFUGAL COMPRESSOR

Figure 12.9 Straight-through compressor casing. *(Dresser-Rand Company, Olean, N.Y.)*

Flexibility of nozzle orientation on a straight-through compressor allows various positions of the suction and discharge flanges. The most common orientation is both nozzles pointed up or down, but in some applications, such as gas recirculators and boosters, the nozzles are located on the side (Fig. 12.10).

Figure 12.10 Side-oriented nozzles on booster compressor. *(Dresser-Rand Company, Olean, N.Y.)*

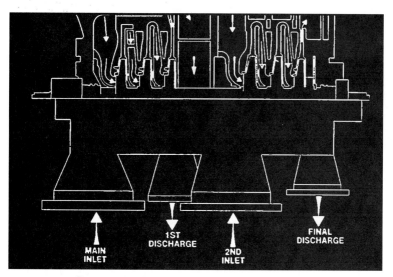

Figure 12.11 Cross-sectional view of compound compressor. *(Dresser-Rand Company, Olean, N.Y.)*

Reviewing the cross section of a *compound compressor,* Fig. 12.11, it can be seen that the flow path is the same as two straight-through compressors in series. That is, the total flow enters at the main inlet of the compressor and is totally discharged at the first discharge connection, is cooled or otherwise reconditioned, reenters the compressor at the second inlet connection, and is totally discharged at the final discharge nozzle.

In applications with high compression ratios, intercooling is desirable to minimize gas temperature and power requirements, as well as meet other process requirements. In many applications, compounding can reduce the number of compressor casings required. Figure 12.12 depicts a string, or train, of casings. Gas discharged from the first casing is led to suitable external heat exchangers before reentering into either the following compressor section or the next casing.

For high pressure ratios, arranging the impellers in a back-to-back configuration results in two inlet and two discharge nozzles, Fig. 12.13. This arrangement balances the axial forces on the rotor and eliminates the requirement for a balance piston, thereby reducing shaft horsepower. Here, the gas flow enters the casing (case) at one end and leaves the case near the middle. The gas is redirected after externally cooling, if desired, in a second section at the opposite end of the case and finally leaves the casing in the center at the required discharge pressure.

Figure 12.12 Compressor string, or casing train. *(Dresser-Rand Company, Olean, N.Y.)*

Two high-pressure compressors operating in series are shown in Fig. 12.14 while undergoing a full-load pressure test. Both of these units utilize a back-to-back configuration with an integral crossover.

In the cross-sectional view of Fig. 12.15, examples of both incoming and outgoing sidestreams are shown. Flow enters the main inlet and is compressed through one impeller to an intermediate pressure level, at which point an incoming sidestream flow is mixed with the main inlet flow in the diaphragm area ahead of the next impeller. The total mixed

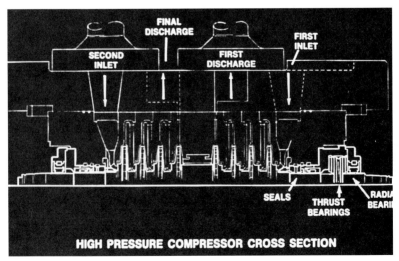

Figure 12.13 Back-to-back impeller orientation. *(Dresser-Rand Company, Olean, N.Y.)*

Figure 12.14 High-pressure compressor in series flow arrangement. *(Dresser-Rand Company, Olean, N.Y.)*

flow is compressed to a higher pressure level through two impellers; a small portion of the flow leaves the compressor through an outgoing sidestream to satisfy a process requirement. The remainder of the flow is compressed through one impeller, mixed with an incoming sidestream, compressed through two stages, and exits through the final discharge.

For refrigeration cycles and other process requirements, the capability to admit or discharge gas at intermediate pressure levels is

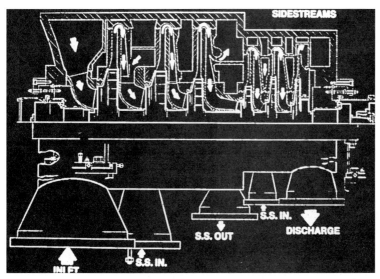

Figure 12.15 Cross-sectional view of sidestream compressor. *(Dresser-Rand Company, Olean, N.Y.)*

Figure 12.16   Sidestream compressor. *(Dresser-Rand Company, Olean, N.Y.)*

required. Compressors provide sidestreams with minimum flow disturbance and provide effective mixing of the main and sidestream gas flows (Fig. 12.16).

In the double-flow configuration of Figs. 12.17 and 12.18, the compressor is divided into two sections. It is effectively operating as two parallel compressors. An inlet nozzle is located at either end of the compressor case. The discharge flow from each section is extracted from a common discharge nozzle at the center of the case. The

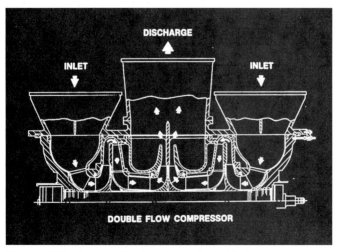

Figure 12.17   Cross-sectional view of double-flow compressor. *(Dresser-Rand Company, Olean, N.Y.)*

Figure 12.18 Double-flow compressor installation. *(Dresser-Rand Company, Olean, N.Y.)*

impellers of one section face in the opposite direction from the impellers in the other section, achieving thrust balance over all operating conditions.

This concept effectively doubles the capacity of a given frame size and has several advantages:

1. *Smaller frame.* For a given capacity, a compressor one frame smaller than the single-flow configuration can be used, thus reducing compressor costs.

2. *Speed match.* In many applications, the flow from the double-flow compressor is discharged to a single-flow compressor of the same frame size. This permits operations at the same speed and allows the use of a single driver or duplication of drivers.

In yet other applications, it is desirable to cool after each stage of compression. The case provides eight nozzle connections allowing intercooling after each impeller. Note that this type of particular design, Figs. 12.19 and 12.20, allows the use of over-the-impeller volute-type diffusers since the stage spacing required for the nozzle allows sufficient room for the required volute area. A typical application utilizing this case and nozzle arrangement would be the compression of oxygen in an air separation plant.

In addition to the multistage centrifugal compressors discussed, single- and multistage, overhung impeller designs are available for low pressure ratio applications. Several of these accessible configurations are shown in Figs. 12.21 through 12.24. Allowable casing pressures

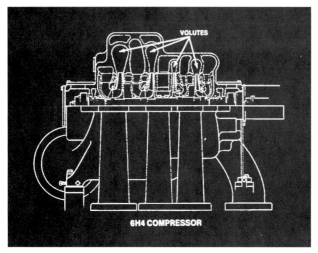

Figure 12.19 Cross-sectional view of over-the-impeller volute-type diffuser. *(Dresser-Rand Company, Olean, N.Y.)*

have been extended up to 2000 psi for this type of casing. Design consideration must be given to start up when the casing is pressurized because of high axial forces on the impeller.

Note the vertical split construction and the location of the inlet nozzle that allows direct inlet flow into the impeller. These units are referred to as direct inlet or axial inlet centrifugal compressors. The fact that the inlet gas enters the impeller without requiring a 90° turn results in lower aerodynamic inlet losses (Fig. 12.23).

Figure 12.20 Compressor with intercooling after each section. *(Dresser-Rand Company, Olean, N.Y.)*

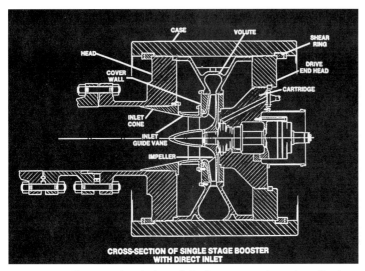

Figure 12.21   Cross-sectional view of single-stage overhung impeller-type compressor. *(Dresser-Rand Company, Olean, N.Y.)*

Head retention by stud bolts used to be common for both vertical and horizontal split overhung impeller machines. More recently, vertically split casings are commonly furnished with segmented shear rings.

On horizontal split casings (Fig. C in the introduction to Part 2), a large number of studs are secured in the lower case flange. The upper case flange has been drilled to receive the studs, and nuts are fastened on the studs when the upper case is properly positioned.

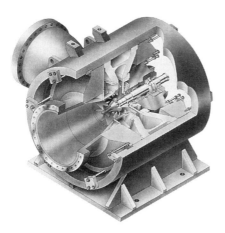

Figure 12.22 Single-stage over-hung compressor for pipeline service. *(Dresser-Rand Company, Olean, N.Y.)*

Figure 12.23 Single-stage overhung compressor installation for pipeline service showing side nozzles. *(Dresser-Rand Company, Olean, N.Y.)*

In earlier designs, vertical split casings of either between-bearing or overhung design are similarly fitted with studs that pass through the heads, and the heads are secured by the stud nuts (Fig. 12.1).

In contrast, the current use of segmented shear-ring head retention designs for between-bearing as well as overhung design vertical split centrifugal compressors offers the benefits of greater strength and faster disassembly. Proper assembly is assured by elimination of pre-

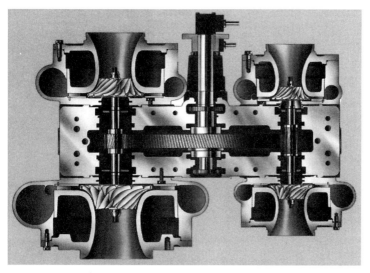

Figure 12.24 Multistage overhung impeller compressor. *(Mannesmann-Demag, Duisburg, Germany)*

cise torque requirements of bolted heads. In shear-ring designs, the head or end closures are retained by segmented shear-ring members positioned in an annular machined groove in the case. The incorporation of an O-ring assures positive sealing.

Even on multistage overhung impeller compressors (Fig. 12.24) the shear-ring head retainer is occasionally used to advantage. A small secondary spacer ring assembly allows removal and installation of the primary shear ring. As illustrated in Figs. 12.2 and 12.6, the most advantageous use of shear-ring head retention is with very high pressure casings. It is difficult to properly design bolted heads for casings with pressure ratings over 5000 psi because of the required physical size of the studs.

Moving to Fig. 12.25, we note how surprisingly many external connections are required to enter the case through the heads. As shown in the photograph, virtually all available space is used in a small high-pressure head by the various external support systems such as lube and seal oil supply and drains, vents, and reference pressure lines.

As will be discussed later, compressor casing materials consist of cast iron or cast steel, fabricated steel, or forged steel. Horizontal split centrifugals commonly are produced from castings or fabrications, whereas vertical split casings use all three types.

The cast-steel or cast-iron casings are more commonly used in horizontal split cases since case pressure ratings are relatively low. Case patterns allow flexibility in the number of stages but result in fixed-stage spacing as opposed to variable-stage spacing.

Advanced manufacturing and welding techniques are also being used in the production of compressor casings. Welded construction has

Figure 12.25 Piping connections entering compressor head. *(Dresser-Rand Company, Olean, N.Y.)*

Figure 12.26   Reinjection compressor with forged steel vertically split casing. *(Dresser-Rand Company, Olean, N.Y.)*

supplemented the high-capacity line of cast cases and provides the option of purchasing horizontally or vertically split compressors for high-capacity, low- and medium-pressure service.

Manufacturing techniques used for welded casings are basically the same whether the compressor is horizontally or vertically split.

Forged steel vertically split casings, Fig. 12.26, are required for very high pressure application. Case pressure ratings over 10,000 psi have been successfully tested for gas reinjection application with units operating in the field well over 7000 psi.

12.5.2   Flowpath

Having completed our review of the compressor casing, we will now turn our attention to the stationary flowpath. The stationary flowpath is contained within the casing and is aerodynamically matched to the inlet and outlet passages and the impellers. Figure 12.27 depicts this stationary flowpath in a vertically split compressor cross section. Note the casing components that include the heads.

Moving to a more detailed examination of a typical centrifugal compressor, we observe in Fig. 12.28 the stationary flowpath and rotor, removed from the casing. This part of the compressor is referred to as the *bundle* and contains guide vanes, diaphragm, return bend, diffuser, and discharge volute. It is evident that on a vertically split compressor, the complete bundle is removed as a single piece. To remove the rotor, the bundle is usually designed with horizontal splits that allow removal of the top half. Some compressor designs employ one-piece rather than split bundle stationary components that are stacked onto the rotor shaft at the same time the impellers are installed.

Figure 12.27   Stationary flowpath in a vertically split compressor. *(Dresser-Rand Company, Olean, N.Y.)*

Note also that the inlet wall is the first diaphragm and completes the inlet channel. The gas passage formed by the inlet wall directs the gas into the first impeller. The opposite side, inboard, of inlet wall forms the diffuser area from the first impeller. After the diffuser, the gas enters the return bend or crossover, which turns the gas streams 180°. The return channel guides the gas into the next impeller. The diaphragm is the stationary element between two stages that forms half of the diffuser channel stage, the return bend, and half the return channel.

Figure 12.29 shows how the inlet guide vanes direct gas flow into the impeller "eye." One method of controlling the stage performance char-

Figure 12.28   Stationary flowpath and rotor. *(Dresser-Rand Company, Olean, N.Y.)*

Figure 12.29 Inlet guide vanes, stationary (fixed) type. *(Dresser-Rand Company, Olean, N.Y.)*

acteristics is through the use of different inlet guide vane angles. Guide vanes can direct the flow into the impeller against rotation, radially, or with impeller rotation. The guide vanes shown are of fixed predetermined angles.

If we want to change the compressor operating performance by means other than speed reduction, movable inlet guide vanes can be incorporated. They are more effective on single-stage compressors (see Fig. A in the introduction to Part 2), with diminishing effect as stages are added. It is extremely difficult from a mechanical standpoint to install and operate movable inlet guide vanes (Fig. 12.30) in any but the first stage of a centrifugal compressor.

Next, in Fig. 12.31, the return bend is clearly shown in the internal portion of the stationary flowpath. After the gas leaves the last-stage diffuser, it is collected in a discharge volute, Fig. 12.32, and then directed to the discharge nozzles. The figure on the left shows a typical scroll-type over-the-impeller volute, whereas on the right is a parallel wall diffuser followed by a more conventional spillover volute.

The scroll-type over-the-impeller volute (Fig. 12.19), as well as the spillover volute, are complex shapes from a manufacturing point of view and are usually a casting. The over-the-impeller volute does offer low-diffusion losses over a wide operating range but requires a larger case diameter and stage width than does the parallel wall diffuser-spillover volute. In horizontal split compressors the stationary flowpath components are designed so they are removed with the top half. The rotor, resting in the lower half, is then readily accessible for inspection and/or removal. This is shown in Fig. 12.33.

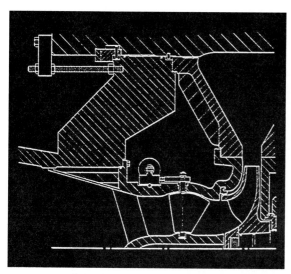

Figure 12.30 Movable inlet guide vanes. *(Dresser-Rand Company, Olean, N.Y.)*

12.5.3 Rotors

Having examined the compressor case and stationary flowpath, we now turn our attention to the rotating flowpath, or rotor. A thorough understanding of rotor nomenclature is necessary to understand some important design considerations. Rotor nomenclature is explained in Fig. 12.34.

Figure 12.31 Internal return bend flowpath. *(Dresser-Rand Company, Olean, N.Y.)*

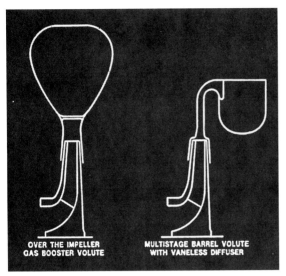

Figure 12.32 Discharge volutes—over the impeller and spillover types. *(Dresser-Rand Company, Olean, N.Y.)*

There is a step in the shaft at the bearing area. On the thrust bearing end, a precision ground spacer that butts against the shaft shoulder is installed. This spacer locates the thrust disk that, in turn, will locate the rotor in the compressor.

The major components of a centrifugal compressor rotor are:

1. Shaft
2. Impellers

Figure 12.33 Compressor rotor resting in lower half of stationary flowpath components. *(Dresser-Rand Company, Olean, N.Y.)*

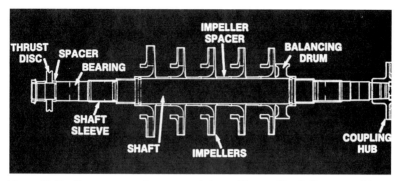

Figure 12.34 Compressor rotor nomenclature. *(Dresser-Rand Company, Olean, N.Y.)*

3. Balancing drum (if required)

4. Impeller spacer

5. Thrust disk

6. Coupling hub

In the seal area, sleeves are provided to protect the shaft. Under labyrinth seals, the sleeves are stainless steel; under oil film seals, the sleeves are often monel with a hard colmonoy or similar overlay to protect against scratches from dirt particles in the oil or gas.

Figure 12.35 shows a complete rotor being installed in a horizontally split centrifugal compressor. All rotor components have been attached to the shaft and the rotor balanced prior to installation.

Figure 12.35 Compressor rotor installed in horizontally split centrifugal compressor. *(Dresser-Rand Company, Olean, N.Y.)*

The shaft is precision machined from an alloy steel forging. This solid rotor shaft design ensures maximum parallelism of rotor components. Impellers and balance pistons are normally forged steel, SAE 4330, with stainless steel available for corrosive gas applications. Impeller spacers are typically machined from a 400 series stainless steel.

Between each impeller is a spacer sleeve, Fig. 12.36. In addition to the function of locating the impellers on the shaft, sleeves also protect the rotor shaft in the event of contact with the labyrinths.

12.5.4 Impellers

A cross section of the impeller, Fig. 12.37, reveals the three components—blade, disk, and cover. The blade increases the velocity of the gas by rotating and causing the gas to move from the inlet, impeller eye, to the top or outside diameter. The impeller disk or hub is attached to the shaft and drives the blade. The cover is attached to the blades and confines the gas to the blade area.

To provide flexibility to meet the many process requirements, several types of impellers are used. These include closed impellers that have a blade mounted between a disk and cover, the cover being the inlet side of the impeller; and open or semiopen impellers that, as the names imply, consist of a disk and blade but have the cover removed. Refer to Fig. 12.38.

A number of manufacturing methods are used in the production of impellers. These include riveted construction, where the disks and cov-

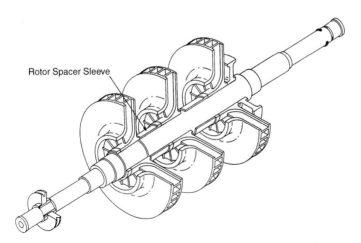

Rotor Spacer Sleeve

Figure 12.36 Spacer sleeves for centrifugal compressor rotor. *(Dresser-Rand Company, Olean, N.Y.)*

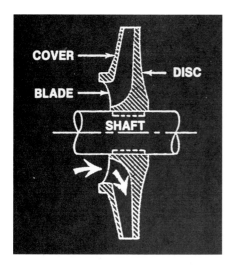

Figure 12.37 Cross-sectional view of impeller. *(Dresser-Rand Company, Olean, N.Y.)*

ers are joined to the blades by rivets; cast construction; electrolytic machining; five-axis milling; and welding.

Over the history of the centrifugal compressor, the most universally used manufacturing method had been riveted construction. This method is rarely used today and has been replaced by welded impellers in modern installations.

Sand and die-casting techniques are available to fabricate various types of impellers. Advanced casting methods are applied in the manufacture of open impellers (Fig. 12.39).

Riveted impellers, Fig. 12.40, are fabricated by riveting the blade to both the cover and the disk. Two types of riveted construction exist. As

Figure 12.38 Impeller types found in centrifugal compressors. *(Dresser-Rand Company, Olean, N.Y.)*

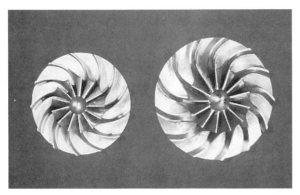

Figure 12.39 Cast, three-dimensional open impeller. *(Dresser-Rand Company, Olean, N.Y.)*

shown, the blade sides are rolled over at a 90° angle and attached by rivets through the rolled-over portion. To decrease the susceptibility of the riveted impeller to corrosive and erosive failure, an integrally riveted impeller was developed. Integrally riveted impellers require a thicker blade since the rivets attach to the blade edge, thus eliminating the requirement to bend the blade 90° on both sides (Fig. 12.41).

Welded impellers are structurally homogeneous. Welded construction, Fig. 12.42, also allows maximum flexibility to alter aerodynamic

Figure 12.40 Riveted impeller, "Z"-type blading. *(Dresser-Rand Company, Olean, N.Y.)*

Figure 12.41 Integrally riveted impeller. *(Dresser-Rand Company, Olean, N.Y.)*

designs. Pattern arrangements required for cast impellers and tools associated with electrochemical milling methods are not necessary to modify the aerodynamic design.

In impellers of high specific speed, where the three-piece and open construction techniques are employed, three-dimensional blade shaping provides optimum aerodynamic geometry.

Varying manufacturing techniques are used to produce the different types of welded impellers including:

- Three-piece construction
- Open impeller construction

Figure 12.42 Structurally homogenous welded impeller. *(Dresser-Rand Company, Olean, N.Y.)*

- Two-piece construction (tip width greater than ⅝ in [16 mm])
- Two-piece construction (tip width less than ⅝ in [16 mm])

In the three-piece construction, Fig. 12.43, the disk and cover are machined from forgings. The die-formed blades are then tack welded to the cover with the use of locating fixtures. The final welding is a continuous fillet weld between the blade and cover. Subsequently, the blade cover assembly is joined to the disk by a continuous fillet weld between the disk and blade.

Impellers of open construction consist of a disk and blades (Fig. 12.39). The cover is eliminated. This type of impeller is characterized by an inducer section that directs the gas flow into the eye of the impeller. The blades are either die formed or precision cast. The welding procedure is the same as for the three-piece construction with the final weld being a continuous fillet weld between the disk and blade.

In two-piece type construction, Fig. 12.44, the blades are machined on either the disk or cover forging. The impeller is completed by a continuous fillet weld to the mating piece (disk or cover) around the entire blade interface. This type of construction is used for impellers with a relatively low tip width-to-diameter ratio, i.e., low specific speed.

Observing Fig. 12.44, we note that the welding techniques previously described are limited to impellers with a channel width of more than ⅝ in (16 mm). Increasing numbers of applications are now requiring the advantages of welded construction for impellers with channel width of less than ⅝ in (16 mm). To meet this need, an advanced welding technique provides a method of welding the blades to the disk from

Figure 12.43 Three-piece welded impeller. *(Dresser-Rand Company, Olean, N.Y.)*

Figure 12.44 Two-piece welded impeller. *(Dresser-Rand Company, Olean, N.Y.)*

the outside of the disk. The method produces an impeller with greater strength than a riveted impeller because of the continuous weld along the entire length of the blades. Any blade contour may be designed without affecting the weldability of the impeller, thus minimizing compromises in aerodynamic design. Using this technique, welded impellers can be manufactured to the smallest practical aerodynamic width.

Figure 12.45 illustrates the wide range of impeller sizes required to serve the widely varying process industry applications. The large rotor is from a compressor with flow to 180,000 cfm (5100 m³/min) and oper-

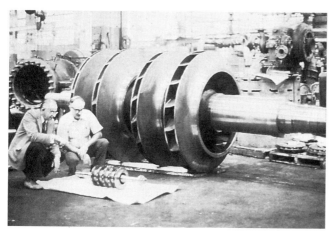

Figure 12.45 Rotor sizes found in centrifugal compressors. *(Dresser-Rand Company, Olean, N.Y.)*

ates to a speed of 4000 r/min. The small rotor is for a compressor that has a flow capability to 3500 cfm (100 m³/min) and operates at speeds to 20,000 r/min. This wide range of welded impeller sizes and types requires a variation of manufacturing techniques.

Balancing drums, Fig. 12.46, are employed to modify or adjust the axial thrust developed by compressor rotors. These drums are typically required when all impellers are facing in the same direction. A balancing drum is mounted behind the last stage impeller, as shown in Fig. 12.47.

Impellers are mounted on the shaft with a shrink fit with or without keyways, depending on the frame size. Prior to rotor assembly, impellers

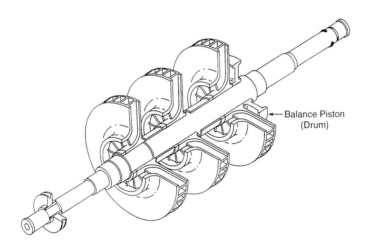

BALANCE PISTON FUNCTION

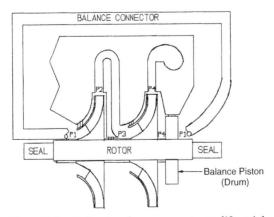

Figure 12.46 Balancing drum serves to modify axial thrust developed by differential gas pressures in compressors. *(Dresser-Rand Company, Olean, N.Y.)*

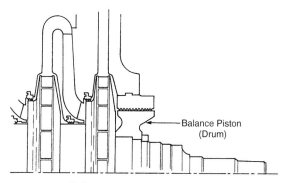

Figure 12.47   Balancing drum mounted behind last-stage impeller. *(Dresser-Rand Company, Olean, N.Y.)*

are dynamically balanced and oversped. Impellers are mounted in pairs beginning at the center of the shaft; successive pairs of impellers are added, one from each end, until the rotor is complete. The rotor is dynamically balanced after the addition of each set of impellers. At each balancing operation, balance correction is done only on the newly added components. Figure 12.48 illustrates a balance operation.

As the flow requirements of a centrifugal compressor application increase, the use of radial flow impellers may be restricted because of low efficiency. The development and use of mixed flow impellers, Fig. 12.49, results in acceptable efficiencies since the gas is allowed to flow through the impeller channels at angles less than 90°. As the name implies, a mixed flow impeller is neither radial flow or axial flow but somewhere between the two extremes.

Figure 12.48 Centrifugal compressor rotor being balanced. *(Dresser-Rand Company, Olean, N.Y.)*

Figure 12.49 Mixed flow impeller. *(Dresser-Rand Company, Olean, N.Y.)*

12.5.5 Axial blading

On very high flow application at low suction pressure, such as atmospheric air, the use of axial bladed compressors, Fig. 12.50, is attractive. Because the gas flows axially through the rotating flow path, turning losses are minimized and thus high efficiency is attainable. Limitations exist in using axial bladed compressors in high-density gas streams due to resulting high rotor thrust loads. Further consideration must be given to the fact that axial compressor performance generally results in less stability and reduced overload characteristics compared to typical radial impellers with backward bent blading.

Figure 12.50 Axial bladed compressor rotor. *(Dresser-Rand Company, Olean, N.Y.)*

12.5.6 Seals

The compressor industry has developed a complete range of seals for all types of applications. Four basic types of seals are offered:

1. Labyrinth
2. Contact
3. Oil Film
4. Gas Seals

Labyrinth seals are suitable for use as main casing seals for compressors operating at moderate pressures. These seals are available with ports for injecting inert gas and/or educting process gas as required depending on the process. This type of seal, Fig. 12.51, has been used for over 25 years in air and oxygen compressors. It is almost identical to the labyrinth seals typically found in steam turbines.

Modern labyrinth seals are often made of a honeycomblike material, Figs. 12.52 and 12.53. Honeycomb seals provide an order-of-magnitude more direct damping, lower whirl frequency ratios, and reduced leakage when compared to conventional labyrinth seals. The whirl effects are causing aerodynamic instabilities or *rotating stall*. Countermeasures are sometimes based more on experimentation than solid computer-generated analytical predictions. Nevertheless, both honeycomb seals and shunt holes (Fig. 12.54) will reduce both gas whirl (or swirl) risk and intensity. This abatement action is sometimes called *reduced cross-coupling,* although purists will assign minor differences to the respective definitions of the two terms.

Published nondimensional data on honeycomb seals are given in Figs. 12.55 and 12.56; these are of primary interest to designers of

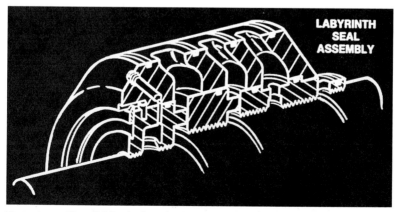

Figure 12.51   Ported labyrinth seals. *(Dresser-Rand Company, Olean, N.Y.)*

Figure 12.52 Special anti-instability labyrinth, or honey-comb seal. *(Nuovo Pignone, Florence, Italy)*

high-pressure compressors since swirl effects and the attendant aerodynamic instabilities are usually associated with high differential pressures.

The mechanical contact seal, Fig. 12.57, was developed in the early 1950s. The major benefit of this type of seal is its ability to maintain a positive seal when the compressor and oil systems are shut down. Mechanical contact seals are suitable for intermediate pressures of 450 psi (32 bar) and are particularly popular in refrigeration applications.

The primary components of this seal are:

- A spring-loaded stationary seal ring
- A floating carbon ring

Figure 12.53 Honeycomb seal segment for high-pressure centrifugal compressor. *(Nuovo Pignone, Florence, Italy)*

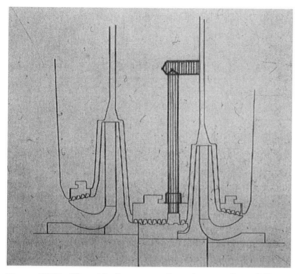

Figure 12.54 Shunt holes connecting the vaneless diffuser of the last impeller with the first grooves of the labyrinth. This minimizes inlet gas swirl and thus cross-coupling action. *(Nuovo Pignone, Florence, Italy)*

- A rotating seal ring
- A spring-loaded shutdown piston
- An oil pressure breakdown ring
- A labyrinth with provision for buffer gas injection

The contact seal design is unique in that it provides a separate sealing surface in the shutdown condition. In the event of a failure of the carbon ring, a standard contact seal would be inoperative. The design will still maintain a positive seal at shutdown conditions; consequently, it provides a fail-safe feature.

When the compressor is stopped under pressurization, the shutdown piston is held closed by this gas pressure. Sealing is accomplished by means of an elastomeric ring sealed against the rotating shaft ring. For operation, oil is introduced at a pressure of 25 psi (1½ bar) above the gas pressure. This oil pressure overcomes the gas pressure and spring tension on the shutdown piston, causing the piston to open and admit oil flow to the carbon ring seal. As the compressor is started, the carbon ring floats between the rotating ring and the stationary ring. The carbon ring seeks its own rotational speed, approximately one-half of shaft speed. The seal is loaded by springs, forcing the stationary ring against the carbon ring.

Honeycomb Seals Improve Rotor Stability

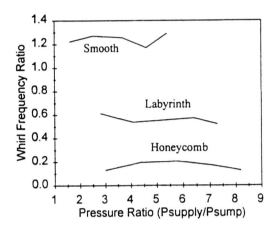

Honeycomb Seals Leak Less

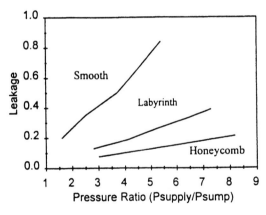

Figure 12.55 Rotor stability and leakage of honey-comb seals. *(Rotordynamics-Seal Research, North Highlands, Ca.)*

A small amount of oil flows across the carbon ring. This oil is prevented from contacting the gas stream by means of a labyrinth seal with an oil slinger on the shaft and is drained to a trap assembly. The majority of oil flows across the entire seal assembly, thus cooling the seal. It is then returned to the reservoir. A small amount of oil flows across a floating ring on the outboard end into the bearing chamber. This floating ring provides the orifice that maintains oil pressure in the seal area.

Honeycomb Seals Produce More Damping

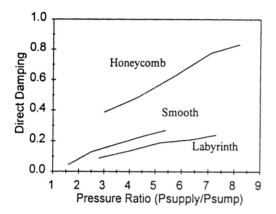

Honeycomb Seals Have Positive Stiffness

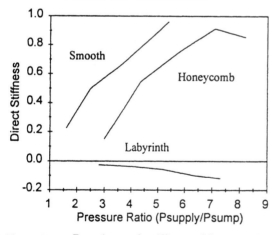

Figure 12.56 Damping and stiffness of honeycomb seals. *(Rotordynamics-Seal Research, North Highlands, Ca.)*

It was the development of the oil film seal, Fig. 12.58, that made possible the application of centrifugal compressors in high-pressure applications for hazardous gases. This seal design was introduced in gas transmission service in 1948. The major benefits of the oil film seal are:

Simplicity. The oil film seal is simple in concept and does not involve rotating or contacting parts. This provides minimum service and maintenance requirements.

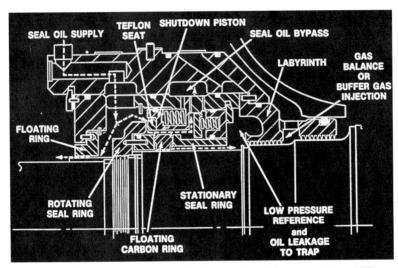

Figure 12.57   Mechanical contact seals. *(Dresser-Rand Company, Olean, N.Y.)*

High pressure.   The oil film seal has higher pressure capability than any other type seals, and, with continued development, its capability may be virtually unlimited.

Fail-safe.   In the event of damage to the seal rings, oil consumption will increase, but a seal will be maintained as long as sufficient oil is supplied, allowing continued operation of the compressor.

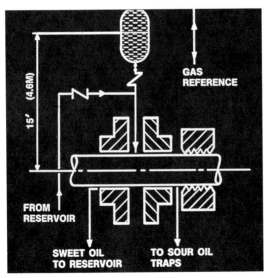

Figure 12.58 Oil film seal assembly schematic. *(Dresser-Rand Company, Olean, N.Y.)*

As further illustrated in Fig. 12.59, the oil film seal consists of two or more babbitted rings that do not rotate but are free to float radially to follow shaft movement. Oil from an overhead tank (see Fig. 12.58) is introduced between the rings at a pressure slightly above the reference gas pressure applied to the top of the tank.

The oil flows across these rings to the internal and external drains. Oil flow is quite small across the inner ring because of the low 5 psi (0.35 bar) differential pressure, which is controlled by the height of the tank above the compressor centerline. Oil flowing across the internal ring opposes the outward flow of gas, thus effecting a positive seal. This small amount of oil is insufficient for cooling the inner ring. The major flow of oil passes through the outer ring or rings and takes the full pressure drop between the oil supply pressure and the atmosphere drain to the reservoir. This oil flow passes by the back of the inner seal ring and provides cooling of this ring.

An optional labyrinth with provision for buffer gas injection can be provided, when required, to keep the seal oil drain separate from the lube oil drain. The basic oil film seal, as previously described, consists of a high-pressure or inner ring and a low-pressure or outer ring. As seal pressures began to exceed the 1000 psi (70 bar), it became necessary to modify the seal to add another ring, allowing the pressure breakdown from seal supply pressure to atmosphere to be accomplished over two rings.

To properly seal a centrifugal compressor, careful attention must be given to gas pressure within the casing. A series of connecting ports allows the designer to seal against a known and predetermined inter-

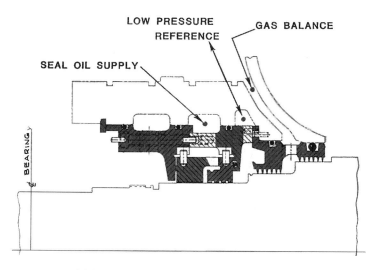

Figure 12.59   Oil film seal details. *(Dresser-Rand Company, Olean, N.Y.)*

nal pressure. The most significant element is to provide a pressure-flow bleed connection between the gas balance port and the seal reference area. While the recompression of this balance piston chamber gas requires additional shaft horsepower, it results in sealing against an internal gas pressure slightly above suction pressure and not the high discharge pressure. It also balances and thus equalizes the gas pressure on the seals at both the suction and discharge ends of the compressor, thereby simplifying the seal oil system controls. Figure 12.60 explains essentials.

Trapped bushing seals, Fig. 12.61, are often incorporated in compressors made by the A-C Compressor Corporation. Under static conditions the trapped bushing seal operates like any other bushing-type seal. A sealant (normally the oil from the lubrication system) is supplied at about 10 gal/min to the seal at a positive pressure above process gas. The sealant flows three ways:

1. A relatively small portion of the sealant buffers the process gas at 1 to 2 gal/h in the clearance area between the inner portion of the stepped dual bushing (4) and the impeller (2). At this point the sealant enters the inner drain and is ultimately bled down to atmospheric pressure by way of a trap.

2. A larger portion of the sealant takes a pressure drop to the atmospheric outer drain between the outer portion of the stepped dual bushing (4) and the impeller (2). This flow rate is a function of the process gas pressure level.

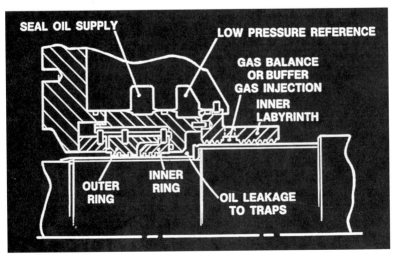

Figure 12.60 Balance piston chamber and seal reference area relationship. (*Dresser-Rand Company, Olean, N.Y.*)

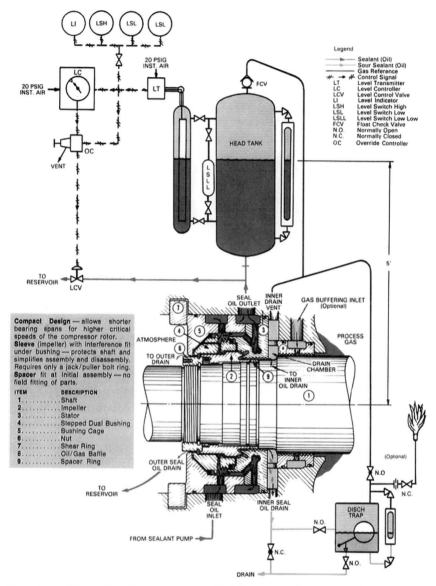

Figure 12.61   Trapped bushing seal system. *(A-C Compressor Corporation, Appleton, Wis.)*

3. The excess sealant passes through the stator (3) cooling passages and out of the seal, where a 5-ft head of sealant is maintained above the process gas pressure by a level control. This level controller (LC) modulates a level control valve to provide the pressure drop for the sealant's return to the reservoir.

While operating normally, the trapped section of the trapped bushing seal performs like a pump during dynamic operation. All of the sealant flowpaths are exactly the same as during static operation except that the sealant buffering the process gas is whirled by the trapped portion of the seal, which consists of two principal parts:

1. The step within the stepped dual bushing (4) works in combination with the outer shoulder on the rotating impeller (2) as a pump, to ensure that the entire clearance area between the inner portion of the stepped dual bushing (4) and the impeller (2) is a positive pressure level with respect to the process gas. Pressure patterns in this clearance area are similar to those in a lightly loaded journal bearing.

2. The portion of the stator (3) that meshes with the inner portion of the rotating impeller (2) acts as a dead-end pump without a suction source. The pumping action just balances the head of the pumping action of the outer shoulder of the impeller (2) plus the 5-ft head maintained in the head tank; hence, there is no pressure drop through the clearance between the stepped inner portion of the dual bushing (4) and the impeller (2). This pressure drop being zero, the inner sealant flow is considerably less than an untrapped bushing seal.

The fifth major type or configuration of compressor seal is the so-called *gas seal*. Developed in the late 1970s, this relatively new sealing device is of sufficient importance to be given special coverage later in this text.

12.6   Bearing Configurations

12.6.1   Radial bearings

Radial bearings, sometimes referred to as *journal bearings,* support the compressor rotor.

Technology in bearing design has increased significantly over the years to meet the increasing demands on the equipment. The original centrifugal compressors were furnished with plain sleeve bearings. The typical fully concentric straight-sleeve bearing was discovered to be inadequate as rotor speed dramatically increased in the late 1940s. In the 1950s, a pressure dam bearing design was developed to increase resistance to half-frequency whirl. The bottom half bearing liner is the same as the straight-sleeve bearing, but the top half is relieved (Fig. 12.61). Thus, an area of high pressure is generated where the relief slot terminates. This increases the bearing loading. Continuing development led to improvement in the stability of lightweight rotors operating at high speeds. This work culminated in the multishoe tilting pad radial bearing, Figs. 12.63 and 12.64. This bearing design has been so successful that it is now considered standard throughout the industry.

In tilting pad radial bearings, the pad surface in contact with the bearing housing is radiused, allowing it to pivot against the bearing housing. As the shaft rotates, a hydrodynamic film is formed between the journal and each pad. The oil enters the clearance between each pad and the shaft, tending to force the pad leading edge away from the shaft. Since the pad can pivot or tilt in its housing, the clearance at the trailing edge of the pad is reduced and the clearance at the leading edge of the pad is increased. This results in a wedge-shaped clearance between the pad and the shaft. This wedge-shaped clearance produces hydrodynamic pressure in the bearing. By making design adjustments in the shape of the pads and bearing clearance, bearing stiffness and damping characteristics can be controlled.

The tilting pad bearing shown in Fig. 12.64 is a five-shoe bearing that has one pad located on the vertical centerline in the lower half of the bearing housing. This ensures that the shaft is supported properly when it is at rest. Other tilting pad bearings may have three or four shoes, and not all rotors are designed for load-on-pad, i.e., load-between-pad orientation may be necessary to obtain desired rotor behavior at high speeds.

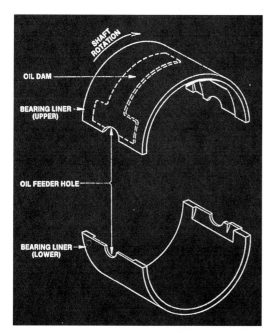

Figure 12.62 Pressure dam bearing. *(Dresser-Rand Company, Olean, N.Y.)*

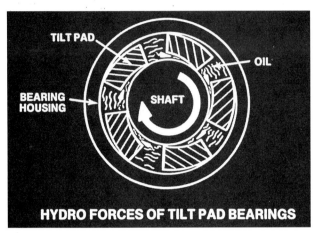

Figure 12.63 Tilt pad bearing, schematic representation. *(Dresser-Rand Company, Olean, N.Y.)*

12.6.2   Thrust bearings

One of the most critical components of a centrifugal compressor is the thrust bearing, Fig. 12.65. Axial thrust is generated in a centrifugal compressor by the pressure rise through the impellers. The major portion of the thrust load is compensated by either a balancing drum or by placing the impellers in a back-to-back arrangement, whereby the thrust generated by one set of impellers opposes the thrust generated by the other set of impellers. In either case, the relatively small residual load is carried by the thrust bearing. The thrust bearing must also be suitably designed to withstand additional load and thrust reversals that may occur during normal operating conditions.

The pressure environment surrounding each impeller creates an axial unbalance force on the impeller and thus the rotor. The total axial

Figure 12.64   Radial tilt pad bearing assembly. *(Dresser-Rand Company, Olean, N.Y.)*

BASE RING   RUNNING
             DISC

PADS

TILT PAD THRUST BEARING ASSEMBLY

BEARING PAD

Figure 12.65 Thrust bearing assembly. *(Dresser-Rand Company, Olean, N.Y.)*

unbalance on the rotor is calculated and is first compensated by the installation of a balance piston as part of the compressor design. The balance piston is generally sized to compensate for approximately 100 percent plus 10 percent of the total unbalance force generated by the impellers at the design operating parameters. The remaining or resulting unbalance force or thrust on the rotor is absorbed by the casing through the thrust bearing.

Figure 12.65 depicts the actual bearing parts including the base ring, running disk, and bearing pads. This bearing design has been extremely successful as evidenced by long-term operation in thousands of centrifugal compressors in the process industry. It has exhibited excellent load-carrying capabilities and even proven ability to withstand reverse rotation without damage to the bearings.

As an option, self-equalizing bearings, Fig. 12.66, can be provided on large frame compressors. This option can be applied when there is concern for potential thrust disc misalignment with the shaft. The self-equalizing bearing provides a uniform distribution of load on the thrust shoes over a wider range of thrust disc misalignment than can be accommodated by standard bearings. The disadvantage of this optional arrangement is that a greater shaft overhang is required which can become a limiting factor in critical speed analysis or rotor dynamics studies.

12.6.3   Flexure Pivot™ tilt pad bearings

Up to now, our overview dealt exclusively with rocking pivot tilt pad bearings. The insert in Fig. 12.67 allows us to compare these more conventional tilt pad bearings to a relative newcomer, the flexure pivot tilt

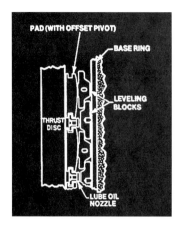

Figure   12.66 Self-equalizing thrust bearing. *(Dresser-Rand Company, Olean, N.Y.)*

pad bearing. The one-piece construction of this particular tilt pad bearing eliminates the multipiece construction typical of conventional tilt pad bearings. This also significantly reduces the manufacturing tolerances, as could be visualized from Fig. 12.68, which, incidentally, depicts a split design flexure pivot tilt pad bearing.

ROCKING PIVOT TILT PAD

FLEXURE PIVOT™ TILT PAD

Figure 12.67 Radial Flexure Pivot™ tilt pad bearings. *(KMC, Inc., West Greenwich, R.I.)*

Figure 12.68 Radial Flexure Pivot™ tilt pad bearing, split design. *(KMC, Inc., West Greenwich, R.I.)*

Thrust-type flexure pivot bearings have been supplied with a number of compressors, and Fig. 12.69 shows this simple, yet effective, design.

A recent development is the hydraulic fit thrust runner disk, Fig. 12.70. On centrifugal compressors, the thrust disk must be removable from the shaft to enable the inner seals to be removed from the shaft during maintenance. To prevent fretting of the shaft material under the thrust disk, the common method of attachment of the thrust disk is by shrink fit. Shrink fit by heating of the disk is unacceptable because of the obvious hazards of applying open flame heat to the disk at plant site.

The hydraulic fit thrust disk shown here allows removal of the disk without heating the rotor. This design features a thin sleeve that is mounted to the shaft with a loose fit. The outside surface of the sleeve is cone shaped or tapered to allow the thrust disk to be mounted by hydraulic pressure onto the sleeve.

To remove the hydraulically fitted thrust disk, one proceeds as shown in Fig. 12.71. Oil at high pressure from pump no. 1 is supplied through

Figure 12.69 Thrust-type Flexure Pivot™ bearing for small compressor. *(KMC, Inc., West Greenwich, R.I.)*

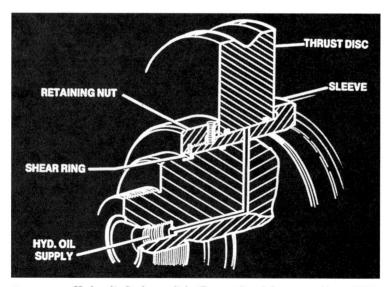

Figure 12.70 Hydraulic fit thrust disk. *(Dresser-Rand Company, Olean, N.Y.)*

holes in the center of the shaft and sleeve to expand the thrust disk over the sleeve. A pusher pump no. 2 pushes the disk on the sleeve. Releasing the hydraulic pressure on pump no. 1 allows the thrust disk and sleeve to shrink to the shaft. The high 1.5 mils/in (1.5 μm/mm) of shaft diameter shrink fit ensures a tight fit on the disk and sleeve to the shaft.

The coupling hubs of most modern centrifugal compressors are hydraulically mounted. The resulting interference fit allows the transmission of torque from the driver to the compressor without the use of keys or keyways. The hydraulic fit method of coupling attachment involves a tapered shaft end, a matching taper in the coupling bore and a method of introducing high-pressure oil between the shaft and the coupling hub. A high-pressure oil fitting is provided in the end of the shaft. The coupling hub is installed on the end of the shaft, and high-pressure oil is introduced to expand the hub that is then moved axially along a shaft taper to a measured distance in relation to the shaft end. Once the coupling is in place, oil pressure is released, and the coupling assumes a tight shrink fit on the shaft end. Refer to Fig. 12.72.

An alternative method of hydraulically mounting coupling hubs involves the injection of high-pressure hydraulic fluid through a threaded port machined into the coupling hub, 90° to the coupling or shaft axis. A hydraulic ram then pushes the expanded hub axially up the tapered shaft.

Centrifugal compressors require one or more major support systems to supply oil to lubricate the bearings, contact seals, and monitoring or

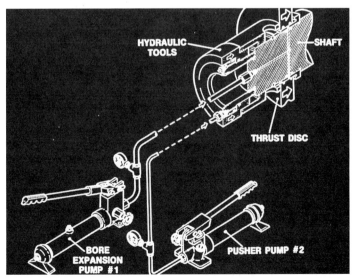

Figure 12.71 Hydraulic fitup procedures. *(Dresser-Rand Company, Olean, N.Y.)*

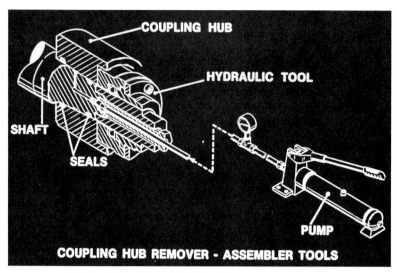

Figure 12.72   Hydraulic fitted coupling hub. *(Dresser-Rand Company, Olean, N.Y.)*

control system. A typical lubricating system furnishes clean, cool oil to the journal and thrust bearings. The lube system can be separate or combined with the seal oil system. This important subject is dealt with later in the text. Also, to properly monitor, control, and protect a complex and expensive piece of machinery such as a high-speed centrifugal compressor, a variety of systems are available. A well-designed control system, including surge control, is a key element in the successful, reliable operation of the centrifugal compressor. Again, this will be discussed separately.

12.7   Casing Design Criteria

The requirements for particular capacity and pressure capabilities of a centrifugal compressor design are determined by a manufacturer based on an analysis of the markets being served. Once a market need is determined, conceptual design for the equipment can begin. The casing receives much initial attention as it represents the major pressure-containing structure.

The maximum-capacity requirement determines the areas required both inside the compressor casing and in the attached nozzles. The pressure requirement influences choices as to whether the casing, often called *case* for short, can be horizontally or vertically split; whether it can be cast, formed from plate steel, or must be forged; what approximate thicknesses are required; and what form the nozzle connections must take. ASME formulas, from Section VIII, Division 1 of the ASME Code, can be used to rough out a minimum case thickness to start. API Standard 617 requires that all cases be designed such that hoop-stress values comply with ASME Code maximum allowable stress values.

As an example, a case for a large-capacity propylene refrigeration compressor was made from steel plate, ASTM-A-516, Grade 65 (minimum tensile strength of 65,000 psi). ASME Code permits a maximum allowable stress of one-quarter of the minimum tensile strength; thus, 16,250 psi. Casing inside diameter was sized at 112 in for the large-volume flow. For a case rating of 350 psig, with a requirement for 100 percent radiograph inspection of the welds, the minimum case wall thickness is:

$$t = \frac{PR}{SE - 0.6P} = \frac{(350)\,(56)}{(16250)\,(1.0) - (0.6)(350)} = \quad 1.222 \text{ in}$$

$$\text{Corrosion allowance} \quad \underline{+0.125}$$

$$\text{Minimum allowable wall thickness} \quad 1.347 \text{ in}$$

where t = thickness, in
P = pressure, psig
R = radius, in
S = allowable stress, psi
E = joint efficiency

It should be pointed out that the actual case wall thickness for the above compressor was 3.75 in, or almost 3 times the minimum required. Compressor case thicknesses are usually quite conservative compared to code thickness requirements, since the primary design criterion is the mechanical rigidity of the case. Casing hoop-stress levels, based on a simplified formula of Stress = $P \times R/t$, at maximum rated pressure, typically fall in the following ranges:

3000–5000 psi	Horizontally split, cast iron
5000–7000 psi	Horizontally split, cast steel
5000–7000 psi	Horizontally split, fabricated steel plate
6000–9000 psi	Vertically split, fabricated steel plate
10,000–16,000 psi	Vertically split, forged steel

Account must be taken of the location and support of end closure heads and internal pieces, as required. Deflections at these positions must be held within predetermined limits. Consider, for example, an O-ring on the outside diameter of a vertically split case bundle. O-rings are commonly used between the bundle and case to prevent recycle of gas in a compressor section. The case growth under pressure must be considered at the O-ring land location to ensure that the O-ring is not extruded in operation. O-rings in the outside diameter of the heads at the case ends present the same problem. Neither would it be desirable to have a horizontally split case deflect to such a degree as to allow opening of stationary internal splits, again causing excessive recycle flow. Deformation at the case rail fits must be held to reasonable levels to control split gaps and interstage labyrinth seal clearances.

Horizontally split cases create an additional concern over the pressure capability of the split joint. Split flanges must be rigid enough to maintain their shape and proximity at pressures of 1½ the case rating, or hydrostatic pressure. This usually requires very thick flange sections, held together with numerous, large, highly stressed studs. Casing contours of horizontally split cast cases often contain bulges and indentations at nozzle and volute sections and around studs. A balance must be achieved between the aerodynamic requirements on the inside and the mechanical and maintenance requirements on the outside. This gives rise to difficult shapes and resultant stress concentrations that must be recognized and accommodated.

Typical materials used in the construction of horizontally split cases are listed in API 617 and include:

	Material	Application range
Cast iron	ASTM A278	−50–450°F
Ductile iron	ASTM A395	−20–500°F
Cast steel	ASTM A216	−20–750°F
	ASTM A352	−175–650°F
Fabricated steel	ASTM A516	−50–650°F
	ASTM A203	−160–650°F

Lower temperature services often use high nickel content materials, such as ASTM A296.

Typical bolting materials for split flanges include ASTM A307 Grade B for cast-iron casings and ASTM A193 Grade B7 for cast- and fabricated steel casings. Nuts are typically supplied per ASTM A194 Grade 2H. For low temperatures, ASTM A320 bolting is usually supplied.

The most difficult area of design for a vertically split case is usually the overhang outboard of the head shear ring. The pressure inside the case tends to swell the case in the middle, causing the overhung ends to deflect inward. At the same time, the pressure on the head enclosure results in a force on the shear ring that tries to deflect the overhung end outward. The stress concentration effect of the angular shear-ring groove aggravates the condition. These considerations usually make necessary a finite-element analysis of the area before final design details are established.

The casing heads are also typically subjected to a finite-element analysis; it is obviously important to control deflections. Efforts to apply ASME formulas intended for simplified flat head design can lead to erroneous results because these formulas are just not applicable to typical compressor head shapes.

A manufacturer has to have confidence that casing and head deflections can be accurately predicted; the manufacturer must be certain that the compressor design will perform as intended. Excessive or unexpected movements can distort the case, cause tilting of a head that can lead to excessive runout between the thrust bearing and thrust disk, and cause gas leaks both internal to the machine (recycle) and/or through a case seal. This concern leads to strain gauge and dial indicator measurements of various cases and heads during pressure testing. These data are compared with predictions in a continuing effort to update and improve the analytical tools. As an example, Fig. 12.73 shows the location of dial indicators on key areas of a medium-size propane compressor and compares predicted (based on finite-element analysis) and actual deformations, thus serving to verify the analysis.

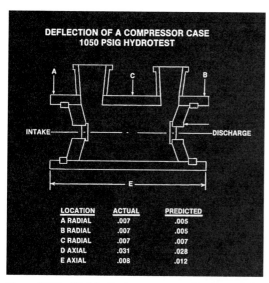

Figure 12.73 Dial indicators measure key areas of a compressor on the manufacturing floor. *(Dresser-Rand Company, Olean, N.Y.)*

Typical materials for vertically split casings include:

	Material	Application range
Welded	ASTM A516	–50–650°F
	ASTM A203	–160–650°F
Forged	ASTM A266	–20–650°F

Again, lower temperatures are possible with high nickel materials. Pressure containing heads are designed from:

	Material	Application range
Cast steel	ASTM A216	–20–750°F
	ASTM A352	–175–650°F
Plate	ASTM A516	–50–650°F
Forged	ASTM A266	–20–650°F
	ASTM A350	–150–650°F

Shear and retainer rings are usually from similar material.

A generalized pressure and flow chart is shown in Fig. 12.74. Comparison with Fig. 12.3 will demonstrate that specific capacity and pressure steps vary from manufacturer to manufacturer. The pressure capability of horizontally split designs is limited to approximately 1000 psi because of the practical problems of sealing a split at higher pressure levels with associated increased distortion.

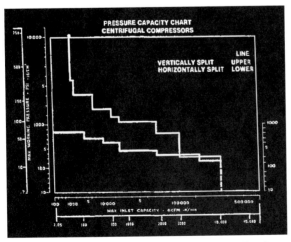

Figure 12.74   Generalized pressure and flow capacity chart. *(Dresser-Rand Company, Olean, N.Y.)*

High-pressure compressors may be defined as those operating over 1000 psi, and these fall in the vertically split category. Equipment of this type is used in applications such as ammonia and methanol synthesis gas, CO_2 for urea, as well as natural gas storage, gas lift, and reinjection. Figure 12.75 shows how these ratings have grown over the years as the market changed. Reinjection now accounts for the highest pressure installations in the world. To date, centrifugal compressors have been proven at 10,500 psi, and even this can be extended when the design is needed.

Figure 12.13 earlier illustrated a typical high-pressure compressor. This particular one has eight impeller stages in a back-to-back arrangement. This design provides essentially a balanced thrust force condition that, under most circumstances, eliminates the need for a balance piston. The casing and heads are steel forgings, and the heads are retained by shear rings. Process pipe connections at inlet and discharge connections are made by machining flat areas on the casing and attaching flanges by means of stud bolts mounted in the case.

As discussed in "Design Overview," Sec. 12.5, the choice between a straight-through design and a back-to-back arrangement for high-pressure applications is one of design philosophy. The back-to-back design is subject to the adverse effect of section mismatching. High-thrust loads can occur during operation at excessively high flow, or stonewall, because of loss of second section pressure ratio. But this is an off-design operating condition problem that can be prevented by

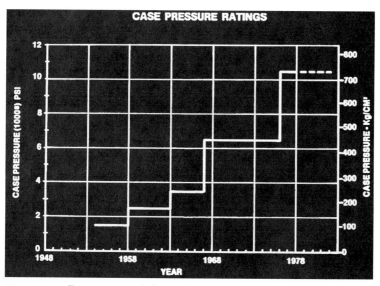

Figure 12.75   Pressure growth chart. *(Dresser-Rand Company, Olean, N.Y.)*

controls or proper operation. Compressor thrust bearing failures caused by balance piston seal deterioration in high-pressure, straight-through machines point to a design or operating condition problem; these are more difficult to monitor and prevent.

The advantages of a back-to-back arrangement in a high-pressure case include:

- The mentioned elimination of potential thrust bearing failure due to failure of the large-diameter balance piston labyrinth. (Balance piston labyrinths in straight-through designs are required to withstand differential pressures as high as 5000 psi.)

- Reduction of recirculation losses in the compressor since:

    The pressure led to the seals and balanced to the compressor suction is an intermediate pressure as opposed to full discharge pressure on straight-through flow designs.

    Seals balanced to suction can have much smaller diameters (and, therefore, much smaller flow clearance area) than balance piston labyrinths on straight-through flow designs.

    If the low-pressure section feeds contaminated gas into a scrubber or other purifier and the *cleaned* gas is subsequently entering into the higher pressure section, internal gas leakage will go from the clean section to the contaminated section: The contaminated gas cannot leak into the clean gas.

High-capacity compressors may be defined as those operating at over 35,000 cfm inlet capacity. Two major petrochemical plants that use

Figure 12.76   Large refrigeration compressor. *(Dresser-Rand Company, Olean, N.Y.)*

these large-capacity machines are ethylene plants and liquified natural gas production facilities.

Figure 12.76 shows a typical large refrigeration compressor. These compressors handle gases such as propylene and propane. Fairly stringent requirements influence the selection of equipment for this service. Cases are usually designed with materials suitable for −40 to −50°F.

The ethylene compressor is still relatively small, even for the largest-size ethylene plants. It is usually handled in a medium-size case. The gas is sweet and clean and the pressures are moderate. A key design consideration is the selection of materials for very low temperature, approximately −150°F.

Traditionally, for pressure ranges that offer a choice, the market has preferred the horizontally split machine from the viewpoint of maintenance. Most often, with steam turbine or electric motor drivers, the compressor and driver will be mounted on a mezzanine. Nozzles are directed downward, and the upper half of the case can be removed relatively easily, exposing all internals. The horizontally split case is still the preferred choice for throughdrive applications. Here, room does not normally exist at either end of the case to extract a bundle. In recent years, there has been a trend toward vertically split compressors (Fig. D in the introduction to Part 2) for high-capacity applications. This is especially true with gas turbine drivers, where the compressors and gas turbines are located at grade level. The gas connections for the compressors are then located at the top of the case. If horizontally split units were selected, it would be necessary to dismantle the gas piping before removing the upper half. This is not necessary with a vertically split case because the internals are withdrawn axially from the end of the case (Fig. 12.77).

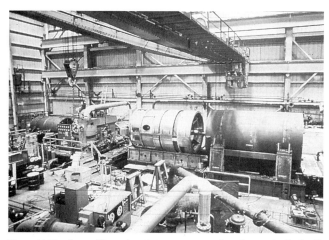

Figure 12.77   Assembly and maintenance procedure for vertically (radially) split compressor casing. *(Dresser-Rand Company, Olean, N.Y.)*

When lifting capabilities are limited and crane capacity does not allow removal of the upper casing half, vertically split compressors are also preferred. After the internal assembly is removed from the casing, the upper half of the bundle assembly can be dismantled by removing throughbolts and removing one section at a time (Fig. 12.78), thus dramatically reducing crane requirements. Alternatively, the bundle top half can be removed in a single lift (Fig. 12.79).

The typical vertically split casing design incorporates many features to ease the installation and removal of the internal assembly. The case internal diameter is machined (Fig. 12.80) in a series of steps, the largest diameter being at the maintenance end. It is designed so that when the bundle is installed, it is only in the last inch or two of travel that the bundle rises onto its locating fits and O-ring lands. Ramps are machined into the case for guiding these pieces. For very corrosive services, stainless steel lands are welded into the case inside diameter at all locations of contact between bundle and case. The intent here is to reduce corrosion and to facilitate disassembly.

For medium- and large-size compressors, adjustable rollers (Fig. 12.81) are designed into the bundle outside diameter at several positions along its length to permit easier travel along the case and support cradle. For down-connected cases, with large holes in the bottom of the case, ribs are provided across the nozzle openings in the line of roller travel. This design feature prevents the bundle rollers from losing contact.

Cast compressor casings are usually provided with integral feet, two at the suction end and either one centered at the discharge end or two located at the sides (Fig. 12.81). The feet must firmly hold the case, yet

Figure 12.78 Vertically split compressor being dismantled. *(Dresser-Rand Company, Olean, N.Y.)*

Figure 12.79 Removal of bundle top half of vertically split compressor. *(Dresser-Rand Company, Olean, N.Y.)*

permit thermal expansion. The case is typically doweled at the suction end feet, which, incidentally, is the end of the case where the thrust bearing is usually located. This allows case and rotor to thermally grow and expand in the same direction. If the discharge end is supported solely in the middle, the support may take the form of a *wobble plate* (see Fig. 12.97). These plates are kept thin to permit deflection without excessive stress. As supported weights increase, additional thin plates are added, parallel to the first. When the weight capacity of this design is exceeded, it is customary to provide two discharge end supports, one at each side of the case, called *sliding feet*. These feet typically rest on a PTFE (Teflon™) pad with clearance provided around the bolting to permit travel in the axial direction.

The dowels and bolting anchoring the case are designed to withstand all reasonable piping forces and moments and torque requirements. Case and nozzle thicknesses are rarely affected by piping loads. On U.S.-built compressors, these sections are very conservatively stressed, as shown previously.

12.8   Casing Manufacturing Techniques

The typical line of cast cases being offered by world-scale manufacturers includes cast-iron and cast-steel designs to satisfy market needs.

Figure 12.80   Compressor casing bore. *(Dresser-Rand Company, Olean, N.Y.)*

Figure 12.81   Adjustable rollers facilitate bundle installation. *(Dresser-Rand Company, Olean, N.Y.)*

The choice of materials for a given application is a function of several variables, including pressure and gas characteristics, as pointed out in API 617. Cast iron and cast steel each require their own supply of patterns, because shrinkage rates vary significantly between the materials. The decision to offer a complete line of machines can prove expensive for a manufacturer because of the quantity and complexity of patterns that must be built and maintained. Case variations are numerous, such as cast iron versus cast steel, standard-flow models versus high-volume reduction stepped cases, straight-through and double-flow configurations, nozzle connections up or down, various sizes of sidestream connections up or down connected at various axial locations in the case, and different size compound connections with different orientations. All of these special items require different patterns or pattern sections.

The majority of cast cases are horizontally split as shown in Fig. 12.82. Bearing chambers may be cast integral with the case; the case would provide integral bearing supports. This design ensures permanent, built-in alignment, providing the case is machined in one piece. Other arrangements involve bolted-on bearings.

Once the case is released from the foundry, the typical manufacturing sequence requires rough machining of the split surfaces to within a fraction of an inch of finish. Next, the sequence progresses to layout and rough machining of the inside contours and fits, welding on the case drains (cast steel only), complete stress relief, finish machining of the splits, and finish boring of the internal diameters. Return bends are generally machined integral with the case. Split line bolting may

Figure 12.82 Centrifugal compressor with cast casing. *(Dresser-Rand Company, Olean, N.Y.)*

be somewhat indented between the return bends to minimize the distance off the centerline.

Casing integrity is verified by hydrostatically testing to 1.5 times the casing design pressure. Casing splits are a metal-to-metal fitup and generally sealed with the aid of room-temperature vulcanizing joint compound. Split line bolting, stressed to prescribed values, pulls the casing halves together.

Fabricated case construction, an example of which is shown in Fig. 12.83, is used on the larger-capacity medium- and low-pressure vertically split and horizontally split compressors. Advanced manufacturing and welding techniques, which can vary somewhat from manufacturer to manufacturer, are used in the production of these cases. Steel plate and forgings are used.

The nozzles may be cast or may be formed from plate. The fabricated design is such that the transition from a rectangular shape (for the connection to the case) to a cylindrical shape is done by straight-line brake bending (Fig. 12.84). Very large nozzles are made from four separate pieces which are welded together (Fig. 12.85). The concept of straight-line seams is used to simplify the joining, allowing automatic submerged arc welding (Fig. 12.86). Forged steel flanges are welded to these nozzle bodies (Fig. 12.87) to complete the assembly. This again is done by the automatic submerged arc welding method.

The case cylinders (Fig. 12.88) are rolled from steel plate and welded along the longitudinal axis. The nozzles are then positioned and welded in place (Fig. 12.89). For long or stepped cases, two cylinders, one for the inlet end and one for the discharge end, are joined with a girth

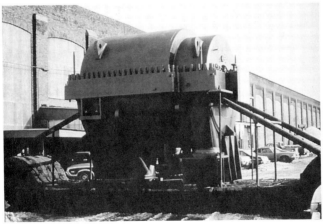

Figure 12.83 Centrifugal compressor with fabricated casing. *(Dresser-Rand Company, Olean, N.Y.)*

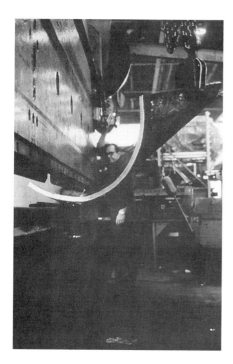

Figure 12.84   Straight-line brake bending produces cylindrical casing contour. *(Dresser-Rand Company, Olean, N.Y.)*

Figure 12.85   Nozzle welding in progress. *(Dresser-Rand Company, Olean, N.Y.)*

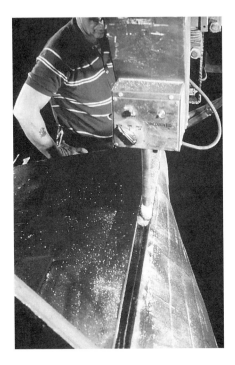

Figure 12.86 Automatic submerged arc welding. *(Dresser-Rand Company, Olean, N.Y.)*

Figure 12.87 Forged steel flanges being welded to nozzle bodies. *(Dresser-Rand Company, Olean, N.Y.)*

Figure 12.88   Casing cylinder. *(Dresser-Rand Company, Olean, N.Y.)*

weld. To make this weld, the case is rotated on a special machine, and the submerged arc welding fixture is held stationary (Fig. 12.90).

The procedure thus far is the same for either a horizontally split or vertically split case. If the case is horizontally split, the next step is to split the case longitudinally into two halves (Fig. 12.91). A split flange is burned out of thick plate (Fig. 12.92) and welded to each half. Return bend sections may be formed from steel plate and attached to the case

Figure 12.89   Nozzle positioning and welding. *(Dresser-Rand Company, Olean, N.Y.)*

Figure 12.90 Case rotation fixturing and girth welding. *(Dresser-Rand Company, Olean, N.Y.)*

by continuous structural welding to create a rib effect that provides additional stiffness to the case (Fig. 12.93).

The casings, assembled as described, are then rough-machined, heat-treated for stress relief (Fig. 12.94), and finish-machined (Fig. 12.95). Figure 12.96 shows a completed horizontally split fabricated compres-

Figure 12.91 Longitudinal splitting of compressor casing. *(Dresser-Rand Company, Olean, N.Y.)*

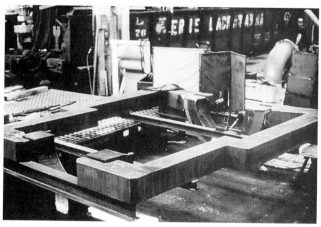

Figure 12.92 Flange production from thick plate. *(Dresser-Rand Company, Olean, N.Y.)*

sor with the top half removed. Note the variation in impeller stage spacing. The large spaces allow for large-capacity sidestream entries.

Figure 12.97 shows a completed compressor.

During manufacturing, the fabricated steel cases undergo quality assurance inspections such as:

1. All pressure welds are subject to 100 percent magnetic particle inspection of:

a. Plate edges prior to welding

Figure 12.93 Attaching return bend section to compressor casing. *(Dresser-Rand Company, Olean, N.Y.)*

Figure 12.94 Heat treatment of welded casing. *(Dresser-Rand Company, Olean, N.Y.)*

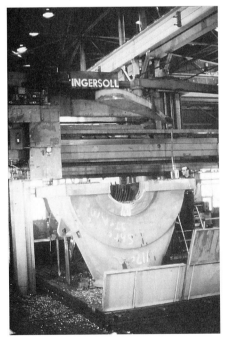

Figure 12.95 Finish-machining in progress. *(Dresser-Rand Company, Olean, N.Y.)*

Figure 12.96 Horizontally split fabricated compressor, lower casing half. *(Dresser-Rand Company, Olean, N.Y.)*

b. Weld root pass

c. Finish welds before and after stress relief

2. Structural welds are subject to 100 percent magnetic particle inspection of:

a. Internal parts before stress relief

b. External parts before and after stress relief

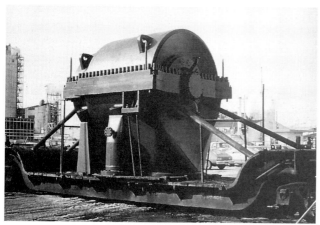

Figure 12.97 Completed fabricated compressor. *(Dresser-Rand Company, Olean, N.Y.)*

3. For design temperatures below −20°F (−29°C), impact tests of base material and weld metal are taken as part of the weld procedure qualification.

4. Hydrotest is done to 1.5 times the maximum working pressure.

Small high-pressure vertically split cases are often manufactured from forgings. This becomes a requirement when thicknesses become too great to roll from plate. Some of these cases are 10 to 12 in thick to contain the high pressures. Nozzles are sometimes cast separately and welded to the case, but for very high pressure applications the process pipe connections are made by machining flat areas on the casing and installing studs for flanges.

Typical manufacturing techniques follow those established for the other casing types, i.e., rough machine, stress relief, and finish machine. Inspection procedures closely follow those for welded steel cases.

12.9   Stage Design Considerations and Manufacturing Techniques

12.9.1   Stage design criteria

The most important element in a centrifugal compressor is the impeller. It is by way of the impeller that the work introduced at the compressor shaft end is transferred to the gas. The treatment the gas receives both before and after the impeller is also important. Typically, a guide vane directs the flow to the impeller, and a diffuser return bend

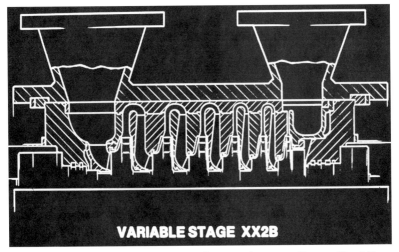

Figure 12.98   Stationary gas passages making up six stages. *(Dresser-Rand Company, Olean, N.Y.)*

and return channel (diaphragm) are downstream of the impeller (Fig. 12.98). Collectively, these parts make up a compressor *stage*. Some of the design considerations involved with these components will be discussed in this section.

Individual impeller designs are related and evaluated by a parameter called *specific speed,* which classifies impellers based on similarity of design, such as angles and proportions. Although this was explained earlier in this text, a bit of amplification may be of interest to the reader.

Specific speed.   One aspect of similarity is called *kinematic similarity,* where velocity ratios between two respective points within two separate designs are the same. For example, if a particular location velocity V as compared to the tip velocity U is constant between two different impellers and this is satisfied throughout the flow field, kinematic similarity is satisfied. Substituting for the velocity V, the capacity Q divided by the diameter D squared, and substituting for the tip speed U, the speed N times the diameter D, results in Q/ND^3, referred to as the *flow coefficient.* This is also referred to as a *capacity coefficient.* It represents the velocity similarities between two designs.

$$\frac{V}{U} = \frac{Q/D^2}{ND} = \frac{Q}{ND^3} = \text{constant} \Rightarrow \text{flow coefficient} \qquad (12.1)$$

The second aspect of similarity is called *dynamic similarity.* Designs are considered to be similar if the forces and the pressure fields are proportional to each other. This similarity is expressed by the pressure coefficient, which is defined as head H divided by tip speed U squared.

$$\frac{H}{U^2} = \text{constant} \Rightarrow \text{pressure coefficient} \qquad (12.2)$$

These two elements are combined to form specific speed. As can be seen, only operating parameters remain: speed, volume, and head.

$$N_S = \int\left(\frac{Q}{ND^3} ; \frac{H}{U^2}\right)$$

$$= \int\left(\frac{Q}{ND^3} ; \frac{H}{N^2D^2}\right)$$

$$= \frac{\left(\dfrac{Q}{ND^3}\right)^{1/2}}{\left(\dfrac{H}{N^2D^2}\right)^{3/4}}$$

where N_S = specific speed
Q = actual flow, ACFS
N = speed, r/min
H = head, ft-lb/lb
D = diameter, in
U = tip speed, FPM

$$= \frac{N(Q)^{1/2}}{H^{3/4}} \qquad (12.3)$$

The specific speed can be used in two ways. First, in application of equipment, it points out where the impeller is operating relative to previous experience. The second important use is the correlation between stage efficiency and specific speed. Such a correlation exists, as shown in Fig. 12.99, based on test data. Also shown is the approximate shape of the impellers in their respective specific speed area.

For any particular compressor frame size, there must exist a range of available impellers to satisfy the flow and head requirements. They may typically cover a range of specific speeds from approximately 20 to over 200. Designs on the order of 200 to 300 have been appearing in recent years because of a trend in gas transmission services of decreased head requirements and increased capacity levels. As stated before, impeller geometry changes with specific speed. One can see the pattern of change in straight-through compressors of numerous stages.

It is worth looking at some general considerations, such as radial versus backward bent blading, blade entrance angles, and impeller blade shapes.

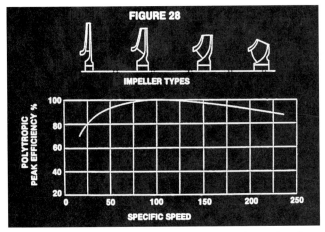

Figure 12.99   Efficiency and specific speed correlation based on shop tests. *(Dresser-Rand Company, Olean, N.Y.)*

Radial bladed impellers produce the most head but exhibit a reduced range of operation with reduced stability, compared to that achievable through backward leaning of the blades. It is for this reason that most centrifugal impellers have backward-leaning blades, generally at angles of approximately 40 to 50°. This inclination has been found to produce a good balance between head and operating range.

The design flow for a particular impeller is determined by the blade entrance angle at the eye; this is illustrated in Fig. 12.100. At the inlet, in the stationary coordinate system, the velocity magnitude and direction are represented by C_{M1}. In the relative system, rotating with the blade at its velocity, the flow approaches in the direction and magnitude of W_1. The design point of a stage represents optimum efficiency; it occurs when the vector W_1 is in line with the blade angle. Since velocities across the impeller inlet eye can vary because of the turning of the flow, the blade angle must also vary across the eye to accommodate this requirement.

As the throughflow velocity C_{M1} increases or decreases, the efficiency drops, primarily because of the positive or negative angle of incidence at the leading edge. Selections are usually made within 5 percent of the design flow.

The level and shape of the impeller pressure coefficient curve is a function of velocity triangles at the discharge of the impeller. The more backward bent the blade, the lower the pressure coefficient and the higher the rise to surge.

There are several types of impeller blading that can be chosen for a particular design requirement and some are shown in Fig. 12.101. At top left is a circular arc blade that is oriented in the axial direction. At

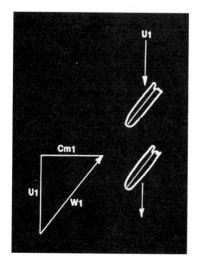

Figure 12.100 Impeller inlet diagram. *(Dresser-Rand Company, Olean, N.Y.)*

top right is a three-dimensional blade shape generated on the surface of an inclined cylinder. At bottom left, we depict a three-dimensional blade shape generated on a conical shape. Finally, at bottom right is a completely arbitrary blade shape. The complexity and manufacturing difficulty of these shapes increases in the order mentioned. The choice is made on the basis of satisfying the aerodynamic requirements. If more than one type of blade is satisfactory, then the least complex design will be selected.

Blade thicknesses are determined by a stress criterion, because the blades must hold the cover to the disk for the typical closed design at the frame speeds desired. Moreover, the blades have to be structurally sound.

The largest flow impellers for a particular frame size are of the open type, with no attached cover, as shown earlier in Fig. 12.39. Radial blading and elimination of a cover on this particular design provide for a maximum amount of flow and head in one stage, yet keep stress levels sufficiently low, even in large diameters, to allow acceptance of maximum material yield point criteria as specified for H_2S service. Note that API 617 stipulates the use of 90,000 psi maximum yield strength, Rc 22 hardness (235 BHN) steels. For ease of manufacturing, the minimum yield strength of carbon steels generally applied in H_2S service is 80,000 psi.

Open impeller blades can be backward curved if desired, with some sacrifice to head and special consideration to blade stresses. The typical open impeller has an almost axial inlet section, referred to as an inducer section. This impeller is typically used in the first stage and

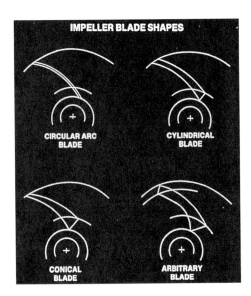

Figure 12.101 Impeller blade shape choices. *(Dresser-Rand Company, Olean, N.Y.)*

requires more axial space than the smaller designs. The primary advantages of this design are higher flow and pressure ratios, with some sacrifice to stability and efficiency. The impeller is designed to run with generous clearances (approximately ⅛ in for a 40-in-diameter impeller) between its periphery and the stationary shroud. It is industry practice to provide one or two stages of backward-curved impellers behind this radial flow impeller to produce a rise to surge performance characteristic in a given section.

A variation of the open impeller design is to equip it with a cover. Some gain is made in efficiency, but the attached cover increases the stresses and significantly lowers the maximum speed capability.

The typical welded-closed impeller used in the vast majority of applications is one of three designs: three-D welded, three-piece welded, or two-piece (milled-welded) construction.

The three-D welded (three-dimensional) has a blade shape that is a portion of a rolled conical or cylindrical surface, or may be a combination of the two. An example is shown in Fig. 12.102. These blades are positioned on the disk or cover at a predetermined inclination and location. The blades then form a three-dimensional contour with respect to the cover or disk.

The choice of cone or cylinder size, location, and inclination to satisfy the aerodynamically required angles at the leading and trailing edges

Figure 12.102 Three-dimensional, welded impeller. *(Dresser-Rand Company, Olean, N.Y.)*

of the blade is difficult. In the past, it was not unusual for a design drafter to spend several weeks trying different combinations before arriving at a satisfactory geometry. Use of the computer has reduced this time to minutes, which allows the designing of more than one blade for review and detailed analysis.

The basic advantage of three-dimensional impellers is better performance, with higher efficiency. However, material, tooling, and welding requirements are higher than with other closed impellers. They also require more axial space in the machine.

The three-piece design (Fig. 12.103) is the next step down in complexity. This type of impeller also has three basic components, i.e., blades, cover, and disk, welded together. This construction allows more freedom for aerodynamic design than the milled-welded impeller design that follows. The blades, however, do not form a true three-dimensional contour. Instead, they are rolled into a circular shape. The impeller requires less axial spacing than the three-D type discussed earlier and is less complex to manufacture.

The two-piece impeller (Fig. 12.104) has the blades milled onto the disk or cover and therefore requires welding on one side only. This type of impeller also has good efficiency and requires a minimum of stage spacing. The two-piece impeller, therefore, combines the features of good performance and ease of manufacturing.

Figure 12.105 gives an idea of the range of impeller sizes available. The large rotor is from a large refrigeration machine, and the small rotor is typical of high-pressure services such as gas injection.

Once the flow channel through the impeller is set, the blade, disk, and cover contours and thicknesses must be developed consistent with

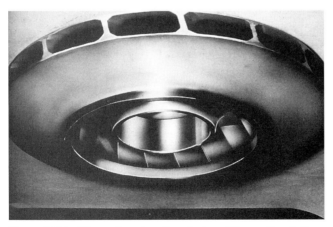

Figure 12.103 Three-piece impeller design. *(Dresser-Rand Company, Olean, N.Y.)*

Figure 12.104 Two-piece impeller with integrally milled blades. *(Dresser-Rand Company, Olean, N.Y.)*

the anticipated speed. Welded impellers are sufficiently complex to mandate computer-based stress and deflection analyses. A manufacturer's stress programs are typically backed up by extensive testing of prototype and/or production impellers using strain gauges, proximity probe measurements of elastic deformations, stress coat and photo stress techniques for studying stress patterns, and overspeed-to-destruction tests.

From both the experimental and analytical investigations, the minimum strength requirements of welded impellers are determined, taking into account appropriate factors of safety. All impellers are heat treated per specifications developed through an engineering depart-

Figure 12.105 Range of impeller sizes typically available. *(Dresser-Rand Company, Olean, N.Y.)*

ment, with acceptable physical property ranges targeted to cover yield and tensile strength, hardness, minimum elongation, and minimum reduction in area.

Typical strength levels used in impellers are as follows:

Material	Yield strength range, lb/in^2
Alloy steel	80–145,000
403–410 stainless	70–135,000
17–4 PH stainless	125–170,000

We can now look at some considerations involved with the design of the other pieces that contribute to the stage, such as the guide vane, diffuser, return bend, and return channel, as depicted earlier in a six-stage compressor (Fig. 12.98). The efficiency of a given stage is a function of the friction and diffusion losses through the stage components. These loss mechanisms can be used to explain the shape of the efficiency versus specific speed curve, Fig. 12.99. Peak efficiency decreases as the specific speed is reduced or increased from an optimum range. This comes about as a result of the combined friction and diffusion losses reaching a minimum value.

Friction losses in a straight pipe are proportional to the velocity squared. The larger the pipe area, the lower the losses. Diffusion losses for areas that are changing are proportional to the velocity ratio $V_{entrance}/V_{exit}$ to the fourth power. Furthermore, losses are incurred by any bends and are proportional to the degree of turning and the tightness of the turn. The lesson in all this is that unnecessary diffusion and bending should be avoided in high-performance compressors.

Characteristically, friction losses increase at the lower specific speeds. This is due to the increased wetted surface and smaller hydraulic channel diameters. Diffusion, on the other hand, increases at the higher specific speeds, reflecting the effects of larger gas capacities being turned in tight bends. When the two losses are added, an area of minimum loss and maximum efficiency results, shown in Fig. 12.99.

The internal geometry of a compressor is more complicated than that of a bent or diffusing pipe. To calculate the velocities within a machine, a manufacturer may use advanced numerical solutions, whereby the geometry would be defined and the velocity fields that must satisfy radial equilibrium and continuity at every spacial point in the flow field would be iteratively calculated. Typically, for a compressor stage analysis, the velocity field from upstream of the impeller to upstream of the next stage impeller is analyzed.

To optimize the efficiency, the geometry of each component is reviewed in terms of velocity and velocity gradients. Unnecessary accelerations are eliminated and diffusion losses are minimized. The return bend radii are generous, the return channel contours are specially shaped, and the areas through the channel and guide vane are closely specified.

The velocity distributions for a final-stage geometry are shown in Fig. 12.106, from the trailing edge of one impeller to the leading edge of the subsequent impeller. The velocity decreases from the first impeller trailing edge through the vaneless diffuser in a uniform manner.

This is followed by the return bend, where the maximum velocities on the inner wall and the minimum velocities at the outer wall are proportional to the severity of the bend and the local surface curvatures. The diffusion losses on both surfaces have been minimized.

Following the return bend, the return channel velocities are shown. The mean velocity through the channel is represented by the middle line; the upper and lower velocities are those along the vane surfaces.

The return channel vanes are optimized. The vane shape, thickness distribution, and number of vanes are analyzed in conjunction with the channel axial height to arrive at the final coordinates.

Following the return channel is the guide vane, which turns the flow from radially inward flow to axial flow. This turning results in acceleration along the inner contour. Unnecessary accelerations and decelerations are carefully scrutinized and minimized.

Next is the impeller, which must be designed consistent with the approach velocities leaving the guide vane and, in turn, impart the

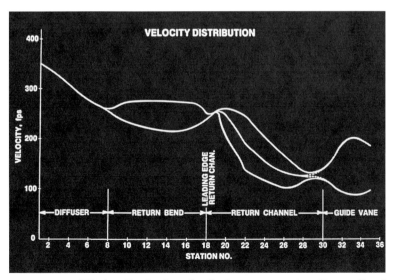

Figure 12.106 Velocity distributions for final stage geometry. *(Dresser-Rand Company, Olean, N.Y.)*

energy increase to the fluid. The process repeats itself for as many stages as there are in the unit.

In summary, the velocity decreases in the diffuser downstream of the impeller, it is then turned from radially out to radially in; the tangential velocity or swirl is removed in the diaphragm or return channel, and finally the flow is turned from the radial to the axial direction and enters the next impeller.

Inlet guide vanes, shown earlier in Fig. 12.29, also provide one method of controlling stage performance, because they may be used to direct the flow into the impeller at different angles: against impeller rotation, radially, or with impeller rotation. The influence of various guide vane angles on a given impeller head characteristic is shown in Fig. 12.107.

How this effect is caused is shown in the impeller inlet diagram, Fig. 12.108. When inlet tangential velocity or swirl exists because of turning guide vanes, the inlet triangle is modified. For a particular value of inlet tangential velocity, C_{U1}, there will be a throughflow velocity, C_{M1}, which will result in a good approach angle to the impeller blading. This is where the peak efficiency will occur.

12.10   Impeller Manufacturing Techniques

As mentioned earlier in "Design Overview," Sec. 12.5, the riveted impeller has been around for many years. There are two basic types of construction, one being the two-piece with milled vanes, and the other is the three-piece with separate disk, cover, and blades.

Large-capacity impellers require the three-piece construction (Fig. 12.103). The cover and disk are fabricated from forgings, usually either

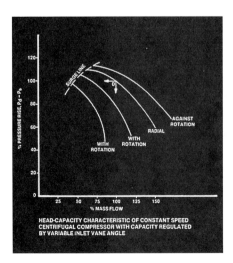

Figure 12.107 Guide vane angles vs. impeller head characteristics. *(Dresser-Rand Company, Olean, N.Y.)*

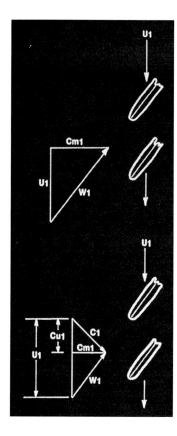

Figure 12.108 Impeller inlet diagram. *(Dresser-Rand Company, Olean, N.Y.)*

of alloy steel or a 400-series stainless. The blades are usually die-formed from stainless steel and are attached to the disk and cover by short stainless steel rivets extending through the disk or cover material and the flange provided along the blade by the forming.

Smaller-capacity impellers are of two-piece construction (Fig. 12.104), with vanes milled on either cover or disk piece. Disk and cover halves are jointed by longer rivets that pass completely through the cover, the blade, and the disk. Flush riveting is achieved through the use of countersunk rivet holes. The outer surfaces are ground smooth.

For both types of construction, the pieces are typically purchased in the final heat-treated condition, with metallurgical and mechanical properties certified. They are completely machined and inspected, including a magnetic particle check, before joining. After assembly, the impeller typically undergoes balance, overspeed test, and additional final inspection, including magnetic particle and dimensional examination.

Cast impellers have also been used for some time. This technique is limited to designs that have high usage, thus justifying the typically high pattern costs. For this reason, cast impellers are not too common in centrifugal process compressors that are almost always custom designed for the application. The technique has been applied to some small impellers. However, there is obviously a practical limit to the narrowness of the flow channel.

Typically, the impellers can be made of any castable steel, including 400 series and 17-4 PH stainless steels. After the obligatory cleanup, they are heat-treated per the required specification. Bores and other surfaces are then machined as specified. Gas passage areas are essentially left untouched, aside from some possible hand-grinding or vibratory finish, if desired.

Castings can be ultrasonically inspected or radiographed to verify soundness. The typical inspection process involves magnetic particle and dimensional checks before and after the balance and overspeed runs.

Welded impellers now account for the majority of impellers being designed and constructed. They are generally more rugged and will be more able to resist corrosion and erosion than riveted types.

All materials for welded impellers, including the disks, blades, and covers, are supplied in the annealed condition, with appropriate material certifications. Pieces are preliminary-machined in preparation for welding. Location welding, or *tack* welding, is done by the inert gas-shielded, metal-arc process (MIG). The completing welds are done by the same method used in a semiautomatic welding process (Fig. 12.109). The gas-shielded, tungsten-arc (TIG) method is used at times for completing welds around blade ends.

Welding tables provide for proper preheat during the welding process. These tables are provided with heating elements in the table and

Figure 12.109 Impeller undergoing semiautomatic welding. *(Dresser-Rand Company, Olean, N.Y.)*

cover to maintain material temperatures at approximately 600°F. The tables are mechanized to permit rotation of the workpiece under a stationary open flame, or torch.

Final mechanical properties are obtained by heat-treating (normalizing or quenching, with subsequent tempering) after welding. Typically, impellers then undergo final machining, dynamic balancing, and overspeed testing to at least 115 percent of maximum continuous speed, as specified in API 617.

The general manufacturing techniques used to produce the different types of welded impellers vary somewhat and will be looked at now.

For a three-piece construction impeller, the disk and cover are machined from forgings. The die-formed blades are then tack-welded to the cover with the use of locating fixtures. The final welding is a continuous fillet weld between the blade and cover. Subsequently, the blade cover assembly is joined to the disk by a continuous fillet weld between disk and blade.

An open impeller design consists of a disk and blades. The cover is eliminated. This type of impeller, as discussed previously, is characterized by an inducer section that directs the gas flow into the eye of the impeller. The blades are either die formed or precision cast. The welding procedure is the same as for the three-piece construction, with the final weld being a continuous fillet weld between the disk and blade.

For two-piece construction, the blades are machined on either the disk or cover forging. The impeller is completed by a continuous fillet weld to the mating piece (disk or cover) around the entire blade interface.

The welding techniques described are limited to impellers with a flow channel width of more than ⅝ in, to allow insertion of a torch. Increasing numbers of applications require the advantages of welded construction for impellers with channel widths of less than ⅝ in (Fig. 12.110). One method of doing this has been to weld through from the back side (Fig. 12.111). This type of impeller is manufactured from a disk forging and a cover forging, thus being of two-piece construction. Impeller blades, integral with the cover, are formed by removing metal from the inner face of the cover. A matching slot, corresponding to the blade contour, is machined in the disk at each blade location. After machining, the disk is precisely located over the cover with the blades aligned to the slots in the disk. The slot is then filled with a continuous multipass TIG weld (Fig. 12.112). An internal fillet is thus formed at the blade-to-disk junction. Using this technique, welded impellers with the smallest practical channel width can be manufactured. Any blade contour may be designed without affecting the weldability of the impeller, thus minimizing compromises in aerodynamic design. The impellers (Fig. 12.113) are available in the same materials and range of properties as the fillet-welded impellers.

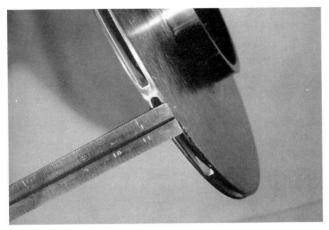

Figure 12.110   Narrow-width impeller. *(Dresser-Rand Company, Olean, N.Y.)*

To ensure quality, the following general inspection procedures are typically carried out on welded impellers:

- Material certification reports are reviewed for compliance with specifications.

Figure 12.111   Slot welding two-piece impeller. *(Dresser-Rand Company, Olean, N.Y.)*

Figure 12.112 Slots in disk being filled. *(Dresser-Rand Company, Olean, N.Y.)*

- Test bars for each forging are serialized and follow the impeller through all heat-treatment cycles.

- Magnetic particle inspection is conducted after each welding cycle, both before and after heat treatment. Final magnetic particle inspection takes place after the overspeed test.

- Dimensional checks are made after initial welding and after completion of the manufacturing and test process.

- Hardness and tensile tests are performed on the test bars to ensure specification compliance. Hardness checks are made of the impeller

Figure 12.113 Finished impeller after slot welding. *(Dresser-Rand Company, Olean, N.Y.)*

cover and disk to verify compliance with specifications and agreement with test bars.

12.11   Rotor Dynamic Considerations

In addition to meeting the aerodynamic expectations, it is equally important that the compression equipment meet the rotor dynamic requirements and be able to operate in a satisfactory and stable manner. Most large equipment today is furnished with noncontacting vibration transducers installed near the bearings. These allow both user and manufacturer to measure rotor vibration amplitudes and frequencies. Maximum acceptable vibration levels at these locations at speed are generally defined in line with applicable API or other industry recommendations. One such guideline essentially allows $(12,000/\text{max. cont. speed})^{1/2}$. In addition, manufacturers have also generally established their own minimum margin requirements between operating speeds and critical speeds. These requirements basically follow those set forth in API 670. The majority of compressors operate with flexible rotors, running between their first and second critical speeds. They exhibit shaft vibration amplitudes as indicated by the GOOD range in Fig. 12.114.

The location of the critical speeds plays a large part in the design of a compressor. A manufacturer has to ensure that the compressor will be able to operate over the full speed range intended; this ensurance comes from rigorous computer analysis of the critical speeds, unbalance sensitivities, and stability of a system. It is generally in the best interest of overall reliability to make the system as stiff as possible, i.e., to minimize the distance between the journal bearings.

The effort to minimize bearing span manifests itself in various ways in compressor design. These include:

- Positioning the bearings as far inboard in the case end enclosures (heads) as possible.
- Minimizing end seal lengths
- Using variable stage spacing

In the past, many casings were designed based on a fixed axial length per stage, usually determined by the largest-size impeller that the manufacturer anticipated would be applied. When smaller impellers were used in these stages, wasted axial space resulted. With variable stage spacing, each impeller gets only the room it requires. A larger number of impellers can thus be accommodated.

The location of the second critical speed can be greatly influenced by overhung weight, which is the weight outboard of the journal bearings

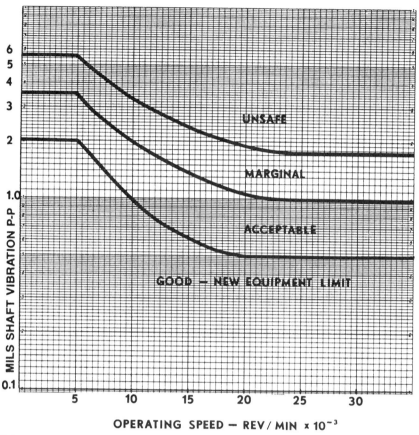

OPERATING SPEED — REV/MIN × 10⁻³

Notes:

1. Operation in the "unsafe" region may lead to near-term failure of the machinery.
2. When operating in the "marginal" region, it is advisable to implement continuous monitoring and to make plans for early problem correction.
3. Periodic monitoring is recommended when operating in the "acceptable" range. Observe trends for amplitude increases at relevant frequencies.
4. The above limits are based on Mr. Zierau's experience. They refer to the typical proximity probe installation close to and supported by the bearing housing and assume that the main vibration component is *1 × rpm frequency*. The seemingly high allowable vibration levels above 20,000 rpm reflect the experience of high-speed air compressors (up to 50,000 rpm) and jet-engine-type gas turbines, with their light rotors and light bearing loads.
5. Readings must be taken on machined surfaces, with runout less than 0.5 mil up to 12,000 rpm, and less than 0.25 mil above 12,000 rpm.
6. Judgment must be used, especially when experiencing frequencies in multiples of operating rpm on machines with standard bearing loads. Such machines cannot operate at the indicated limits for frequencies higher than 1 × rpm. In such cases, enter onto the graph the predominant frequency of vibration instead of the operating speed.

Figure 12.114   Turbomachinery shaft vibration chart. *(Bloch / Geitner, "Machinery Failure Analysis and Troubleshooting," Gulf Publishing Company, Houston, Tex., 1983 and 1994)*

at each end. This typically includes the thrust disk and coupling(s). Relocating the thrust-bearing inboard of the journal bearings on a drive-through unit increases the second critical speed. This permits a higher operating speed or increased bearing span, resulting in operating speeds farther removed from the first critical speed.

A typical rotor dynamic analysis starts with a definition of the rotor configuration. The shaft is divided into sections of a given diameter and length, and the impellers, thrust disk, balance piston, and coupling(s) are represented as weights or forces.

Lateral critical speed calculations are then made, with the aid of a computer. This analysis gives the approximate location of the rigid critical speeds, and, with the input of different support or bearing stiffnesses, criticals are calculated as a function of stiffness. The resulting plot is known as a *critical speed map* (Fig. 12.115). The mode shapes determined by these calculations are of value in determining the location of the unbalance required for exciting a particular critical speed in the response analysis that follows.

There also exist programs that compute the performance characteristics of the bearings, including oil flow, horsepower loss, oil temperatures, effective eccentricity, and load per pad, along with oil film stiffness and damping coefficients in the vertical and horizontal directions as a function of speed. Most often these results are cross-plotted on the critical speed map, as shown in Fig. 12.115. The curve labeled K_y is horizontal bearing stiffness and the curves labeled K_x are vertical bearing stiffness. The K_x curves are for 1, 2, and 3 times the bearing

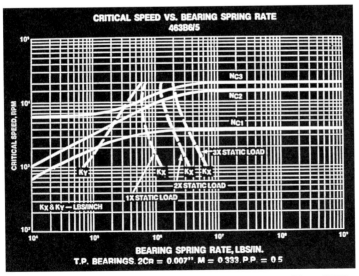

Figure 12.115   Critical speed map. *(Dresser-Rand Company, Olean, N.Y.)*

static load. The 2- and 3-times curves represent 1 g and 2 g dynamic or unbalance loads added to the weight load. Although not shown, dynamic loads would similarly influence the K_y stiffness. The intersection of the bearing stiffness and rotor mode curves are the undamped critical speeds. It is seen that the horizontal and vertical critical speeds can be different and that dynamic or unbalance loads influence the critical speed results. Also, as bearing or support stiffness increases, the critical speeds approach the critical speeds of the rotor on rigid or simple support as a limit.

The next step makes use of a rotor response analysis program that calculates the unbalance response of a rotor in fluid film bearings. It is able to predict rotor synchronous vibration behavior at all speeds for a selected unbalance distribution. Bearing stiffness and damping characteristics are part of the input. The motion of the rotor is treated as two-dimensional. The output gives the major and minor axes of an ellipse formed by the locus of the shaft center. Dynamic bearing forces can also be determined for the bearings. Examples of typical results from such an analysis are shown in Fig. 12.116 (amplitude vs. speed) and 12.117 (bearing force vs. speed). This program has proven to be a most valuable tool for predicting the synchronous vibration behavior of rotating machinery.

The most recent analytical tool, developed during the mid-1970s, is the rotor stability analysis. This technique has been used to increase the understanding of various rotor instability phenomena that became apparent in compressors designed for extreme high pressure applications.

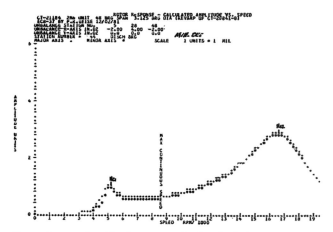

Figure 12.116 Amplitude vs. speed plot (unbalance response plot). *(Dresser-Rand Company, Olean, N.Y.)*

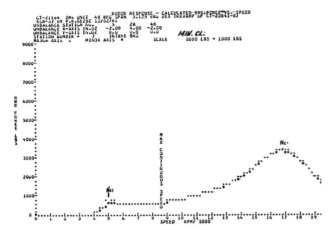

Figure 12.117   Bearing force vs. speed plot (unbalance response). *(Dresser-Rand Company, Olean, N.Y.)*

The stability program integrates rotor geometry with support stiffness and damping to determine system-critical speeds, damped mode shapes, and an exponential representation of system stability. It is complicated in that eight coefficients are used for each bearing: four stiffnesses, normal and cross-coupling, and four damping, normal and cross-coupling. These can be input for the seals if desired. In addition, the program can accept other input such as aerodynamic effects, internal friction, and negative damping.

These analysis methods yield critical speeds and logarithmic decrements together with three-dimensional damped shaft mode shapes. The critical speed and logarithmic or *log* decrement output is plotted in the format illustrated in Fig. 12.118. It displays critical frequency in cycles per minute vs. shaft speed in revolutions per minute so that synchronous excitation will be a 45° line as shown. These curves represent the vertical and horizontal modes of the first and second critical frequencies, respectively. In essence, these curves indicate how the critical frequency varies with shaft speed, and the intersections with synchronous excitation are the critical speeds. Log decrement values offer a relative means in ranking the ability of a system to cope with undesirable excitation. The higher the value, the greater the excitation required to make the system unstable.

Torsional studies represent another type of analysis sometimes performed on a rotor system. The torsional natural frequencies are calculated by dividing the system into a number of mass moments of inertia separated by torsional spring constants for the various shaft sections. An example is given in Fig. 12.119. Results are obtained by use of a computer program.

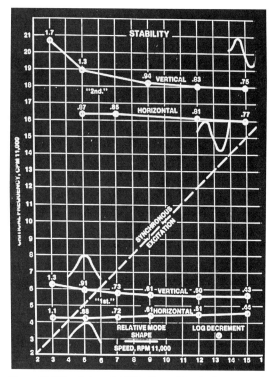

Figure 12.118   Rotor stability (logarithmic decrement) plot. *(Dresser-Rand Company, Olean, N.Y.)*

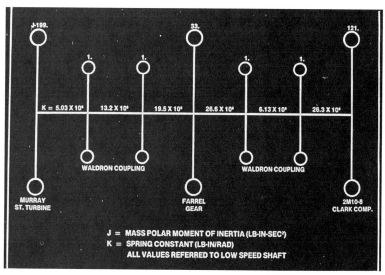

Figure 12.119   Torsional natural frequency plot. *(Dresser-Rand Company, Olean, N.Y.)*

For the majority of systems, only shaft rotational frequency is considered as potential excitation. A system is considered satisfactory if the torsional natural frequencies are 10 or more percent away. If closer than 10 percent, the system is tuned. An interference diagram is shown in Fig. 12.120. The sloped line represents excitation. Note how the natural frequencies are well removed from the operating range in this example.

While steady-state torsional problems are rare, a potentially troublesome transient torsional problem exists in conjunction with synchronous motor drivers during start-up. Widely different motor start-up characteristics are possible, and their effect on the system must be analyzed. Computerized procedures that can accurately predict alternating and peak torque levels have been developed. System design integrity can thus be ascertained.

Steady-state torsional problems are somewhat more likely to occur in gear-driven compression systems. Here, an appropriate torsional analysis takes on added importance.

12.12   Fouling Considerations and Coatings*

Process compressors are required to run at or near peak efficiencies for long periods of time. It is on this basis that two solutions to the poly-

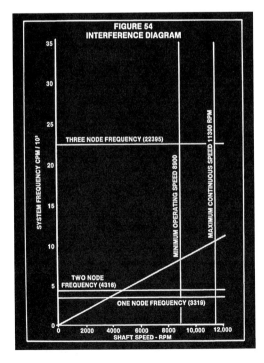

Figure 12.120 Interference diagram. *(Dresser-Rand Company, Olean, N.Y.)*

merization problem, flush liquid injection and the application of coatings, have been implemented.

The introduction of liquid into the gas stream is always process-dependent and is thus not considered within the scope of this text. However, modern coating technology should be reviewed within the context of centrifugal compressor design.

12.12.1 Polymerization/fouling

The chemical mechanism that takes place to generate polymerization is not well understood, as it applies to compressor fouling. However, what is known is that hydrocarbons inherent in the process gas, or formed during the compression process, can bond tenaciously to components and lead to significant performance loss (see Fig. 12.121). Deposits of this type have been found in compressors used for hydrocarbon processing, coke gas blowers, and other units where the gas contains sufficient amounts of hydrocarbons under the right conditions of pressure and temperature.

Factors which have been found empirically to be critical to the fouling process are:

1. *Temperature*—Polymerization occurs above 194°F (90°C).

2. *Pressure*—The extent of the fouling is proportional to pressure.

3. *Surface finish*—The smoother the surface, the less apt the component is to foul.

---

* SOURCE: R. Chow, B. McMordie, and R. Wiegand, "Coatings Limit Compressor Fouling," *Turbomachinery International,* January/February 1995. Adapted by permission of the publishers.

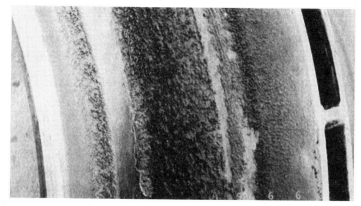

Figure 12.121 Centrifugal compressor impeller, showing fouling deposits. (*Courtesy of Turbomachinery International, Norwalk, Conn.*)

4. *Gas composition*—Fouling is proportional to the concentration of reactable hydrocarbon in the process (inlet) gas.

12.12.2   Fouling and its effect on compressor operation

Component fouling has many detrimental effects on compressor operation.

Unbalance.   One of the most obvious is the build-up of material on the rotor. This can lead to unbalance, which gradually builds until the unit exceeds its allowable vibration limit and has to be shut down to correct the problem. In addition, operation with significant rotor unbalance can lead to fatigue loading and a possible reduction in component life.

Abrasive wear.   Deposits have also been known to reduce both the axial and radial clearances between the rotor and the stationary components. This clearance reduction has led to abrasive wear, which has severely damaged numerous impellers and labyrinth seals.

Unbalance and abbrasive wear are progressive, with the costs associated with correction showing up after longer periods of operation.

Loss of efficiency.   When considering fouling that affects unit performance, the losses and associated costs are revealed very quickly, typically only months after the unit is started. This has been confirmed by actual operation.

The case study which follows describes a 6 percent decrease in efficiency after only 17 months of operation. In addition, it was found elsewhere that the most intensive growth of the deposition layer occurred during the first 50 to 200 hours of operation. These examples reinforce the fact that fouling degradation occurs early, and can cause significant losses, making it increasingly important to take corrective action from the start to assure optimum efficiency.

Fouling affects efficiency through three basic loss mechanisms:

1. Friction losses
2. Flow area reductions
3. Random changes of pressure distribution on the blade

These mechanisms affect both the stationary flow path and the rotating element. However, in the past, attempts to correct the fouling problem were limited to the diffuser and return channels, which were considered to be the most susceptible. The rotating element is less likely to foul due to the dynamic force applied to the deposits because of the dislodging effect of rotation. In addition, by design, the stationary flow paths have slightly rougher surface finishes than the rotating element. Today, however, fouling of both stationary and rotating flow path components needs to be addressed.

The Elliott Company has modeled the effect of fouling on compressor performance using their compressor performance prediction program. This work simulated the effect of deposits on stationary flow path surfaces on the performance of an Elliott 38M9 centrifugal compressor, operating under the following conditions:

1. Compressor speed = 6150 r/min

2. Inlet pressure = 228 psia

3. Inlet temperature = 142°F

4. Gas containing a mixture of various hydrocarbons

The study evaluated the effects of diffuser passage width and surface finish on head and efficiency. It showed that losses of 10 percent or more could result from passage width restrictions of 10 percent or greater and surface finishes of 500 rms or higher (Figs. 12.122 and 12.123).

It should again be noted that compressor washing, using either water to lower gas temperature, or a hydrocarbon solvent to dissolve the deposits, has been used to reduce the extent of fouling.

12.12.3   Coating case study

Novacor Chemicals' Ethylene 2 Plant is a world-scale petrochemical plant located in Joffre, just east of Red Deer, Alberta, Canada. The main feedstock, ethane, is cracked to make ethylene and other hydrocarbon by-products. Approximately 1.8 billion pounds of ethylene is produced by Ethylene 2 annually.

Ethane is cracked in furnaces, and then compressed by the cracked gas compressor string for finish processing and separation of products. The cracked gas compressor string consists of Elliott 88M, 60M, and 46M compressors driven by an Elliott NV9 steam turbine (Fig. 12.124). Historical performance data shows that the compressors foul during operation. The formation of polymers in the 46M compressor is greatest, as this has the highest pressure and temperature in the compressor string.

To increase reliability and improve the run time of the equipment, coating of the compressor parts was considered. A coating would have to provide three benefits:

1. A *nonstick surface,* so that fouling could not form on the surfaces and degrade performance.

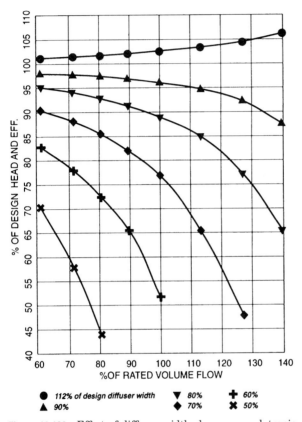

Figure 12.122 Effect of diffuser width change on polytropic head and efficiency. (*Courtesy of Turbomachinery International, Norwalk, Conn.*)

2. An *erosion barrier* to the current wash oil injection practice in the compressors.

3. *Corrosion protection,* to maintain the finish on aerodynamic surfaces.

A risk analysis of the coating showed minimal impact to the process and equipment, should the coating not function as designed. The only concern was that unstacking the spare rotor was necessary to coat the compressor wheels. At the next shutdown of the plant, the rotor wheels were coated by Sermatech using the SermaLon coating system.

12.12.4 SermaLon coating

The SermaLon coating system was developed by Sermatech, and was designed for wet, corrosive environments. It combines the benefits and features of the three types of coatings generally used to combat corro-

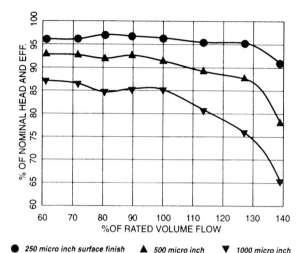

Figure 12.123   Effect of diffuser finish on head and efficiency for 80 percent diffuser width. (*Courtesy of Turbomachinery International, Norwalk, Conn.*)

sion of metal components: barrier coatings, inhibitive coatings, and sacrificial coatings.

The outermost layer of the coating system is a high-temperature resin film, which is a barrier against corrodants in the environment. Barrier coatings prevent corrosion by sealing the substrate from environmental effects. But once this seal is broken, corrosion proceeds unchecked at the point of the breech in the coating.

The intermediate layer of the SermaLon coating system is a durable, inhibitive coating. Inhibitive coatings contain pigments (e.g., chromates or complex metalo-organic compounds) which prevent corrosion

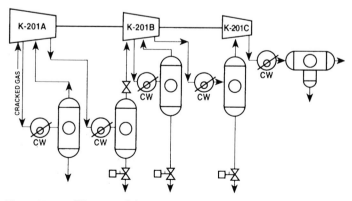

Figure 12.124   Diagram of the compressor string at Novacor Chemicals; K-201A = Elliott 88M, K-201B = Elliott 60M, K-201C = Elliott 46M. (*Courtesy of Turbomachinery International, Norwalk, Conn.*)

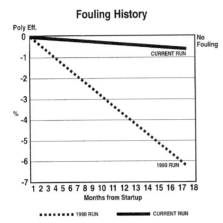

Figure 12.125 Fouling history of the compressor at Ethylene 2. (*Courtesy of Turbomachinery International, Norwalk, Conn.*)

by modifying the chemistry of environmental corrodants contacting the coated surface. These reactions change the pH, reactivity, and even the molecular structure of the corrodants. Inhibitive coatings are very effective as long as the film remains intact and inhibitive pigments remain reactive. When the coating is breeched, corrosion of the exposed substrate is slowed until active pigments are depleted.

The foundation of the coating system is a tightly adherent layer of a sacrificial aluminum-filled ceramic. Sacrificial or *galvanic* coatings prevent corrosion of structural hardware by corroding instead of the substrate. They are made of a more *active* metal which, when placed in contact with a less-active (more *noble*) metal, will be entirely consumed by the environment before the more noble material begins to corrode.

12.2.5 Results

As of 1995 (17 months of operation), compressor performance monitoring has shown that the efficiency has remained virtually constant. For comparison, there was a 6 percent reduction in efficiency during the previous run, without coatings, in the same amount of time (Fig. 12.125). The success in maintaining the performance of the compressor is attributed to the coating and washing system. However, it is impossible to quantify the effect of each alone. It is estimated that the payback for the coating of the rotor was two months.

Dry Gas Seal and Magnetic Bearing Systems*

In the 1970s, a forward-looking Canadian company, Nova, initiated a program to identify, research, and implement technologically and economically sound means of eliminating the many problems associated with oil systems on their centrifugal compressors. The purpose of this program was to improve safety and reduce maintenance and operating costs.

In the years that followed, Nova worked closely with mechanical dry gas seal (dry seal) and magnetic bearing manufacturers on the development, application, and installation of these technologies for centrifugal compressors. By 1978, a fully functional dry seal was installed successfully. By 1985, the first oil-free compressor using both dry seal and magnetic bearing technologies was operational.

Development of the technologies continues, with recent advancements in thrust-reducing seals and magnetic bearing control system enhancements. Other manufacturers have become active in the field and dry seal technology has advanced to the stage where, in the late 1980s, the industry began actively discussing specification standards.

13.1 Background

Internal studies conducted by Nova's Alberta Gas Transmission Division in the early 1970s indicated that a significant portion of the down-

---

* Developed and contributed by Stan Uptigrove, Paul Eakins, and T. J. Al-Himyary of Revolve Technologies, Calgary, Canada. Revolve is an employee-owned company specializing in rotating equipment and the design and application of dry seal and magnetic bearing systems. Revolve was spun out of Alberta, Canada–based Nova Corporation in an effort to further develop and implement these technologies.

time on their 79 centrifugal compressors related to problems with the seal or lube oil systems. The problems ranged from failures in auxiliary systems, such as motors and pumps, to leaks and failures in pressure piping resulting from vibration.

By 1978, Nova and a seal vendor developed a dry seal that reliably replaced the conventional seal oil system in centrifugal compressors. Dry seal development with other vendors continued, with the recent application of alternative face materials, groove patterns, and thrust-reducing configurations—all with successful results.

In 1985, an active magnetic bearing system was installed in one of the centrifugal compressors that had been retrofitted earlier with dry seals. As of 1994, this compressor, an Ingersoll-Rand (IR) CDP-230 driven by a General Electric LM 1500 gas generator and an IR GT-51 power turbine rated at 10.7 MW (14,500 hp), had accumulated 9 years of operation since the bearing retrofit. It also served as a test bed for several other pilot installations of new developments in this technology.

In 1988, Nova retrofitted magnetic bearings onto the power turbine of this package. Operating history since then indicates that this higher temperature environment poses no limitations to magnetic bearing technology. This makes it possible to eliminate the lube oil systems commonly serving the gas compressor and the power turbine, a feature found frequently in aircraft derivative gas turbine compressor packages.

A system expansion undertaken by Nova in the late 1980s has seen the procurement of over 40 new compressors equipped with one or both of these technologies, along with the increasing involvement of equipment vendors in the development of these technologies. Other user companies, such as Alberta Natural Gas, Marathon Oil, Shell Canada, and Exxon Chemicals, have moved in the same direction.

13.2   Dry Seals

13.2.1   Operating principles

Dry seal design is based on the gas film technology used successfully in other applications, such as air bearings in high-precision machining and measurement equipment. The heart of the sealing mechanism is comprised of two seal rings (Fig. 13.1). The mating ring has a groove pattern etched into a hard face and rotates with the shaft. The primary ring has a softer face and is restrained from movement except along the axis of the shaft.

Springs are located to axially force the primary and mating ring faces toward one another. When the compressor is shut down and depressurized, the spring forces result in contact of the faces. As the compressor is pressurized, the balance of static pressure forces on the seal mechanism allows a minute volume of gas to leak past the faces.

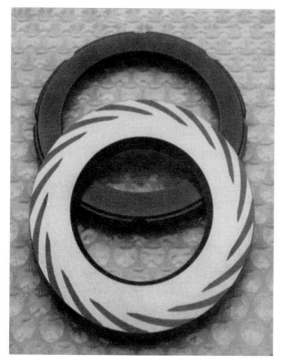

Figure 13.1 Dry seal rings. *(Revolve Technologies, Calgary, Canada)*

When the compressor is running, the combination of process gas pressure (hydrostatic forces) and the pumping pressure provided by the spiral grooves (hydrodynamic forces) results in noncontact seal face equilibrium. Increased clearance reduces gas film pressure, and hydrostatic pressure behind the faces tends to reduce clearance. The noncontacting nature of the gas seal film means there is virtually no mechanical wear.

Depending on the process application, the seal mechanism can be used by itself in a single (Fig. 13.2), double, or tandem arrangement. The tandem arrangement (Fig. 13.3) is most common in natural gas pipeline applications. Each stage is capable of sealing against full process gas pressure, up to approximately 10,350 kPa gauge (1500 psig).

In normal operation, the primary stage seals against full process gas pressure, while the second stage is unloaded. The secondary stage only sees process gas pressure in the event of a failure of the first-stage seal. This provides backup for safe shutdown. For ease of installation, the entire seal assembly is encapsulated so that it can be installed and removed as a complete unit (Fig. 13.4).

A monitoring and control system ensures that the seals are provided with a clean gas supply to prevent potentially dirty process gas from

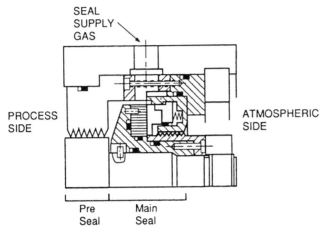

Figure 13.2 Single dry seal configuration. *(Revolve Technologies, Calgary, Canada)*

entering the seal (Fig. 13.5). Although other sources can be used, seal gas supply is typically taken from the compressor discharge piping and filtered. A coalescing 0.1-μm filter is used in case liquids are present.

Seal gas is monitored through a small flow meter and sent to the seal gas supply, where the majority reenters the process cavity across a labyrinth seal. Only the volume of leakage gas pumped by the grooves passes across the seal faces; then it drops across the smooth dam area of the mating ring until it reaches the pressure in the first-stage leakage port.

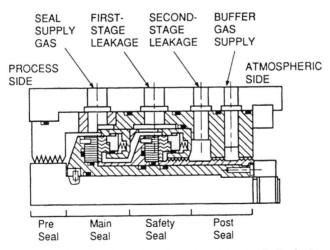

Figure 13.3 Tandem dry seal configuration. *(Revolve Technologies, Calgary, Canada)*

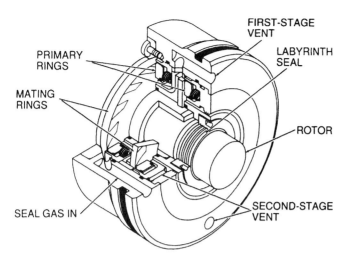

Figure 13.4 Dry seal cartridge. *(Revolve Technologies, Calgary, Canada)*

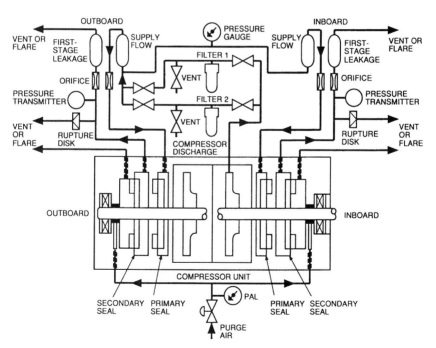

Figure 13.5 Dry seal monitor and control system. *(Revolve Technologies, Calgary, Canada)*

Leakage volume is measured at this point because it provides a diagnostic measure of seal operation. An increase or decrease in leakage volume beyond predetermined levels results in an alarm or shutdown of the compressor. A second flow meter gives visual indication of leakage volume, enabling operators to gauge the cleanliness of the gas leaving the seal. Leakage volumes vary with size and speed but are typically below 80 L/min (3 scfm).

The seal can also be contaminated by oil from the adjacent bearing cavity. Nitrogen or instrument air is provided as a buffer to prevent oil contamination.

13.2.2 Operating experience

Nova is one of the largest users of dry seals. From the initial pilot installation, the dry seal retrofit program has grown to include over 30 units, ranging in shaft size from 45 to 255 mm (1.75 to 10 in), with pressures up to 10,880 kPa gauge (1600 psig) and speeds up to 27,060 r/min.

The program was expanded to include the specification of dry seals on newly procured gas compressors. As of 1994, the total exceeded 75 units with dry seals. Nova's lead has been followed by numerous other companies, and dry gas seals should be considered for the majority of centrifugal compressors. With dry seals, gas compressor safety has increased, and operating and maintenance costs have decreased. Reliability is almost always better than that of oil systems. Dry seals are now an accepted design standard at many cost-conscious and reliability-minded user companies.

13.2.3 Problems and solutions

There were several technical problems encountered over the course of the program. One problem that continues to occur intermittently is static pressurization of the compressor casing at the beginning of the start-up cycle. Working with the seal vendors, users have minimized problems by changing the spring rate of the primary ring carrier springs and by improved quality control of dynamic O-ring and retainer.

In the past, seal gas contamination caused problems because of the extreme flatness tolerances required on the seal faces. Even minute particulates can damage the soft primary ring enough to disrupt seal film stability. Filtration system improvements, including a second coalescing filter upstream of the original, have resolved most of these problems. When a new compressor package is being commissioned, a separate source of seal gas, usually nitrogen, should be provided. This protects the seals from contamination from debris that may be present in recently constructed compressor systems or mainline pipe. Normally, such debris does not cause a problem beyond the first several hours of operation.

Another problem concerned the explosive decompression of O-rings. During operation at pipeline pressures of 4500 to 11,000 kPa gauge (650 to 1600 psig), gas becomes entrapped in the elastomeric O-ring materials. Upon shutdown and compressor casing depressurization, this gas would blister or burst the O-rings as the gas attempted to escape. Using higher density elastomers and changing the compressor control logic to depressurize only in the event of an emergency have largely corrected this problem.

13.2.4   Dry seal upgrade developments

In 1986, Nova began working with a seal vendor on the design of a dual hard face dry seal. Hard faces enable a face geometry to be devised that allows complete static separation of the faces, taking the noncontact concept one step further. The use of hard silicon carbide primary ring provides this capability. It also enables dry seal technology to be used in applications as high as approximately 20,000 kPa gauge (2900 psig). User companies have been applying this type of seal since 1988 with very good results.

More recently, additional groove patterns are being applied successfully. In 1991, the first bidirectional T-groove seals (Fig. 13.6) were installed. Unlike the spiral groove and other one-directional mating ring designs, the bidirectional seal can be installed on either end of the

Figure 13.6 Bidirectional dry seal faces. *(Revolve Technologies, Calgary, Canada)*

compressor, which reduces spares inventory requirements. Also, if the compressor rotates in reverse, the seal is not damaged. Operationally, this seal has proven to be as effective as spiral groove seals and offers substantially lower leakage (Fig. 13.7). Since most dry seals in natural gas pipeline applications vent their seal leakage to atmosphere, this may be of interest in light of increasing attention to environmental emissions. As of 1994, Nova had over 30 compressors operating with the bidirectional seals.

13.3   Magnetic Bearings

13.3.1   Operating principles

Magnetic bearings for gas compressor applications are used in both radial and axial configurations, performing the same tasks as their hydrodynamic counterparts. Each individual bearing consists of a rotor and stator, position sensors, and an electronic control system (Fig. 13.8).

The rotor of a radial magnetic bearing consists of a stack of circular laminations pressed onto a sleeve that can be fitted to the compressor shaft. Used to reduce eddy current losses, these laminations are selected from a material with high magnetic permeability for higher magnetic flux conductance. The radial magnetic bearing stator is similar to that of an electric motor, with a stack of slotted laminations about which coils of wire are wound (Fig. 13.9). The stator is divided equally into four distinct electromagnetic quadrants, each with pairs of north and south poles. In horizontal rotor applications, quadrant center lines are oriented at 45° to the vertical, so that forces due to gravity

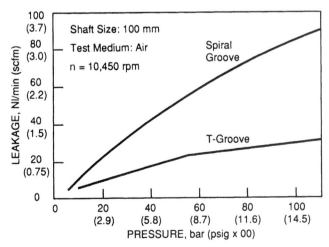

Figure 13.7   Groove shape and seal leakage rates. *(Revolve Technologies, Calgary, Canada)*

AXIAL SENSOR ROTOR
(End Shaft)

AXIAL BEARING STATOR

RADIAL
SENSOR
STATOR

RADIAL
BEARING
STATOR

AXIAL
SENSOR
STATOR

AXIAL
BEARING
ROTOR

RADIAL SENSOR ROTOR

RADIAL BEARING ROTOR

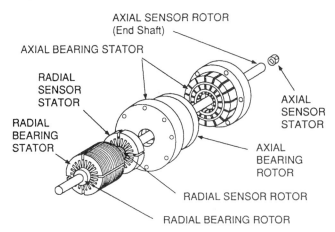

Figure 13.8   Magnetic bearing construction. *(Revolve Technologies, Calgary, Canada)*

are reacted by the upper two adjoining quadrants. This increases load capacity and stability.

The rotor of an axial magnetic bearing is a solid ferromagnetic disk secured to the compressor shaft. The axial bearing stator (Fig. 13.10) is made from solid steel wedges within which coils are wound in annular grooves to form electromagnetic windings. Laminations of highly permeable material are placed in between the wedges to decrease eddy current losses. By positioning a stator on both sides of the rotor disk, a double-acting thrust bearing is created.

The position sensors provide feedback to the control electronics on the exact position of the rotor. Among the well-proven sensor types, we find

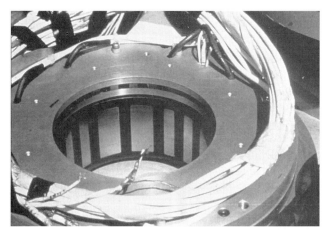

Figure 13.9   Radial magnetic bearing stator. *(Revolve Technologies, Calgary, Canada)*

Figure 13.10 Axial magnetic bearing stator. *(Revolve Technologies, Calgary, Canada)*

those that form an inductive bridge. As the air gap increases or decreases, the inductance varies. When the bearing rotor is centered, the position error signal is zero. A shift in the shaft location results in a corresponding change in inductance that alters the position error signal.

The position signal from the sensors is sent to the control electronics and compared to a reference signal that indicates where the compressor shaft should be. Any difference between these two signals generates an error signal. This error signal is processed by the control system. The output of the controls is used to vary the current supplied to the appropriate electromagnet via power amplifiers. The dc voltage for the amplifiers is stepped down for use in the control logic circuitry (Fig. 13.11) by dc-dc converters.

Also located in the control electronics is a monitoring and security system that can initiate alarms and shutdowns to protect the unit from damage. A battery backup system is provided to maintain operation in the event of electrical power failure.

The bearing rotor and stator surfaces are ground smooth to minimize mechanical runout and variation in forces. For this reason, it is important that these surfaces do not come into contact during operation or any other time. To prevent this, an auxiliary landing system is provided, consisting of rolling element bearings located in a removable bearing holder. The clearance between the shaft and auxiliary bearings is normally half the clearance between the magnet rotor and stator surfaces. When the system is deenergized (either in motion or at rest), the shaft coasts down on or remains at rest supported by the auxiliary bearings. Use of a removable sleeve fitted to the shaft, with predeter-

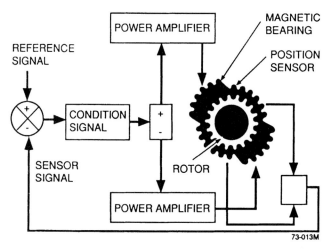

Figure 13.11 Control loop for magnetic bearings. *(Revolve Technologies, Calgary, Canada)*

mined radial and axial dimensions, provides a sacrificial means of maintaining the desired clearances. Auxiliary bearings are typically rated for three emergency coastdowns from full load and speed.

Finally, purge air is supplied to the bearing cavities to meet electrical code requirements. Since instrument air is already available at virtually every compressor location, this requirement is satisfied easily.

Compressor start-up begins with the supply of air to the bearing cavities, followed by static levitation of the shaft. The casing is then pressurized, and the unit valves that isolate the compressor from the piping system are opened. The driver start sequence then begins. Normal shutdowns involve closing of the unit valves upon detection of no compressor shaft rotation. The casing is left pressurized, but the shaft is delevitated after a predetermined period of time.

13.3.2 Operating experience and benefits

Magnetic bearings have been applied by a number of users, and their acceptance is clearly on the increase as run hours are accumulated. At the Alberta Gas Transmission Division of Nova, retrofit programs included three compressors that have the shaft supported by bearings on either side of the impeller (beam-type), one compressor that has the shaft supported by bearings on one side of the impeller (overhung-type), and one overhung power turbine. These installations include a range of shaft sizes from 100 to 175 mm (4 to 7 in) at the journal diameter, weights from 320 to 1360 kg (700 to 3000 lb), and rotational speeds from 5000 to 11,500 r/min. Newly procured gas compressors were specified by Nova with magnetic bearings. As of 1994, more than 30 of these had been installed at that company. Other turbomachinery installations elsewhere exceeded a total of 200 machines.

The benefits of magnetic bearing systems include increased efficiency due to elimination of the parasitic shear losses associated with the oil system. Power consumed by the magnetic bearing system averages 3.5 kW (5 hp). This power represents losses in the bearing coil windings and control and amplifier electronics, as the energy is transferred back and forth between the two. No other power is consumed. On a compressor or power turbine package with two magnetic bearing systems, the energy savings amount typically to just under 3 percent of package output power.

Another benefit is the increase in safety of the installation due to elimination of the oil system. Insurance company statistics indicate that, within the rotating equipment user industries, 80 percent of all equipment fires are oil-related.

The signals from the magnetic bearing system (current and position) provide a source of diagnostic information useful for machine condition and system operation monitoring. This is helpful in detecting negative trends and taking preventive action. Assignment of alarm and shutdown set points to these signals further improves the operating safety and risk of damage to the machine.

To date, conditions detected by magnetic bearing systems include improperly installed or damaged inlet and exit guide vanes, balance line blockage, and incorrectly sized balance pistons, as well as buildup of debris on the rotating parts of the aero-assembly, causing imbalance. Many of these situations would remain hidden with hydrodynamic bearings, resulting in higher loads and shorter bearing life spans and, ultimately, would manifest themselves as bearing failures.

When evaluating the application of magnetic bearings, substantial credit may be given to significant reductions in weight and space. This may be of special importance in offshore installations using turboexpanders and canned motor centrifugal pumps. The latter would eliminate mechanical seals as well.

13.3.3 Problems and solutions

Several technical problems encountered with magnetic bearings are more closely related to design and methods of manufacture than to the technology itself. One bearing rotor lamination failure caused damage to the radial bearings. This failure was traced to manufacturing procedures and material selection. After making appropriate modifications, there were no other failures associated with the bearing mechanical hardware itself.

Failures have occurred with commercially procured subcomponents in the control and power systems. These include problems with axial position sensors, dc-dc converters, and power amplifiers. After the vendor replaced the sensors and converters with improved designs, no failures occurred. Most amplifier problems are related to loose connections within the amplifier, and preventive tightening of connections corrects these.

Establishment of stable control loops spanning the range of disturbance frequencies encountered is achieved by means of physically changing electronic components on printed circuit boards. The time required to complete this activity cannot be predicted with great precision. Depending on operational constraints, this can be construed as a problem. Experienced companies such as Revolve that manufacture magnetic bearings have focused on this area for further development. The use of digital technology has made significant improvements possible.

13.4 Development Efforts

Some of the areas where further development would improve industry acceptance of magnetic bearing technology include auxiliary landing systems, higher permeability of materials, sensor developments, standardized electronic hardware, and software tuning capabilities.

The auxiliary landing systems used by Nova have typically included only rolling element bearings. However, these bearings are not specifically designed for this application and have only a limited life span in such demanding service. Alternative systems, developed without rolling elements, rely on the passive friction between the rotating element and the stationary element. Materials that have low coefficients of friction and that are sized to allow dissipation of the energy removed from the shaft in the form of heat have yielded encouraging results. To date, they have not been used on shafts within the heavier range of major compressors. Nevertheless, with careful attention applied throughout the design phase, such a landing system could be implemented successfully.

Materials with higher magnetic permeability would enable bearings to be made of smaller physical size. This has advantages not only in terms of overall physical size but also in minimizing the colocation distance between inductive sensors and the center of the load-bearing portion of the magnetic bearing. These considerations can point to certain rotordynamic constraints that make good designs more difficult to achieve.

The phenomenon of colocation involves physical separation of the point on the shaft where position is sensed and where the centerline of the reactive force is applied to the shaft. Depending on the rotordynamic characteristics of the shaft, the displacement of the shaft at the bearing and sensor locations may be opposite. A signal to correct the rotor position could actually cause the shaft to move farther away from the desired position rather than closer. As one of the major users of magnetic bearings, Nova has no evidence that this has been a problem in any of its magnetic bearing installations. The use of different types of sensors that are colocatable with the centerline of the load-bearing portion of the bearing is a recent development that will ensure that this does not become a cause for concern.

Standard electronic hardware would eliminate the need for spares specific to each installation. Currently, this is not the case with many of the systems in use, which are tuned by changing out hardware components on printed circuit boards.

Software-tuning capabilities would greatly reduce the time taken for hardware tuning of bearing systems. It would also make them more user-friendly and, thereby, effect greater industry acceptance of the technology. This feature involves the development of digital control systems that are responsive and robust enough to handle the envelope of loads imposed during compressor operation. Their introduction has been delayed by limitations in hardware speed, combined with the complexity of the necessary control algorithms.

A number of users, including Nova, have recently put into operation an enhanced digital magnetic bearing control system that not only incorporates standard electronic hardware and a more user-friendly interface but also takes better advantage of the diagnostic abilities of magnetic bearing technology.

13.4.1   Thrust-reducing seals

To take advantage of current dry seal and magnetic bearing technology, Revolve is focusing considerable attention on thrust-reducing seals for both overhung- and beam-type compressors. The first thrust-reducing seal for overhung compressors was installed in 1988, making possible the application of magnetic bearings in this type of compressor. Overhung compressors typically undergo much higher thrust loading upon start-up than do beam-type compressors because their pressure forces are not balanced. Without the thrust-reducing seal, the size of the axial magnetic bearing required would have been prohibitive. The seal enabled reduction of the bearing to a size easily incorporated into the housing.

The compressor selected for this application, originally of radial inlet design, was retrofitted to an axial inlet configuration (Fig. 13.12). Basically, a tube encloses a volume projecting from the eye of the impeller, roughly equivalent to the cross section of the shaft. A dry seal located at the eye of the impeller is used to isolate this volume from the compressor process cavity (Fig. 13.13).

Figure 13.12 Axial inlet compressor being retrofitted with magnetic bearings. *(Revolve Technologies, Calgary, Canada)*

Thrust control begins on start-up with the tube at atmospheric pressure. As head builds across the compressor and impeller, a net thrust results in the direction of the flow into the compressor, thereby loading the Z2 axial magnetic stator (Fig. 13.14). The increasing current signal is processed through an electropneumatic transducer that allows discharge pressure seal supply gas into the tube. This pressure increases, counteracting the thrust load on the shaft, and reduces axial bearing current until equilibrium is attained.

For beam-type compressors, a thrust-reducing seal was installed in 1991. This seal acts upon the current signal from the active stator of

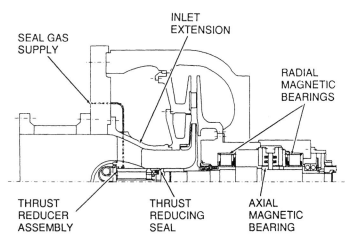

Figure 13.13 Axial inlet thrust reducer. *(Revolve Technologies, Calgary, Canada)*

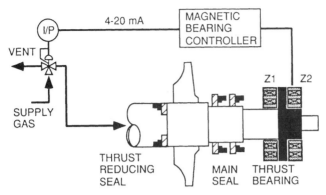

Figure 13.14 Control for overhung thrust reducer. *(Revolve Technologies, Calgary, Canada)*

the axial magnetic bearing (Fig. 13.15) in much the same way as the system for overhung compressors. Seal stages of different diameter, combined with an active pressure regulator and discharge pressure seal supply gas, provide an envelope of variable axial load. This is used to counteract the net thrust on the shaft imposed by pressure differential across the impeller and gas momentum forces.

These loads are normally counteracted by a balance piston. As was discussed in Sec. 12.5 a balance piston, or balance drum, is a cylinder of predetermined size fitted onto the shaft adjacent to the discharge side of the impeller. The pressure differential across the impeller and resultant net thrust is balanced by a reversal of the same pressures across this cylinder. This is done by fitting a labyrinth seal to the outside diameter of the cylinder and by allowing a stream of gas from the

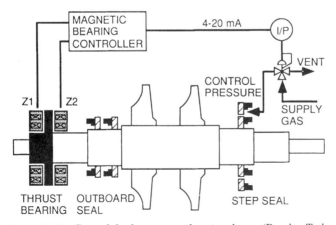

Figure 13.15 Control for beam-type thrust reducer. *(Revolve Technologies, Calgary, Canada)*

discharge side of the impeller to flow across the seal and then through a "balance line" back to the suction side for recompression. Use of a thrust-unloading seal can minimize or completely eliminate the need for a balance piston leakage and, thereby, increase the overall efficiency of the compressor. Balance piston leakage can be as high as several percent of the flow through the compressor.

13.5   Integrated Designs Available

Magnetic bearing technology yields information previously unknown, such as the effect of internal aerodynamic design on bearing loads. The uncertainty associated with this design factor is one reason magnetic bearing technology has not gained greater acceptance. By combining the operational knowledge gained to date with recent advances in numerical analysis hardware and software, a predictive model for bearing loads could be devised that would reduce or eliminate this uncertainty.

To date, the retrofit of these technologies has served to illustrate the practicality, robustness, and limitations of various configurations. This does not mean that the optimum has been reached for all factors affecting operational efficiency and serviceability. Thrust-reducing seals provide an excellent example. The value of magnetic bearing signals in active control loops governing other aspects of turbomachinery performance has been demonstrated with thrust-reducing seal applications. Of the many factors that influence turbomachinery design, Revolve and such large-scale users as Nova Corporation believe these technologies can assist in attaining greater compressor efficiencies when incorporated at the conceptual stage.

Dry seal and magnetic bearing technologies offer significant advantages over the systems they replace. The vast majority of problems encountered have been solved and are not related to the technologies themselves. Recent advancements in magnetic bearing control technology and thrust-reducing seal applications take greater advantage of the benefits offered by these technologies. Where they have been incorporated into the design concept stage of turbomachinery development, they have proven to lead to interesting and advantageous designs.

Two of these advantageous designs are embodied in the Sulzer **mo**tor **pi**peline **co**mpressor (Mopico) and Sulzer-Acec **H**igh-speed **o**il-free **i**ntelligent **m**otor (Hofim) compressors.

The Mopico gas pipeline compressor features a high-speed, two-pole, squirrel cage induction motor. Motor and compressor are housed in a hermetically sealed, vertically split, forged steel casing (Figs. 13.16 and 13.17). The center section contains the motor and bearings, and each of the end casing sections houses a compressor wheel, a fixed vane diffusor, and the inlet and discharge flanges.

Mopico compressors can be operated in series or parallel (Fig. 13.18). Magnetic radial bearings and a double-acting magnetic thrust bearing maintain the runner in position. The motor is cooled by gas metered from the high-pressure plenum of one the compressor housings. Hence the Mopico runs completely oil-free.

The speed and thus the discharge rate of the Mopico unit is controlled by a thyristorized, variable-frequency drive. This drive was developed by Ross Hill Controls Corp., Houston, Texas. It uses thyristors that can be switched out. These enable pulse-free run-up without current peaks and an operating speed range of 70 to about 105 percent.

Refer to Fig. 13.19 for an overall installation schematic and note that the following conditions can be complied with through the new combination of elements:

- Low installation, maintenance, and energy consumption costs
- Broad operating range at high economic performance
- Compatibility with existing compressors
- Unattended remote control
- Emissionless and oil-free
- Can be installed outdoors

Based on cost per installed horsepower, the cost of the Mopico compressor is only about two-thirds that of a gas turbine unit and less than half that of a low-speed recip. It is some 10 percent less than a conven-

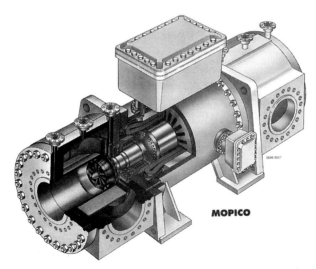

Figure 13.16 Sulzer Mopico motor pipeline compressor, incorporating magnetic bearings. *(Sulzer, Ltd., Winterthur, Switzerland)*

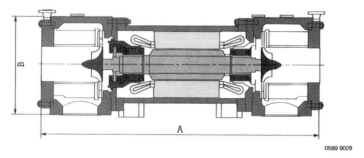

Dimensions and weights

Frame size	Power kW	A mm	B mm	Weight kg
RM 28	2000	2500	900	9 000
RM 35	3500	2900	1000	11 000
RM 40	6000	3300	1200	13 000

Figure 13.17 Dimensions, weight, and simplified cross section of Mopico compressor. *(Sulzer, Ltd., Winterthur, Switzerland)*

tional centrifugal with dry seals, magnetic bearings and a direct drive, high-speed induction motor with variable-frequency drive.

On the other hand, a conventional centrifugal with motor gear drive and variable inlet guide vanes is less expensive. This system is, however, unacceptable for pipeline application because of poor efficiency at low-pressure ratio conditions.

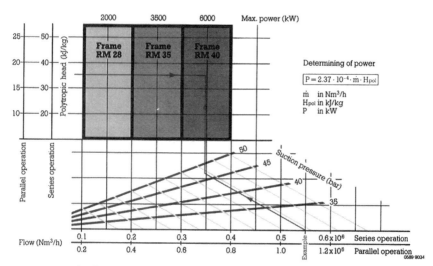

Figure 13.18 Application ranges for Mopico compressors. *(Sulzer, Ltd., Winterthur, Switzerland)*

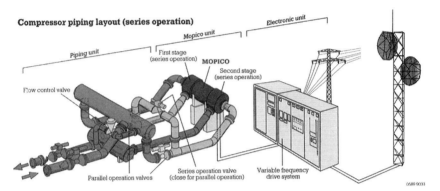

Figure 13.19 Installation schematic for Mopico compressor. *(Sulzer, Ltd., Winterthur, Switzerland)*

The second, equally promising, new design concept was mentioned as Sulzer-Acec's Hofim. This very recent high-speed oil-free intelligent motor compressor is shown in Fig. 13.20. It features separate motor and compressor directly coupled, with both machines supported on magnetic bearings. The prototype application is in a natural gas storage facility, for which the key parameters are:

Motor nominal speed	20000 r/min
Motor nominal power	2000 kW
Compressor inlet pressure	5000 kPa (725 psia)

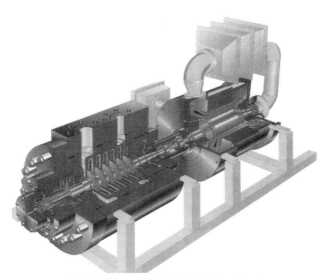

Figure 13.20 Hofim high-speed oil-free intelligent motor compressor. *(Sulzer, Ltd., Winterthur, Switzerland)*

Compressor discharge pressure   15,500 kPa (2250 psia)

Compressor speed range   70–102%

Compressor flow   38,000 Nm^3/h

The motor is an asynchronous squirrel cage induction machine, driven by a solid state variable-frequency drive. The compressor is a six-stage barrel machine of fully modular design. Dry gas seals are used to minimize internal and external leakage. An active balance system controls residual thrust of the whole unit to a level compatible with the capacity of the axial magnetic bearing. The unit was developed jointly by four companies in conjunction with the European BRITE program.

Couplings, Torque Transmission, and Torque Sensing

14.1 Coupling Overview*

High-reliability and low-maintenance requirements are almost always among the key demands of process machinery users. This explains why coupling selection deserves considerable attention.

Major requirements for high-performance couplings are high-torque and high-speed capacity, low weight and small envelope size, low overhung moment, and low residual unbalance. The coupling must be capable of transmitting the system design torque at maximum continuous speed for extended periods. It must be able to handle speed and load transients at defined misalignment conditions with minimum reactions on the drive system.

There are several major factors that, in combination with each other, determine the continuous duty rating of a coupling. These factors are:

1. Speed of rotation: High speeds may limit the coupling diameter. This in turn sets the tooth pitch diameter and tooth loading of gear couplings, Fig. 14.1, or bolt pitch diameter of nonlubricated couplings, Fig. 14.2 through 14.4.

2. The torque that can be transmitted by gear couplings of a given pitch diameter and tooth length is a function of allowable contact pressure and the relative sliding velocity between hub and sleeve teeth.

---

* Based on application summaries provided by Michael Calistrat & Associates, Missouri City, Tex.; product information courtesy of Lucas Aerospace, Utica, N.Y., and the various contributors listed as sources of illustrations.

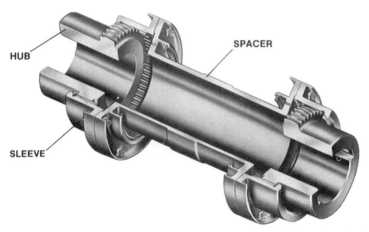

Figure 14.1   Gear coupling requires lubrication. *(Zurn Industries, Erie, Pa.)*

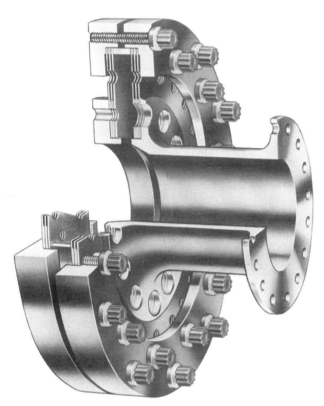

Figure 14.2   Nonlubricated "Flexxor" coupling. *(Coupling Corporation of America, York, Pa.)*

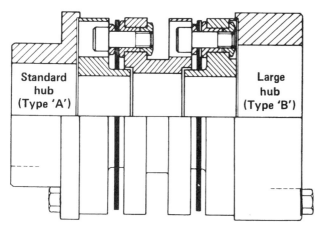

Figure 14.3   Disk pack coupling. *(Flexibox, Inc., Houston, Tex.)*

3. The torque that can be transmitted by metal disk or diaphragm-type high-performance couplings is a function of allowable stresses in the flexing members.

4. Running misalignment determines the relative sliding velocity between the gear teeth for a given pitch diameter and speed of rotation. Similarly, running misalignment affects the stress levels in flexing members of nonlubricated couplings.

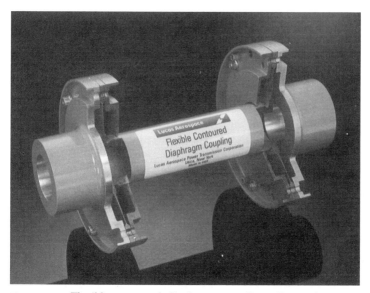

Figure 14.4   Flexible contoured diaphragm coupling. *(Lucas Aerospace Company, Utica, N.Y.)*

5. Tooth hardness determines the allowable contact pressure. Heat-treated alloy steel is suitable for many applications, but increased surface endurance may be obtained by suitable hardening procedures, regardless of coupling type.

14.1.1 Low overhung moment

The effect of coupling overhung moment is felt not only in machine bearing loads but in shaft vibrations. The advantage of a reduction in overhung moment is not only to reduce bearing loads but to minimize shaft deflection, which results in a reduction of the amplitude of vibration. The reduction of the coupling overhung moment produces an upward shift in shaft critical speeds. This change in natural frequencies results in an increase in the *spread* between natural frequencies. For many applications, reduced overhung moment is an absolute necessity to enable the system to operate satisfactorily at the required operating speed.

Low overhung moment is generally achieved with the conventional gear coupling configuration that consists of gear meshes between a shaft-mounted hub and sleeve-spacer assemblies (Fig. 14.1).

Diaphragm couplings can achieve similar results with geometries as illustrated in Fig. 14.5. In either case, the heavy coupling sections are placed as far back as possible so that the resultant gravity force due to the weight and center of gravity of the hub and the distributed weight of the sleeve-spacer-sleeve assembly applied at the centerline of the hub teeth is the least distance from the centerline of the machine bearings.

The diaphragm safety, or noncontacting gear-type coupling configuration, Fig. 14.6, usually consists of a spool or distance piece extension

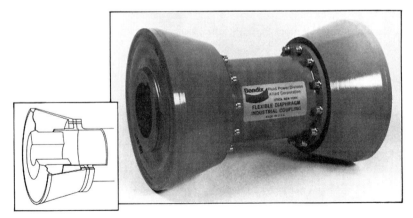

Figure 14.5 Low-moment-type diaphragm coupling. *(Lucas Aerospace Company, Utica, N.Y.)*

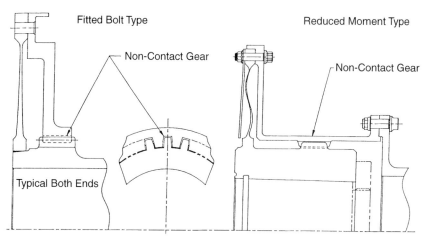

Figure 14.6   Contoured diaphragm safety backup coupling. *(Lucas Aerospace Company, Utica, N.Y.)*

having internal gear teeth spaced between the teeth machined on the periphery of shaft-mounted coupling hubs.

The disadvantage of the concentration of the weight of the spool piece at points between the connected shaft ends can often be offset by using shorter extensions of the shaft from the machine bearing and by using a smaller and lighter weight coupling because of the inherent large bore capacities of rigid hubs.

14.1.2   Low residual unbalance desired

The gear-type high-performance coupling consists of a number of components that are fastened and hinged together by the tooth meshes. The minimum number of components for a double-engagement coupling is three. Angular and offset misalignment can thus be accommodated (Fig. 14.7). Three-piece couplings include the continuous sleeve, or flangeless coupling, and the marine, or spool, configuration. Five components—two hubs, two sleeves, and a spacer—are used in most other configurations.

In diaphragm couplings, Figs. 14.4 through 14.6, torque is transmitted from the driving shaft hub to the diaphragm rim through the body-fitted bolts and then through the contoured profile to the rigid diaphragm hub electron-beam welded to the spacer tube. From here, the torque passes to the opposite diaphragm hub and through the contoured profile to the integral rim and again through the rim body-fitted bolts to the driven hub and output shaft.

Contributing factors to the total residual unbalance of a coupling are:

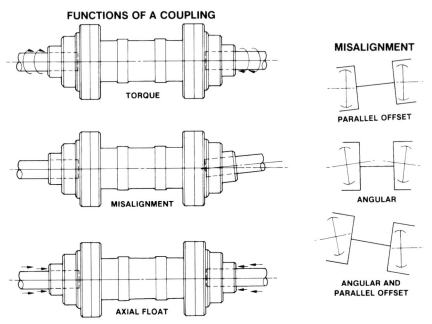

Figure 14.7 Functions of a coupling. *(Zurn Industries, Erie, Pa.)*

1. Balance correction tolerance.

2. Balance machine accuracy: machine sensitivity and driver error.

3. Arbor assembly unbalance: mandrel and bushing unbalance.

4. Mounting surface runout: mandrel runout, bushing to mandrel clearance, bushing bore to mounting surface runout, and arbor bushing to hub bore (or sleeve rabbet) diametral clearance.

5. Pilot surface runout, i.e., hub bore to hub body pilot OD is a factor if metal must be removed from the pilot OD after balancing to provide assembly clearance.

6. Pilot surface diametral clearance: hub to sleeve pilot clearance and sleeve to spacer diametral clearance.

7. Hardware unbalance: bolts, nuts, retaining rings, etc.

The implication of the aforementioned is that the straightness, concentricities, minimum clearances, and dynamic balance of the tooling are more important than the final correction tolerance in achieving an actual minimum of residual unbalance.

14.1.3   Long life and maintainability

The main mode of failure of a gear coupling is, in most cases, wear or fatigue of the tooth surfaces due to the lack of lubrication, incorrect and water-contaminated lubricant, or excessive surface stress.

Assuming correct lubrication, long life of a gear coupling is attained by proper surface treatment of the teeth. General practice is to make the gear elements of high-performance couplings from chrome-molybdenum steel, or chrome-moly-nickel steel that is heat treated to a core hardness of about 300 BHN.

Diaphragm and disk pack couplings are subjected to potential distress primarily when sensitive surfaces are nicked or scratched, or whenever the flexing metal parts are either pulled apart or pushed together because hub installation errors, unexpected thermal growth, or movement of coupled shafts.

The principal advantage of nonlubricated metal disk and diaphragm-type high-performance couplings is derived from the fact that neither requires lubrication. Gear couplings, on the other hand, will suffer quickly whenever proper lubrication guidelines are violated. As already mentioned, gear-type couplings require lubrication because of the relative sliding motion between the teeth of the hub and sleeve. This sliding motion is alternative and is characterized by small amplitudes and relatively high frequencies.

For example, a gear coupling on a 3-in shaft turning at 10,000 r/min, having an angular misalignment of 2 min, has an alternative motion with a frequency of 167 cycles/s and a peak to peak amplitude of 1.7 thousandths of an inch.

Even with optimum lubrication, such a condition would likely cause fretting corrosion. Fortunately, the load on each tooth is not constant but varies twice per revolution from maximum to minimum. The ratio between the maximum and minimum force on the tooth is a function of misalignment and tooth geometry; above certain conditions the minimum becomes zero, and this means that temporarily there is actual separation between the teeth. On the other hand, as the misalignment decreases, the force on the tooth tends to remain constant, but the amplitude of the oscillation decreases also.

For use in oil-lubricated couplings, additives such as antirust, antifoaming, and antioxidants are not beneficial. On the contrary, in the case of continuous oil flow lubrication, these additives are often retained within the coupling, causing serious problems.

EP additives are not detrimental, but laboratory tests could not prove that they are advantageous in high-performance couplings. It should be noted that these couplings are designed to work under relatively low-contact pressures (less than 4000 psi), and extreme pressures could be developed only during the breaking-in period. For this reason only, EP additives are recommended when lubricating gear couplings.

Very few high-performance couplings are grease lubricated. There are two main reasons for this. One is the fact that grease-lubricated couplings must be serviced more often than continuously lubricated couplings. The other is that most of the greases available today are not resistant to centrifugal forces, and high-speed couplings certainly develop very high centrifugal forces. The 3-in shaft coupling turning at 10,000 r/min would develop a centrifugal force equivalent to 8400 g's. It is worthy of note that industrial centrifuges cannot develop more than 10,000 g.

Under such high forces greases tend to separate into their oil and soap constituents. Unfortunately, the soaps are heavier than oils so that the teeth of the gear coupling, which are at the large diameter, are contacting grease that has an excessive percentage of soap.

14.1.4 Continuous lubrication not a cure-all

Turbomachinery installations that still apply gear couplings are likely to depend on continuous lubrication from the main oil system. The viscosity of this oil is no doubt chosen to satisfy compressor and driver bearing requirements; it is probably too light for optimized lubrication of gear couplings. Perhaps even more damaging is the fact that much of the wear product or water contamination carried by the lube oil ends up being centrifuged out in the coupling. Consider the following.

A coupling requiring an oil flow of 3 gal/min will have a total oil circulation of 1,576,800 gal/y. If we assume a nearly perfect oil purity of only two parts of dirt per million and if the coupling centrifuge effect separates all of this dirt, then in 1 y the coupling would accumulate 3 gal, or approximately 12 L of sludge!

From the foregoing, we may conclude that

1. Grease-packed couplings are acceptable for high-speed machinery only if frequent maintenance downtime can be tolerated

2. Continuously oil-lubricated couplings should be designed so that sludge is not allowed to accumulate in oil retention dams and similar discontinuities

3. The lube oil supply must be virtually free of solid contaminants and, especially, free of water. (Refer to Sec. 15.2 on lube oil systems for a discussion of state-of-the-art water removal methods.)

4. The user should give serious consideration to nonlubricated turbomachinery couplings.

14.1.5 Contoured diaphragm coupling overview

The design of a well-proven high-speed, high-power, nonlubricated coupling is principally centered about the contoured diaphragm (Fig. 14.4 through 14.6). It is this special hyperbolic contouring of the diaphragm that permits the accommodation of torque, speed, axial deflection, and simultaneous angular misalignment while maintaining uniform shear and low tensile stress. Two diaphragm profiles are employed. A straight profile is used on some models (Fig. 14.6, left side). A wavy profile that has the same hyperbolic contour and allows radial freedom of the inner hub and outer rim is used on other models (Fig. 14.6, right side).

As shown in Fig. 14.8, the diaphragm bending stress, resulting from angular and parallel misalignment, is a fully reversing cyclic fatigue stress. This stress occurs at the rate of one cycle per revolution. The diaphragm is designed to distribute this stress across the area of its profile. In addition, the transition from optimal fatigue stress in the profile to outer rim and inner hub should be controlled by generous radii.

Figure 14.9 represents a plot of fatigue data generated by actual life cycle testing of diaphragms. Fatigue life of the diaphragm material AMS 6414 (vacuum melted 4340) is 70,000 psi at 10^7 cycles. Conservative design criteria restrict imposed bending stress to a maximum of only 35,000 psi.

A Modified Goodman diagram is shown in Fig. 14.10. The combined mean stress (steady-state axial or torsional) and combined alternating

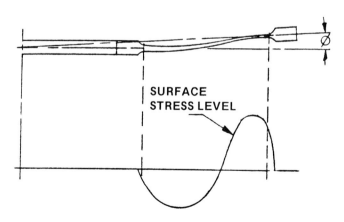

BENDING STRESS DIAGRAM

Figure 14.8 Diaphragm bending stress is a fully reversing cyclic fatigue stress. *(Lucas Aerospace Company, Utica, N.Y.)*

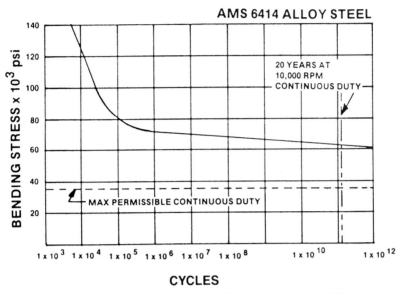

Figure 14.9 Fatigue life curve depicting life cycle testing of contoured diaphragms. *(Lucas Aerospace Company, Utica, N.Y.)*

stress (cyclic axial or torsional and bending) for a given operating condition are plotted on the constant life (Modified Goodman) fatigue diagram. Manufacturers such as Lucas Aerospace require that all continuous and short-term operation must have the plotted operating point fall within the area under the dotted curve. Any point within this area has a minimum cyclic factor of safety of 2.0.

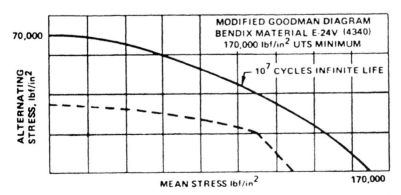

Figure 14.10 Modified Goodman diagram for high-performance contoured diaphragm couplings. *(Lucas Aerospace Company, Utica, N.Y.)*

For corrosion protection each flex unit is coated with multiple layers of Sermetel® "W", which is an inorganically (chemically) bonded aluminum coating offering a sacrificial method of corrosion protection. In this manner anytime an area of base material does become exposed to a hostile atmosphere the Sermetel® coating which is more chemically reactive than steel will be the only surface to corrode. This is referred to as *anodic corrosion protection,* which is highly successful.

Covering this layer are several coats of chemically resistant epoxy paints. Not only does the epoxy provide protection from direct contact with corrosives, but it also is very tough and helps guard the diaphragm from damage due to abrasion.

14.2 Performance Optimization Through Torque Monitoring*

Fouling deposits on blading or nozzles cause performance deterioration of steam and gas turbines. In steam turbines, evidence of these deposits is frequently not discovered until the steam flow increases to a point where no additional steam can be passed through the unit. The turbine can no longer carry an assigned load under these conditions. In gas turbine air compressors, fouling deposits reduce the amount of air available for efficient combustion. Excessive fuel consumption or reduced load-carrying capacity may result. Similarly, process gas compressors often polymerize. The resulting flow impediment can seriously influence process operation and mechanical performance. On-line torque measurement systems provide an easy method of measuring produced power. Comparing fuel consumption, load conditions, and torque enable you to decide whether further steps need to be taken.

Turbomachinery performance can be restored by judicious application of onstream cleaning methods. Abrasion cleaning and solvent cleaning (water washing) are the two principal approaches. Literature, which can be obtained from original equipment manufacturers, provides ample details of suitable procedures and their relative merit. The problem, to date, has been to figure out conveniently and accurately when to initiate the onstream cleaning process.

In pure economic terms, onstream cleaning should commence when the cumulative cost of power lost because of fouling since the last cleaning cycle equals the cost of the cleaning procedure. This is where the on-line torque measuring system comes into play. Installed to mon-

---

* Segment developed and contributed by Bently-Nevada, Minden, Nev. Additional information provided by Torquetronics, Inc., Allegany, N.Y., and by Indikon/Metravib Instruments, Inc., Cambridge, Mass.

itor torque at the coupling, the device also shows related speed and power. Peak torque, speed, and power values are also provided. The indicator can be connected to a computerized monitoring system, strip chart recorders, or tied to a process control computer or programmable logic controller. This additional equipment allows accessing of and correlation with steam or fuel flowmeters and heat rate tables.

Axial compressor fouling continues to be a common and persistent cause of reduced gas turbine efficiency. A 1 percent reduction in axial compressor efficiency accounts for approximately 1½ percent increase in heat rate for a given power output. Even compressor stations not subjected to industrial pollutants or salty atmospheres are frequently prone to fouling. Torque measurements provide an early indicator of changes in efficiency.

Performance deterioration of gas turbines can be detected by combining turbine fuel flow rate with power output. Monitoring systems should incorporate readouts of power. In a computer system, this value can be easily compared to produced power per specific rate of fuel consumed.

Turbine manufacturers provide teststand-verified performance curves. These data are helpful in determining the degree of performance deterioration by comparing actual (fouled) condition and ideal (clean) condition specific fuel consumption rates. For efficiency optimization an operation should monitor the average fuel wastage, or average turbine efficiency, at regular time intervals.

Continuous torque sensing devices can provide valuable information in other areas as well. With torque limitations on one or both of the coupled shafts, a torque indicator can serve as a constraint control. Torque sensing can allow process optimization for computer-controlled compressors where several levels of refrigeration are available. Some turbocompressors can be configured to get desired flow and head by such methods as varying speed, varying stator blade angle, and varying guide vane angle. On large axial compressors, combination of stator blade adjustment and speed variation of about 10 to 15 percent improves the part load efficiency of the compressor and increases the stable operating range. When given sufficient attention, appreciable differences in energy consumption may result, and savings of power may be realized by using torque data.

Increased energy input to the driver due to performance decay of the driver or driven equipment can be detected effectively by torque measurements. The case of a gas turbine driving a centrifugal compressor best illustrates how the issue can be resolved by measuring torque. High driver fuel consumption and high-coupling power shows that the driven machine is more highly loaded, mechanically deficient, or internally fouled. Using the dynamic torque signals and diagnostic and analytical instrumentation, procedures are available to figure out which of these three possible causes is most probable. For example, high driver fuel consumption and normal coupling power would show that the most probable cause of the efficiency decay is turbine fouling. Torque measurements and subsequent action can thus reduce energy waste in compressors incorporating antisurge controls. Equally important: Torque sensing may pinpoint causes of failure.

Although many methods exist for determining how a component failed, torque measurements may show what caused it to fail and provide clues on how to avoid repeat failures. Looking at broken pieces can tell you how something failed. Comprehensive maintenance records can help you predict when something will fail again. Yet, these methods do little to identify the cause of failure or prevent failure recurrences.

It can be difficult to figure out the cause of failure with insufficient accurate information on system loading during machine operation. Is it a running overload problem? Is it a resonance-related frequency or vibration-related problem? These and other questions can be answered by measuring torque on a running system. The data should reveal both the steady state and dynamic torque.

Bently-Nevada's TorXimitor™ torque sensing system is depicted in Figs. 14.11 and 14.12. It consists of two parts: a stationary component that surrounds, but does not make contact with, a rotating system. The

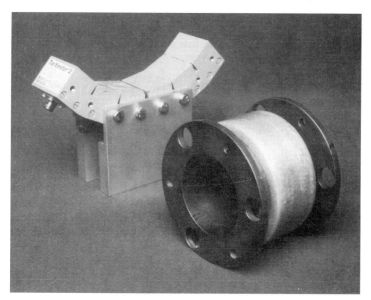

Figure 14.11   TorXimitor torque sensing device. *(Bently-Nevada Corporation, Minden, Nev.)*

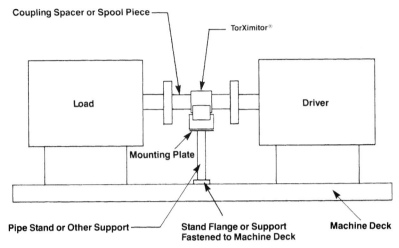

Figure 14.12 Schematic representation of TorXimitor torque sensing device. *(Bently-Nevada Corporation, Minden, Nev.)*

rotating system contains electronic circuitry installed on the coupling spacer or spool piece. The stationary component is installed on a mounting plate that is attached to a pedestal on the machine baseplate or equipment platform.

A similar approach is embodied in Torquetronic's meters. The Torquetronic system measures torque as the shaft twists between a pair of toothed flanges made integral with a coupling spacer. A pair of phase-displaced sinusoidal signals are generated by multiple pickups in the form of internally toothed rings surrounding fully encapsulated circumferential coils that are energized to provide a low-level toroidal flux path. Clearly, the torque required to twist the shaft through one tooth pitch must correspond to exactly 100 percent phase displacement.

The toothed pickup rings are permanently fixed to each end of a rigid stator tube that is supported clear of the rotating shaft from the coupled machines and often replaces the existing coupling guard. Refer to Figs. 14.13 and 14.14. The readout unit, which is a specialized digital phasemeter, measures speed and the phase displacement of the two signals and converts them to torque, speed, and power in engineering units. The integrity of the readout unit can be proved by a complement switch that temporarily crosses the inputs to the phasemeter. If reading + complement add up to 100 percent exactly, all must be correct since it is most unlikely that equal and opposite errors will exist in two measurements that are seen quite independently by the readout.

Our last example describes the Indikon torquemeter system. This instrumentation package can also incorporate on-line continuous

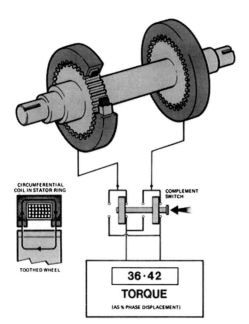

Figure 14.13 Torquetronic torque metering principle. *(Torque-tronics, Inc., Allegany, N.Y.)*

("hot") alignment monitoring. Figures 14.15 and 14.16 illustrate both principles.

Indikon uses a rotary transformer, Fig. 14.17, whose functioning depends on electromagnetic induction between a primary winding and a secondary winding, just like an ordinary transformer. In this case, however, the secondary is attached to the rotating shaft, while the primary is fixed relative to the machine frame. The air gap between primary and secondary is sufficiently large to accommodate worst-case misalignment. Submersion in oil or exposure to oil mist does not affect transformer operation.

An on-shaft calibration circuit periodically generates a millivolt per volt calibration signal that goes through the same shaft electronic circuits as the torque signal. By measuring the ratio of these two signals, the effects of changes in strain gauge bridge voltage, shaft electronic circuit characteristics, and transformer coupling are eliminated, since they affect both signals equally. Zero errors due to shaft displacement are avoided by using different power and signal frequencies.

As in all strain gauge transducers, accuracy is a function of full-scale stress level. This determines whether the magnitude of the full-scale bridge unbalance signal is large enough to reduce to insignificance any residual effect of temperature on bridge balance. In general, a stress level of 15,000 psi permits a system accuracy within the range of 0.20 to 1.0 percent of full scale. Shaft temperatures up to a maximum of 250°F are permitted.

Acceleration levels up to 20,000 g are permissible in those cases where the electronics package can be located along the axis of the cou-

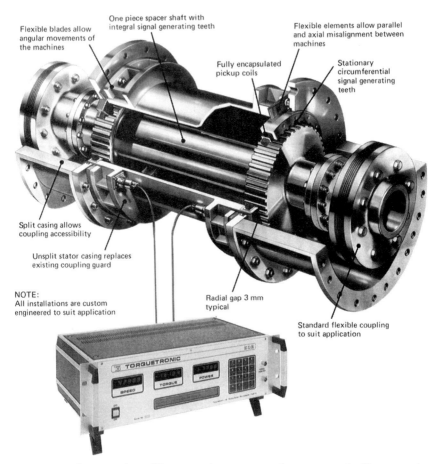

Flexible blades allow angular movements of the machines

One piece spacer shaft with integral signal generating teeth

Flexible elements allow parallel and axial misalignment between machines

Fully encapsulated pickup coils

Stationary circumferential signal generating teeth

Split casing allows coupling accessibility

Unsplit stator casing replaces existing coupling guard

NOTE:
All installations are custom engineered to suit application

Radial gap 3 mm typical

Standard flexible coupling to suit application

Figure 14.14 Cutaway view of Torquetronics torque sensing components. *(Torquetronics, Inc., Allegany, N.Y.)*

pling or torque shaft. In cases where the electronic package must be located on the shaft surface, a maximum of 15,000 g is allowable.

Torque pulsations or fluctuations over the frequency range of 10 to 500 Hz produce an analog output signal for recording purposes or for viewing on an oscilloscope.

Wherever possible, the RT Torquemeter makes use of existing couplings or shafts. If stress levels have to be increased to attain the desired accuracy, this can be accomplished by machining a short reduced section on the coupling or shaft. The resulting change in overall torsional stiffness is usually less than 10 percent.

In new applications, couplings with suitable stress levels can be provided by coupling manufacturers. The preferred location for the shaft electronics package is along the shaft axis. Where this is not possible,

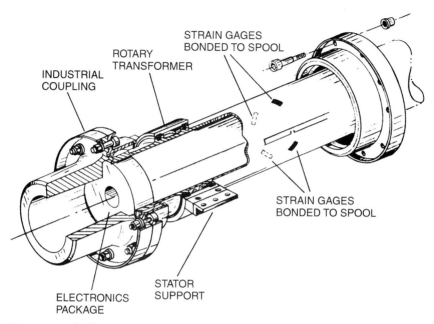

STRAIN GAGES
BONDED TO SPOOL

ROTARY
TRANSFORMER

INDUSTRIAL
COUPLING

STRAIN GAGES
BONDED TO SPOOL

STATOR
SUPPORT

ELECTRONICS
PACKAGE

Figure 14.15 Indikon torquemeter components. *(Indikon Metravib Instruments, Inc., Cambridge, Mass.)*

the circuits are contained in a package distributed around the shaft circumference.

Indikon's hot alignment indicating system measures coupling misalignment under actual operating conditions and displays digitally the X and Y mils/in and mils readjustments required for realignment. It is applicable to both gear-type and diaphragm-type couplings. Solid couplings require a somewhat different approach, using strain gauges.

The basic components of the system are shown in Fig. 14.16. They include inductive proximity probes, rotating with the coupling, which measure the amplitudes of the once-per-revolution variations in gap due to misalignment at each end.

A marker probe at 12 o'clock detects when the shaft is in a reference position, as determined by a slot or a thin metallic target.

By comparing the phase of the misalignment signal with the phase of a reference voltage derived from the marker probe, the X and Y components of the misalignment signal are obtained and displayed digitally on the indicator. Digital displays for either X parallel and X angular misalignment, or Y parallel and Y angular misalignment, are provided, as determined by a selector switch.

Continuous analog outputs for all four quantities are provided to permit the recording of the growth of the machine into its hot aligned condition. This system can also be used in place of rim and face dial

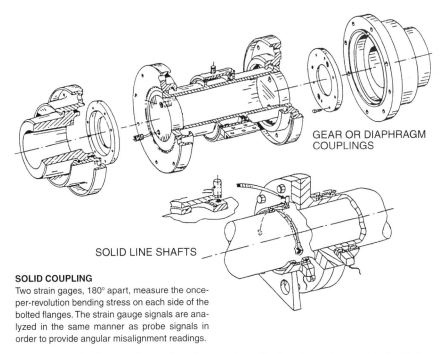

GEAR OR DIAPHRAGM
COUPLINGS

SOLID LINE SHAFTS

SOLID COUPLING

Two strain gages, 180° apart, measure the once-
per-revolution bending stress on each side of the
bolted flanges. The strain gauge signals are analyzed in the same manner as probe signals in
order to provide angular misalignment readings.

Figure 14.16 On-line continuous hot alignment monitoring elements associated with turbomachinery couplings. *(Indikon Metravib Instruments, Inc., Cambridge, Mass.)*

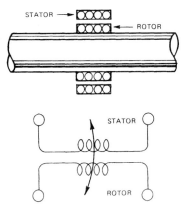

STATOR

ROTOR

STATOR

ROTOR

*Electrical Equivalent
of Rotary Transformer*

Figure 14.17 Rotary trans-
former. *(Indikon Metravib
Instruments, Inc., Cambridge,
Mass.)*

indicators to adjust cold alignment offsets to correct for the misalignment measured under hot conditions.

For gear-coupling applications, an auxiliary system is available to measure mean sliding velocity and to actuate an alarm when conditions develop that could lead to gear surface failure.

The indicator resolves the misalignment signal into its parallel and angular components by making use of two fundamental considerations:

- If the once-per-revolution signals from the probes at both ends are equal, but opposite in phase, the misalignment is parallel. The signal from the probe at one end is then resolved into its X and Y components.

- If the misalignment is angular, the vector sum of the probe signals is not zero, and the difference signal is analyzed to obtain its X and Y components.

Lubrication, Sealing, and Control Oil Systems for Turbomachinery*

15.1 Considerations Common to All Systems

The primary function of an oil system is to provide the proper quantities of cooled and filtered oil at the required regulated pressure levels to the driven and driving equipment. This oil can be used for lubrication, shaft sealing, and/or control oil purposes. The oil system is designed to furnish the oil required at all operating conditions of the equipment. A basic combined lube and seal oil system is described and shown in Fig. 15.1.

A fabricated steel reservoir tank serves to store a volume of oil sufficient for typically 5 to 8 min at normal flow. The tank is fitted with both a dipstick and a sight glass level gauge. Removable heating elements, either steam or electric, are usually provided. These heaters are sized to heat the oil from the minimum site ambient temperature to the minimum required oil temperature required by the turbomachine (usually 70°F) within 12 hours. The tank is furnished with a temperature indicator and with a level switch that activates an alarm when the oil level is below the minimum operating level. The purge and vent connections on the tank provide a means to exhaust any gases that are released from the oil.

---

* Developed and contributed by Roy J. Salisbury, Manager, Customer Service Department, Imo Industries, Inc., DeLaval Turbine Division, Trenton, N.J. Portions derived from Transamerica DeLaval Engineering Handbook, copyright © 1947, by DeLaval Turbine Company.

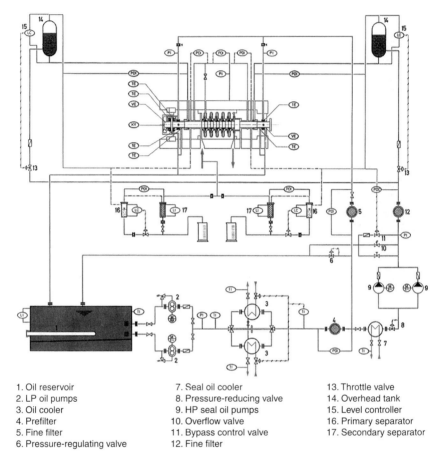

1. Oil reservoir
2. LP oil pumps
3. Oil cooler
4. Prefilter
5. Fine filter
6. Pressure-regulating valve
7. Seal oil cooler
8. Pressure-reducing valve
9. HP seal oil pumps
10. Overflow valve
11. Bypass control valve
12. Fine filter
13. Throttle valve
14. Overhead tank
15. Level controller
16. Primary separator
17. Secondary separator

Figure 15.1   Basic lube and seal oil schematic—forward pressure control. *(Mannesman-Demag, Duisburg, Germany)*

Oil is drawn from the bottom of the reservoir through suction strainers by motor-driven or steam turbine-driven pumps. The main and auxiliary pumps are identical and supply a constant flow of oil. Each pump discharge line has a relief valve that protects the equipment from any overpressure caused by a system malfunction. The relief valves are sized to pass the full pump capacity. A pressure gauge and a check valve are furnished in each pump discharge line. A block valve is placed downstream of the check valve for maintenance purposes.

The flow of oil then passes through a transfer valve that can transfer the flow from one filter-cooler set to the other set without interrupting the flow. Out-of-service units can be opened for cleaning or maintenance while the other units are in service. Two identical coolers, each capable of handling the system's maximum flow and heat load, are furnished. Water flow to the coolers is regulated to maintain the desired oil outlet temperature of typically 120°F (49°C). Two identical filters are usually supplied; additional prefilters are sometimes used. One filter is placed downstream of each cooler to remove particulate material as small as 10 or 5 μm. Filters, with clean cartridges, are sized to handle the maximum system flow and pressure with a pressure drop no greater than approximately 5 psi. A differential-pressure indicator and a differential-pressure switch are placed across the filter-cooler combination to warn of the need to change the filter cartridges.

To avoid oil pressure surging and interruption of the oil supply, the out-of-service filter-cooler combination is filled with oil before it is put on the line. The vent valves on the filters and coolers are used to vent air from the units, while the oil cross-connect line is opened to fill them. This cross-connect line is left open to keep the out-of-service units pressurized. Thermometers are fitted upstream and downstream of the coolers to check unit performance.

A back pressure regulator valve is supplied to establish and control the header pressure after the filter-cooler units. This valve can be either self-operated or pneumatically operated, whichever method is dictated by duty. Oil is taken from a point before the filter-cooler units and bypassed back to the reservoir. The valve is sized to control a wide range of flows, with either one pump or both pumps in operation. In the latter case, the valve would pass a maximum flow of the two pump capacities less the flow required by the system.

Several different pressure levels are often required to carry out the various functions of an oil system. A pressure-reducing valve is used to reduce the header pressure (established by the back pressure regulator) to the required pressure level for lubricating oil, control oil, etc. One valve is used for each required pressure level.

When oil seals are used on a compressor, the flow to the seals is set by a flow control valve. This valve is designed to maintain a constant volume of oil regardless of how the oil pressure may vary.

All control valves—flow, pressure, or differential-pressure—have bypass provisions. In case of malfunction, the control valve can be isolated and the flow or pressure adjusted manually through the bypass globe valve.

The unit lubricating oil line receives oil from the console and delivers it at approximately 20 psig to the various points to be lubricated. Except for modern machinery incorporating such componentry as magnetic bearings, or nonlubricated couplings, or both, the oil supply is fed to the compressor thrust and journal bearings, to the coupling, and to the driving equipment bearings. The line is fitted with a pressure indicator and several pressure switches at the farthest extreme of the header to ensure an adequate oil supply at this outermost point. One switch is set to trigger an alarm, and a second switch starts the auxiliary oil pump, when the lubricating oil pressure falls to, say, 12 psig. The third switch is set to trip the unit at typically 8 psig decreasing.

The lubricating oil drain system collects the oil used by the compressor, coupling, and driving equipment and returns it to the reservoir. Each atmospheric drain line from the bearings and seals is fitted with a sight flow indicator and a thermometer.

15.2   Seal Oil Considerations

Seal oil would be needed for the various contacting face, mechanical or oil film seals on turbocompressors, but not, of course, for labyrinth or dry gas seals. Seal oil, if required, is supplied to each compressor through a separate header. The seal oil pressure is often controlled from the downstream side of the seal but on some applications is controlled upstream (forward pressure control). A differential back pressure regulator valve is often used for mechanical seals, while a head tank is typically used for oil film seals.

When mechanical seals are used, the oil side pressure is set about 45 psi above the gas side. A differential back pressure regulator senses compressor reference pressure or control gas and regulates the seal oil to maintain the proper differential pressure. The seal oil system is fitted with a differential-pressure indicator and several switches. These are connected between the seal oil return line and the control gas connection for back pressure designs, or the seal oil supply line and the control gas for forward pressure control. The switches activate an alarm and start the auxiliary pump when the seal-oil-gas differential falls to approximately 35 psi ΔP.

Oil film seals usually have a head tank that is mounted above the compressor to maintain a 5 psi differential above the compressor reference pressure. Dual head tanks are occasionally found, with a level control valve supplied downstream of the seal oil return lines to maintain the proper level in the head tank. For forward pressure control the level control valve is on the seal oil supply line, as shown in Fig. 15.1. It is usually located approximately 14 ft (4 m) above the centerline of the compressor.

The overhead seal oil head tank arrangement is sized to have the proper capacity and rundown time for emergency operation, coastdown, and block-in. A pneumatic level transmitter sends a signal to a level-indicating controller that operates the level control valve. Level switches and gauges are furnished for monitoring, alarming, and trip functions.

Oil drainers or separators are supplied for both mechanical contact, oil film, and similar seals. They collect the contaminated seal oil and provide an automatic means of discharging this oil for reuse or disposal. An oil drainer is essentially a tank with a float-operated drain valve arranged so that the seal oil leakage can be drained without releasing any gas. Each drainer can be isolated for servicing by closing three valves. The nonoperational drainer is bypassed, and both seals are allowed to drain into the remaining drainer by opening the valves in the crossover lines. A level glass is fitted on the drainer to indicate the operating level.

The majority of the components in the oil system are usually mounted on a preassembled console such as that shown in Fig. 15.2. Consoles of this type are piped and tested prior to shipment; they can generally accommodate a wide selection of valves, flanges, fittings, and other components to meet various specifications. Piping can be entirely carbon steel, entirely stainless steel, or any combination of the two, depending mostly on user preference.

For compressors with gas turbine drivers, compressor oil is usually supplied by the turbine oil system. In such cases, a small seal console that contains seal booster pumps, filters, and the necessary seal oil controls and instrumentation is provided. On some units, the main seal oil pump is driven from the compressor shaft with an auxiliary pump on the seal console. An emergency seal oil accumulator can be mounted directly on the seal oil console, with an external nitrogen supply providing the motive fluid to supply seal oil during plant power failures. These accumulators require special valving to avoid nitrogen ingestion into the system.

In most units the seal oil is combined with the lubricating oil in one system, but separate lubricating oil and seal oil systems can be provided if necessary because of potential contamination of the lubricating

Figure 15.2 Lube oil console package for modern compressor. *(Lubrication Systems Company, Houston, Tex.)*

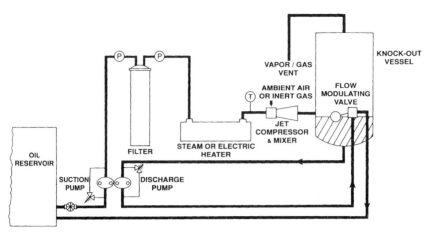

Figure 15.3 Schematic of an onstream lube oil purifier using the air stripping principle. *(Ausdel, PTY, Ltd., Cheltenham, Victoria, Australia)*

oil. Vacuum dehydrators, coalescers, centrifuges, air stripper, and nitrogen spargers are among the devices used to improve equipment reliability and reduce the cost of preventive maintenance. These units are typically designed for permanent installation on critical machinery oil systems for continuous, onstream purification.

A modern, cost-effective lube oil reclaimer or onstream oil purifier schematic is shown in Fig. 15.3. Here, the gear pump forces contaminated oil through a filter to remove particles and corrosion products. The pressurized oil is heated by steam or electric heaters and then enters the jet compressor or mixer where ambient air is induced. The air is humidified by the water in the oil and exits through a vent in the knockout vessel as the dehydrated oil returns to the reservoir. Process conditions are generally at temperatures of 140 to 190°F and atmospheric pressures. The higher the process temperature is, the greater the efficiency will be.

16

Compressor Control

16.1 Introduction*

Every centrifugal or axial compressor has (at a given rotational speed and inlet conditions) a characteristic combination of maximum head and minimum flow beyond which it will surge. Preventing this damaging phenomenon is one of the most important tasks of a compressor control system.

The most common way to prevent surge is to recycle or blow off a portion of the flow to keep the compressor away from its surge limit. Unfortunately, such recycling extracts an economic penalty due to the cost of compressing this extra flow. Hence, the controls must be able to accurately determine how close the compressor is to surging so it can maintain an adequate—but not excessive—recycle flow rate.

This task is complicated by the fact that the surge limit, in general, is not fixed with respect to any one variable such as pressure ratio or the pressure drop across a flowmeter. Instead, it is a complex function that also depends on gas composition, suction temperature and pressure, rotational speed, and guide vane angle.

An understanding of the principles of integrated control and protection systems is thus extremely important to companies and industries operating turbocompressors.

16.2 Control System Objectives

An integrated compressor control system may have any or all of the following objectives:

---

* Segment developed and contributed by Compressor Controls Corporation, Des Moines, Iowa. Special thanks to Dr. Brett Batson.

- **Performance control:** Maintaining a primary process variable (for example, discharge pressure or mass flow rate) at a desired set point level

- **Surge protection:** Preventing surge-induced compressor damage and process upsets without sacrificing energy efficiency or system capacity

- **Limiting control:** Maintaining any process-limiting variables (such as drive motor current and fluid temperature) within safe or acceptable ranges

- **Loop decoupling:** Minimizing adverse interactions between control functions

- **Load balancing:** Distributing the overall compression load among several compressors in a multiple compressor network

- **Event sequencing:** Automating changes in process status by controlling such events as start-up, shutdown, and purging operations.

- **On-line redundancy:** Providing uninterrupted control in the event of a hardware failure

- **Host communication:** Integrating the compressor control system into higher level distributed or supervisory control systems

The purpose of this segment is to highlight the conceptual framework within which these control objectives and methods of meeting them can be understood.

16.3 Compressor Maps

An axial or centrifugal compressor raises the pressure of a gas via energy added to it. If inlet conditions, rotational speed, and guide vane angle are held constant, the amount of energy added per unit mass (polytropic head) of gas will depend only on the inlet volumetric flow rate. Thus, we can construct compressor characteristics in terms of polytropic head and volumetric flow in suction.

The specific mechanical energy of a fluid is defined as

$$e_m = \frac{p}{\rho} + \alpha_{\text{KE}} \frac{V^2}{2} + gz \qquad (16.1)$$

where p = static pressure
   ρ = density
   V = average (or bulk) velocity
   g = gravitational constant
   z = elevation
   α_{KE} = kinetic energy flux coefficient

In most applications dealing with gases, the elevation component is insignificant. Thus we need only be concerned with the first two terms of Eq. (16.1).

Because each of these forms of energy can be expressed as an equivalent elevation (head), they are often referred to as the *pressure head and velocity head*. As explained earlier, they will have units such as $ft \cdot lb_f/lb_m$ *(kJ/kg)* if expressed as specific energies, or simply *ft (m)* when expressed as elevations.

A compressor uses a two-stage process to increase the pressure of its process stream. First, the mechanical energy of the rotor is transferred to the fluid, resulting in its acceleration—and increasing its kinetic energy. Most of the velocity head is then converted to an increase in pressure head by decelerating the fluid through a diffuser.

The work added to the fluid in the compressor depends on the path that the state of the gas takes as it passes from suction to discharge. However, this energy can be characterized by choosing a particular path having the same end points as that of the actual fluid. The path often chosen is a *polytropic compression* path. The overall increase in fluid specific mechanical energy is referred to as the *polytropic head* (H_p). The ratio of the change in mechanical energy divided by the change in stagnation enthalpy (mechanical plus internal energy) is defined as the *polytropic efficiency.*

Unfortunately, it is not possible to measure polytropic head directly—it must be calculated as a function of fluid properties and several measurable process variables. By integrating the thermodynamic relationship for work over a polytropic path, it can be shown that

$$H_p = \frac{Z_a R_u T_s}{MW} \frac{R_c^{\sigma} - 1}{\sigma} \tag{16.2}$$

where $\sigma = \text{exponent} \left(\dfrac{k-1}{k\eta_p} \right)$

η_p = polytropic efficiency
k = ratio of specific heats (c_p/c_v)
MW = molecular weight
p_d = discharge pressure (absolute)
p_s = suction pressure (absolute)
R_c = pressure ratio (p_d/p_s)
R_u = universal gas constant
T_s = suction temperature (absolute)
Z_a = average compressibility factor

The polytropic head developed by a specific compressor will vary as a function of the inlet volumetric flow rate, rotational speed, the position of the guide vanes, and the suction conditions.

Like polytropic head, volumetric flow in suction must be calculated as a function of fluid properties and process variables that can be directly measured. For orifice plates and venturi meters (which are, perhaps, the most commonly used flow measuring instruments), it can be shown that

$$Q_s^2 = \frac{Z_s R_u T_s}{\text{MW}} \frac{\Delta p_{o,s}}{p_s} \tag{16.3}$$

where $\Delta p_{o,s}$ = pressure drop across an orifice plate
     MW = molecular weight
     p_s = suction pressure (absolute)
     R_u = universal gas constant
     T_s = suction temperature (absolute)
     Z_s = suction compressibility factor

Thus, compressor performance is often illustrated by plotting characteristic curves in H_p versus Q_s. For a given speed and inlet guide vane angle, this will produce a single performance curve at constant inlet conditions. By allowing either the rotational speed or guide vane angle to take a series of discrete values, we can generate a family of performance curves, which is called a compressor map. It is important to note that, in the coordinates (Q_s, H_p), the performance curves are only valid for the given inlet conditions.

In a similar fashion, constant resistance curves could be used to plot the amount of energy that must be added to the fluid to sustain a given flow rate. Each of these curves represents some possible combination of gas properties, inlet and outlet piping, valve positions, back pressures, and operating devices.

The shape of the resistance curves depends on the characteristics of each specific application. For flow through pipes, energy required will be approximately proportional to volumetric flow squared. Plotting resistance curves in the coordinates H_p versus Q_s^2 would then yield a series of straight lines radiating from the origin.

In general, however, the exact shape of the resistance curves will defy simple analysis. Fortunately, it is rarely necessary to know their exact shape, so we will represent them as a series of generic curves.

The performance map for a compressor system can be illustrated by superimposing both performance and constant resistance curves (see Fig. 16.1). At any given instant, the value of the compressor's independent variable (such as rotational speed) will determine which performance curve it operates along. Similarly, the resistance of the network at that instant will determine the position of the current line of constant resistance. The intersection of these two curves is called the *operating point*.

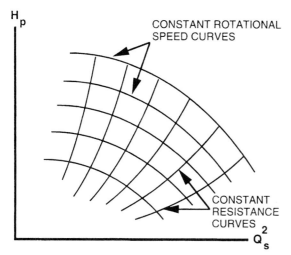

Figure 16.1 Typical compressor performance map. *(Compressor Controls Corporation, Des Moines, Iowa)*

The flow rate at this point is such that the amount of energy added by the compressor is equal to that required to overcome the network resistance. The coordinates of the operating point thus represent the volumetric flow rate through, and polytropic head developed by, the compressor.

16.3.1 Invariant coordinates

As noted above, the compressor performance curves in the coordinate system (Q_s, H_p) are unique for the given suction conditions. In practice, the inlet conditions are not constant, so for the purposes of control, the coordinates used must be invariant to changes in inlet conditions. Performing dimensional analysis on the parameters important to compressors results in several possible coordinates suitable for control. Two of these are presented here.

The first is *reduced polytropic head* (h_r) versus *reduced flow rate in suction* (q_s). (For the same reasons as stated earlier, it is convenient to square the reduced flow rate.) These coordinates are defined as:

$$h_r \equiv \frac{R_c{}^\sigma - 1}{\sigma} \tag{16.4}$$

$$q_s{}^2 \equiv \frac{Q_s{}^2 \mathrm{MW}}{Z_s R_u T_s} \propto \frac{\Delta p_{o,s}}{p_s} \tag{16.5}$$

where $\quad \sigma = \text{exponent} \left(\dfrac{k-1}{k\eta_p} \right)$

$\eta_p = $ polytropic efficiency

$\Delta p_{o,s}$ = pressure drop across an orifice plate in suction
k = ratio of specific heats (c_p/c_v)
MW = molecular weight
p_d = discharge pressure (absolute)
p_s = suction pressure (absolute)
R_c = pressure ratio (p_d/p_s)
R_u = universal gas constant
T_s = suction temperature (absolute)
Z_s = suction compressibility factor

The second coordinate system is *pressure ratio* (R_c) versus *reduced flow rate* in suction (q_s). This combination has been in use for many years and remains a basis for many surge control systems. This has the advantage of requiring only pressure and flow measurements. Often, however, using reduced head, h_r, results in a more invariant surge line.

From the (q_s^2, R_c) coordinate system comes the common surge control system based on Δp_c and $\Delta p_{o,s}$. This is constructed by assuming a surge control line which satisfies

$$R_c - 1 = C\frac{\Delta p_{o,s}}{p_s} \qquad (16.6)$$

The derivation is as

$$R_c - 1 = C\frac{\Delta p_{o,s}}{p_s}$$

$$\frac{p_d - p_s}{p_s} = C\frac{\Delta p_{o,s}}{p_s}$$

$$p_d - p_s = C\Delta p_{o,s}$$

$$\Delta p_c = C\Delta p_{o,s} \qquad (16.7)$$

Note that the gas composition is not required to calculate any of these coordinates. In application, reduced flow rate is calculated using the differential pressure, $\Delta p_{o,s}$, from the flow measurement device divided by suction pressure, p_s so temperature, gas molecular weight, and compressibility are not required.

It should be pointed out that these coordinates are not invariant to variations in isentropic exponent k. However, since k does not vary considerably in most applications, this does not present a problem. Therefore, although the term *invariant* is used, it should be remembered that these coordinates are *nearly invariant*.

Using either of these coordinates for a compressor without inlet guide vanes, the surge limit line is represented by a single curve that

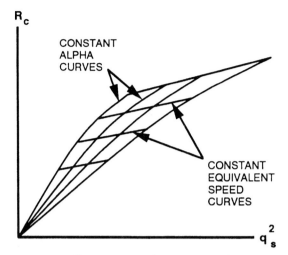

Figure 16.2 Compressor performance map for a compressor with variable inlet guide vanes. *(Compressor Controls Corporation, Des Moines, Iowa)*

is stationary with changing inlet conditions. The rotational speed is not required to determine the relative distance between the operating point and the surge control line.

For compressors with inlet guide vanes, the surge limit is represented by a family of curves that do not depend on suction conditions (Fig. 16.2). In this case, along with the two basic coordinates (reduced polytropic head *or* pressure ratio, and reduced flow rate) another coordinate may be required. This additional coordinate could be either guide vane position, α, or *equivalent speed, N_e,* defined as

$$N_e \equiv \frac{N\sqrt{MW}}{\sqrt{Z_s R_u T_s}} \qquad (16.8)$$

where MW = molecular weight
N = rotational speed
R_u = universal gas constant
T_s = suction temperature (absolute)
Z_s = suction compressibility factor

Note that the gas composition must be known to calculate equivalent speed. This is due to the appearance of molecular weight in the definition.

Although compressor maps are rarely presented in these invariant coordinates, these are the preferred systems for control. This is due, not only to their invariance, but also to the fact that gas composition is not required for their calculation (with the exception of equivalent speed).

Some variations and simplifications are possible to the invariant coordinates. Some other signals that are possible to use to produce invariant coordinates are the differential pressure across the compressor, Δp_c, and the flow measurement in discharge represented by the differential pressure, $\Delta p_{o,d}$.

16.4 Performance Control

One of the basic objectives of a compressor control system is to regulate the output of the compressor to meet the varying needs of the overall process. That objective is accomplished by manipulating a *performance control element*. This control element might be a suction or discharge control valve, guide vane positioner, or rotational speed governor. It would serve to maintain a process pressure or flow rate at a *set point* value.

When illustrating performance control, it is helpful to redefine the compressor map coordinate system. For example, we might plot discharge header pressure against mass flow through the compressor (see Fig. 16.3). The set point will then appear as either a horizontal or a vertical line, depending on whether we want to maintain constant pressure or constant mass flow, respectively.

To achieve this coordinate system transformation, we must define the control element to be part of the compressor system rather than

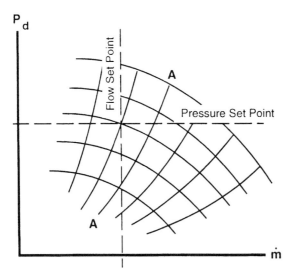

Figure 16.3 Compressor map showing typical performance control set points. *(Compressor Controls Corporation, Des Moines, Iowa)*

part of the network. Thus, repositioning a control valve moves the performance curve rather than the network resistance curve.

Defining the coordinate system in this way implies that any independent process variables (such as fluid composition) have constant values, so that each point in the H_p versus Q_s coordinate system corresponds to a unique point in the new coordinate system. Changes in any of these variables would change the shape and position of the performance and resistance curves.

Within this transformed coordinate system, the discharge pressure and flow rate are determined by the intersection of the performance and resistance curves corresponding to the current control element position and network resistance. The control system must manipulate the control element in response to changes in network resistance and gas properties, so that the operating point is kept as close to the set point line as possible.

For example, assume that the compressor of Fig. 16.4 is operating at point 1, which is on the set point line. If the network resistance decreases, the operating point will move along the performance curve to point 1', where it intersects the new resistance curve. To restore set point operation, the control system must reposition the control element to select the performance curve that intersects the new resistance curve at the new operating point 1''.

Consider the hypothetical situation where all of the assumptions made in transforming the coordinate system were valid, the process and

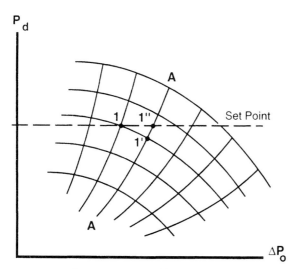

Figure 16.4 Compressor map illustrating typical performance control response. *(Compressor Controls Corporation, Des Moines, Iowa)*

control system were free of dynamic effects, and the equations of the performance and resistance curves were known. Then the control system might (at least theoretically) be able to deduce changes in network resistance from the position of the control element and the value of the controlled variable. At that point the controller could calculate the exact position required by the system to satisfy the control objectives.

In general, none of these conditions can be met. Performance control algorithms must therefore be able to adapt to changes in process parameters that are not (or cannot) be measured. They must be able to overcome dynamic effects and time delays while dealing with the uncertainties in our knowledge of compressor performance and network resistance characteristics. They have to circumvent the difficulties associated with solving even approximate models of these imperfectly understood relationships in real time (usually by *not* solving them at all).

So our hypothetical situation doesn't exist in the real world. As a solution to these problems, we must construct a control system in another way. Very simply, the alternative of choice is a control scheme that always acts in a fashion that reduces the error between the operating point and the set point. This is explained in the following section.

16.4.1 PI and PID control algorithms

The commonly used proportional-integral (PI) and proportional-integral-derivative (PID) control algorithms provide the adaptable method needed to control imperfectly known compression systems.

The basic premise of these algorithms is that the controller output should be a function of the difference (which is called the *error,* ϵ) between the value of the controlled variable (or *process variable,* PV) and its *set point* (SP).

$$\epsilon = \text{SP} - \text{PV} \qquad (16.9)$$

The controller output is calculated as the sum of a constant steady-state value and several bias terms that are functions of the error

$$\text{OUT}_{\text{PI}} = \text{SS} + P + I \qquad (16.10)$$

$$\text{OUT}_{\text{PID}} = \text{SS} + P + I + D \qquad (16.11)$$

where   OUT = value of controller output signal
        SS = steady-state component
        $P \propto \epsilon$ (instantaneous error)
        $I \propto \int \epsilon dt$ (integral error)
        $D \propto d\epsilon/dt$ (derivative error)

To understand how these algorithms work, imagine an application in which the goal is to maintain a steady 100 psig discharge pressure. Assume the compressor is initially operating at steady state with its suction throttle valve open 75 percent. Because the compressor is in steady state and the valve is not at a limit, the error must be zero. Therefore, the instantaneous and derivative errors are zero, and the integral error is constant. Thus, the controller output will not change until the compression system is perturbed or the set point is changed.

Now assume that the downstream process system resistance decreases so that the control valve must be increased to 80 percent open to maintain the desired set point of 100 psig pressure. A 75 percent open valve would then produce a lower than desired pressure. Because the controller is unable to directly calculate the required new valve setting, an intelligent trial and error approach must be taken—simply open the valve a specified amount, monitor the result, and proceed as required from there. This type of approach is known as *closed loop* control.

The relationship between valve opening and discharge pressure is unknown. However, a good starting point is to assume that the additional valve opening should be approximately proportional to the pressure change—a large pressure variation is countered by a large change in the valve position. This assumption leads to the proportional term of the PID algorithm.

Returning to the above example, note what happens if the control were proportional *only* (the integral and derivative terms both set to zero). With the suction throttle valve at 75 percent open, the discharge pressure would decrease, causing the error (and thus the output) to rise. Resetting the output to 80 percent causes the pressure to increase, reducing the error (and thus the output). Note that if the pressure error was to go to zero, the output would go to SS since the control system is proportional only. Therefore, there is an intermediate level at which the error and the output (both nonzero) stabilize, leaving the pressure below its set point.

For example, assume that a valve opening of 78 percent would yield a steady 98 psig discharge pressure. If a 2 psig error produced a 3 percent proportional response (starting at a valve position of 75 percent), the output of the controller at 98 psig would be 78 percent—the exact valve setting needed to maintain that pressure. The discharge pressure would thus stabilize at 98 (instead of the desired 100) psig. The resulting 2 psig disparity is known as a *proportional offset*.

Proportional offsets are eliminated by adding an integral term to the control algorithm. Because this integral will accumulate whenever the error is nonzero, the control action cannot stabilize unless the controlled variable is at its set point. At steady state, then, the proportional part is zero (because the error is zero), but the integral term equals the value required to keep the error at zero.

Therefore, the steady-state controller output is actually the sum of the steady state and integral error terms. As a result, the steady-state (SS) term loses its special significance and is usually merged into the integral term.

In the PID algorithm only, another term is added that is proportional to the first time derivative of the error. The basic premise behind including this term is that the magnitude of the control response should be modulated according to how fast the error is changing.

For example, a fairly large response would be appropriate if the pressure was too low and still falling. In contrast, a smaller response would be warranted if the pressure was too low but rising.

The practical effect of including the derivative term is that it often allows the control response to be accelerated without increasing the risk of instability. However, derivative control will also make the system more sensitive to signal noise. Thus, the simpler PI algorithm is sometimes more appropriate than full PID control.

16.4.2 Stability considerations

Improper tuning can render either the PI or PID algorithm (or even simple proportional control) unstable (see Fig. 16.5). To gain a qualitative appreciation of how this can occur, consider the PI response to a step change in set point for a process initially operating at steady state.

Initially, the integral of the error is constant. So the output changes proportionately to the magnitude of the step change, countering its effect and thus decreasing the error. With purely proportional control,

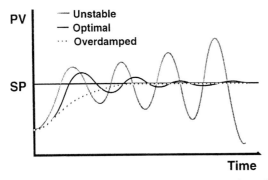

Figure 16.5 System response to improper tuning of PID algorithm. *(Compressor Controls Corporation, Des Moines, Iowa)*

the system will tend toward a new steady state characterized by a proportional offset of the controlled variable relative to its set point.

However, because of the natural inertia of the system, changes in the process will lag behind the control signal. Because that signal will reach its new value before the controlled variable does, the control action will overshoot. As a result, the controlled variable will also overshoot, causing the control action to reverse. The overall process will therefore oscillate about the new steady-state conditions. Proper tuning of the control response will damp out these oscillations.

The magnitude of the overshoot will depend on the value of the control system's proportionality constant (which is known as the *proportional gain*). Increasing this gain will accelerate the control response, reduce the proportional offset, and increase the overshoot. If the proportional gain is set too high, each successive overshoot will be larger than its predecessor, resulting in an unstable system.

Adding an integral control action has the effect of increasing the system inertia and thus exacerbates the risk of instability. The reason is fairly simple—both the error and its integral will lag behind changes in the controller output. The proportional and integral gains must both be compromised to maintain system stability. As mentioned earlier, however, the integral term allows the control system to reduce the error to zero.

On the other hand, adding a derivative term can reduce the risk of instability. This is because the derivative of the error is a measure of how fast the system is responding. If the system becomes unstable, the derivative action will tend to counter the oscillatory action.

16.4.3   Integral or reset windup

A PID controller may encounter situations in which changing the output signal cannot reduce the error to zero. For example, suppose our control objective is to maintain a constant fluid level in a storage tank by manipulating an inlet valve to balance the inlet and outlet flows. Even with the valve fully open, the feed rate might prove insufficient to maintain the desired level. The control system cannot manipulate the controlled variable to eliminate this problem.

In such a situation, the integral error would continue growing indefinitely, eventually reaching a saturation value. Then if the flow rates change so that the level becomes too high, the PID loop must integrate the error (which is now opposite in sign) for some time to reduce its value from the saturation level to a magnitude that allows the valve position to come off of 100 percent open. So the control system is unable to respond rapidly enough to sudden decreases in outflow, possibly causing the tank to overflow.

The breakdown of control under such circumstances is referred to as *integral* or *reset windup*. Compressor control systems that employ PI or PID algorithms must be able to recognize and respond to this condition by disabling the integral portion of the algorithm—setting it to a constant (generally nonzero) value.

16.5 Performance Limitations

As shown previously, a compressor increases the pressure of the process gas by first accelerating the fluid and then converting the resultant kinetic energy into an increased pressure head. Thus, the maximum polytropic head that can be developed is limited by the tip speed of the impeller blades. Polytropic head is observed to be a decreasing function of volumetric flow.

At any given rotational speed and inlet conditions, the performance of the compressor is limited not only by this maximum polytropic head but also by a maximum flow rate. These limitations are illustrated on the typical performance curve shown in Fig. 16.6.

16.5.1 Surge limit

Consider the scenario of a constant-speed compressor with fixed suction conditions. This compressor is initially at steady state, so the operating point is stationary.

If the network resistance increases, the operating point will move along the performance curve to the left. Eventually, a point of minimum stable flow and maximum polytropic head is encountered. Operating the compressor to the left of this point can induce a potentially destructive phenomenon known as surge. This point is known as the

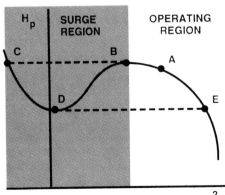

Figure 16.6 Performance curve illustrating typical performance control response. *(Compressor Controls Corporation, Des Moines, Iowa)*

surge limit point. The locus of all such points defines a curve known as the *surge limit line* (SLL), as shown on Fig. 16.7.

The onset of surge can be visualized by considering a simple system consisting of a compressor and discharge valve. If this system were operating in steady state, there would be no accumulation of the process gas (over time) in the plenum between the two elements. In Fig. 16.6, this point is represented by *A*.

If the valve were then closed slightly, the flow rate through it would drop. Process gas would accumulate in the plenum, causing a pressure increase. This would, in turn, cause a decrease in the flow through the compressor and an increase in the flow through the valve until they were once again equal. The new steady-state flow rate would be lower than it was originally, and the new plenum pressure would be higher.

If the valve continued to close in small steps, the compressor would approach its surge limit (point *B* in Fig. 16.6). This approach would be noticeable when a relatively large drop in flow produced a small increase in the final plenum pressure. In other words, the slope of the compressor characteristics tends to reduce in magnitude near the surge limit point.

At the surge limit point, closing the valve further would cause the flow to decrease. As shown in Fig. 16.6, the compressor could not achieve the pressure required to maintain the flow less than that of the surge point. The result is that the flow through the compressor would drop precipitously, typically reversing in approximately 20 to 80 ms (see Fig. 16.8). This is shown as the dashed line *B–C* in Fig. 16.6. The operating point cannot reside on the portion of the curve between *B*

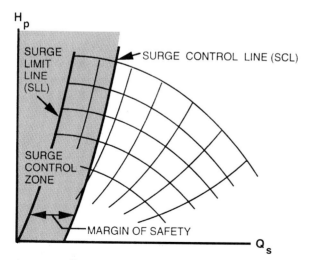

Figure 16.7 Compressor map showing surge limit line, surge control line, and surge control zone. *(Compressor Controls Corporation, Des Moines, Iowa)*

and D. Note that the flow drops off before the pressure. This can be seen in the top plot in Fig. 16.8.

At point C, the flow would be negative, but energy is still being added to the fluid. The plenum would begin to depressurize due to the net efflux of fluid. The trajectory of the operating point would be from C to D in Fig. 16.6.

At point D, the compressor, once again, reaches the region of the curve it cannot reside on. At this point the flow quickly jumps to point E where the compressor is in the "safe region."

With the discharge valve at the position that caused the surge in the first place, if no other action is taken, the compressor will travel from E

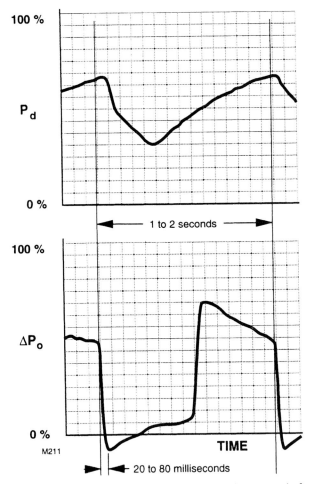

Figure 16.8 Pressure and flow variations during typical surge cycles. *(Compressor Controls Corporation, Des Moines, Iowa)*

to A and then repeat the cycle. The oscillatory pattern will continue until some action occurs to stop it (e.g., opening the discharge valve) or the compressor fails.

During surge, the severe oscillations of flow and pressure create heavy thrust bearing and impeller loads, vibration and rising gas temperatures. If more than a very few surge cycles are experienced, process upsets and severe compressor damage are likely to result.

The past decade has brought on a better understanding of the various modes of instability associated with compressors.

16.5.2 Stonewall

Again, consider a constant-speed compressor with fixed suction conditions. As network resistance decreases, the operating point will move along the performance curve to the right. Eventually, a point of maximum flow and minimum polytropic head is encountered, beyond which further decreases in network resistance will not increase the flow rate. This is known as the *choke point,* or *stonewall.*

The physical phenomenon is that the gas velocity has increased to the local acoustic velocity (therefore Mach 1) at some point in the compressor. When choke occurs, the flow rate cannot increase unless conditions change in the choked region.

Stonewall is not particularly damaging to single-stage centrifugal compressors but can aerodynamically affect the rotors and blades of multistage centrifugal and axial compressors. In such situations, a suitably designed antichoke controller can be used to manipulate an antichoke control valve, thus maintaining sufficient system resistance to prevent choke.

16.6 Preventing Surge

Surge occurs when the network resistance becomes too high for the compressor to overcome. The obvious way to prevent surge is to decrease network resistance whenever the operating point moves too close to the surge limit line. This is accomplished by opening an antisurge valve to recycle or discharge a portion of the total flow.

The chief drawback to this approach is the efficiency penalty that it entails—the energy that was used to compress the recycled gas goes to waste. Thus, the control system should be tailored to open the antisurge valve only when—and only as far as is—necessary. On the other hand, if we do not provide adequate protection against surge, we risk prohibitive repair and downtime costs.

Therefore, accurate and dependable methods of determining the surge limit are required. Antisurge control entails measuring the distance between this surge limit and the operating point and then maintaining an adequate margin of safety without sacrificing efficiency or stability.

The solution is to maintain the operating point on or to the right of a line known as the *surge control line* (SCL; see Fig. 16.7). The distance between the surge control and surge limit lines (the margin of safety) should be just enough to allow the chosen control algorithms to counteract an impending surge.

Whenever the operating point moves into the *surge control zone* (i.e., to the left of the SCL), the antisurge valve must be opened fast enough to keep the operating point from reaching the surge limit line and far enough to return it to the surge control line. On the other hand, when the operating point moves to the right of the SCL, the antisurge valve should be closed as far as possible without moving the operating point into the surge control zone.

16.6.1 Antisurge control variables

Like H_p and Q_s, the distance between the operating point and the surge limit line cannot be directly measured. Nor is there a standard definition relating it to parameters that can be measured. Thus, antisurge protection algorithms can be based on any function of measurable process variables that satisfies the following criteria:

- It should vary monotonically as the operating point approaches the surge limit, so that the required control action is never ambiguous (if not, there must be a means of resolving any ambiguities).

- It must be invariant to any aspect of the process that might change, so that the compressor is adequately protected in all possible situations.

- It must be easily calculated from process variables that can be accurately measured or assumed constant.

- It should be most sensitive to changes that occur when the operating point is near the surge limit.

The obvious possibilities include combinations of the coordinates presented in Sec. 16.3. Other, less obvious candidates, include functions of the compressor drive power or rotational speed—all of which meet these criteria for at least some applications.

The existence of guide vanes may complicate antisurge control. In general, the guide vane angle, α, must be included as part of the variable used for determining the distance between the operating point and the surge control line.

One variable that has advantages of control on flow as well as pressure, is a ratio of a function of the ordinate to the abscissa. From Sec.

16.3, choosing the (q_s^2, R_c) coordinate system, the antisurge control variable, S_s, would be calculated as

$$S_s = \frac{f\,(R_c)}{q_s^2} \qquad (16.12)$$

where $\quad R_c$ = compressor pressure ratio, p_d/p_s
$\qquad q_s^2 = \Delta p_o/p_s$
$\qquad f\,(\cdot) = q_s^2\,|\,surge$

The surge limit line is, therefore a line of $S_s = 1$. As shown in Fig. 16.9, lines of constant S_s comprise a family of curves that emanate from the origin. They are similar in shape to the surge limit line and diverge from the surge limit line.

Using this method, comparing the value of S_s to unity gives a relative distance to surge.

16.6.2 Antisurge control algorithms

In performance control, the goal is to minimize deviations of the controlled variable from its set point—variations to either side are of approximately equal concern. In contrast, the goal of an antisurge controller is to maintain the controlled variable to one side of an absolute limit—deviations to the other side must be prevented at any cost.

Thus, good antisurge control uses a combination of the following types of control responses:

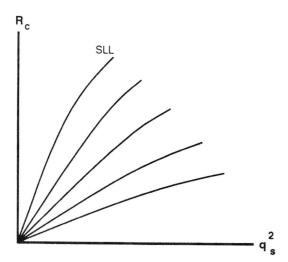

Figure 16.9 Compressor performance map showing lines of constant surge control variables S_s. (*Compressor Controls Corporation, Des Moines, Iowa*)

- Opening and closing the control valve to maintain the operating point on or to the right of the surge control line without allowing deviations to the left of the surge limit line

- Moving the surge control line (relative to the surge limit line) to adapt the margin of safety to changing process conditions

The first of these points may be accomplished by simple closed loop (PI or PID) control if the safety margin is great enough. To reduce the safety margin (thereby increasing the operating range and efficiency of the compressor), rapid increase of the recycle valve set point is expedient.

Another aspect of this issue was noted earlier. The safety margin should be increased if surge occurs under a large disturbance. This reduces the chance that surge will recur under a similarly large disturbance.

Another aspect of this issue is to increase the safety margin *dynamically* when surge is threatening. Basing this adjustment on the derivative of the approach (from the right of the surge control line *only;* per state-of-the-art practice, late 1990s) results in a control scheme that does not compromise stability. It also provides for earlier opening of the recycle valve under fast disturbances.

16.6.3 Controlling limiting variables

The third major responsibility of an integrated compressor control system is to counteract undesirable changes in any process-limiting variables.

The limitation involved might be required to protect the compressor or other process equipment (for example, preventing excessive motor current, bearing temperatures, or excessive pressures). Or it may be necessary to protect the process gas from conditions—such as an excessive discharge temperature—that could chemically degrade or otherwise damage its quality.

In an integrated control system, we are concerned with two categories of process-limiting variables—those that must be controlled by opening the antisurge valve and those that must be controlled by manipulating the performance control element.

Limiting variables that fall into the first category can be controlled by the antisurge controller, provided that it has such a capability. Otherwise, an additional controller must be used, along with a switching device that can dynamically assign control of the antisurge valve to the appropriate controller.

Similarly, if there are limiting variables that can only be controlled by manipulating the performance control element, it is necessary to use either a performance controller with multivariable capabilities or multiple controllers along with an appropriate switching circuit.

Regardless of which approach(es) are chosen, it is necessary to provide protection against integral windup in any control loop using a PI or PID algorithm. When multivariable controllers are used, the operator may or may not be allowed to override automatic control of any or all of those variables.

16.7   Loop Decoupling

The action of the antisurge control system can upset the performance control operation and vice versa. In networks of compressors, the antisurge control actions between the various cases may also require decoupling. The potentially conflicting effects of interacting control loops can be counteracted by implementing a loop decoupling algorithm.

For example, if the performance controller (PIC) in Fig. 16.10 needs to reduce the downstream flow rate, the rate at which it closes its control valve may need to be compromised to avoid destabilizing the system. The optimum response will depend on the proximity of the compressor's operating point to the surge line.

If the compressor was operating on its surge control line, reducing the flow rate would move the operating point into the surge control zone. The antisurge controller (UIC) would respond by opening the recycle valve, which would have the side effect of further reducing the

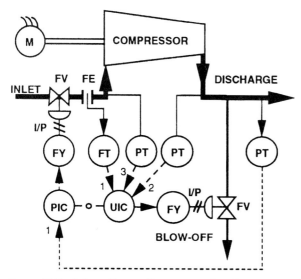

Figure 16.10 Interacting performance and antisurge control loops. *(Compressor Controls Corporation, Des Moines, Iowa)*

downstream flow rate. The performance controller would then need to increase the flow rate, which would (in turn) cause the antisurge controller to reduce the opening of its valve.

These interactions would render both loops more oscillatory and therefore less stable. Stability could only be restored by tuning the controllers less aggressively, thus making them less effective when operating well away from the surge limit. Unless some provision was made for coordinating the two loops, neither could be optimally tuned for all situations.

This problem can be overcome by having the controllers monitor and compensate for changes in each other's outputs. This feed forward control action would allow the performance controller to moderate its actions when the antisurge valve was opening, and vice versa. The loop interactions would thus be decoupled—the actions taken by each controller would then be correct regardless of what the other was doing.

16.8 Conclusion

Compressor control is not only critical (from an economic standpoint); it is a challenging controls problem. Compression systems are typically dynamic, with lags and delays that make it difficult—most of the time impossible—to control them with a simple PID approach.

The antisurge variable used for control should be selected carefully to provide accurate prediction regardless of the inlet conditions of the compressor.

It is important to integrate the various control tasks surrounding the compressor—antisurge, performance, limiting, etc. When the compression system involves multiple compressors, a method of balancing the load between compressors may be required. This method should combine maintaining the performance variable at its set point while not compromising surge control.

17

The Head-Flow Curve Shape of Centrifugal Compressors*

Much has been written concerning what goes on in a centrifugal compression stage. Unfortunately, most such literature has been produced by development people for the benefit of other development people and has been presented mostly in bits and pieces in widely scattered technical papers.

The operating supervisor at the plant level and the equipment specialist at the planning level have a real need for an overall understanding of this subject. Such knowledge can help the operator to better understand the potentials and limitations of his machine. It should help the specialist to determine what can realistically be expected of a centrifugal compressor, and it should help in his analysis of competitive offerings.

It is on this premise that the following paragraphs are written—in the hope that in a small way at least, they can build a bridge of comprehension between the centrifugal compressor investigator and the compressor user.

17.1 The Compressor Stage

This discussion will largely concern itself with the conventional compressor stage, i.e., a radial inlet, closed impeller running at 700 to 900 ft/s (213 to 273 m/s) tip speed, feeding a vaneless diffuser. However, suf-

---

* Segment developed and contributed by Donald C. Hallock, "Centrifugal Compressors: The Shape of the Curve," Compressor Refresher, The Elliott Company, Jeannette, Pa. (Reprint 93). Originally published in Air and Gas Engineering, Volume 1, Number 1, January 1968 and presented through the courtesy of CAGI.

ficient attention will be given to such variations as inducer impellers and vaned diffusers that a general understanding of most combinations of commonly used hardware should result.

Discussion will center on the *impeller* and *diffuser* because these are the two key elements in producing characteristic shape (see Fig. 17.1). Poorly designed inlets, collectors, or return channels can naturally affect performance, but their influence on characteristic shape is usually small and will henceforth be ignored.

This discussion is directly applicable to a single-stage machine and to each stage of a multistage machine. The approach taken will be largely qualitative rather than quantitative, because it is not our purpose to produce a design manual, but rather to produce understanding. A general familiarity of the reader with centrifugal equipment is assumed. Certainly, the preceding pages of this text will be quite helpful in this regard.

17.2 Elements of the Characteristic Shape

Any discussion of characteristic shape must, like ancient Gaul, be divided into three parts. We have a *basic slope* of head vs. flow, upon which we must superimpose a *choke* or "stonewall" effect in the overload region, and a minimum flow or *surge* point in the underload region. The resulting overall characteristic will then be the basic slope as altered and limited by choke at high flow and as limited by surge at low flow in Fig. 17.2. We will discuss each of the three parts in turn.

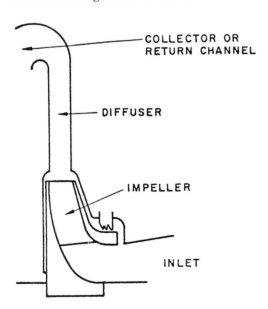

COLLECTOR OR
RETURN CHANNEL

DIFFUSER

IMPELLER

INLET

Figure 17.1 Impeller and diffuser geometry influence compressor performance curve. (*Elliott Company, Jeannette, Pa.*)

17.2.1 Basic slope

To understand *basic slope,* it is necessary to look at what is going on at the impeller tip in terms of velocity vectors. In Fig. 17.3, V_{rel} represents the gas velocity relative to the blade. U_2 represents the absolute tip speed of the blade. The resultant of these two vectors is represented by V, which is the actual absolute velocity of the gas. (By vector addition, $U_2 + V_{rel} = V$.) It can be seen that the length of the vectors and the magnitude of the exit angle α are determined by the amount of backward lean in the blade, by the tip speed of the blade, and by gas velocity relative to the blade, which is in turn dictated by tip-volume-flow rate for a given impeller.

Having the magnitude and direction of the absolute velocity V, we now break this vector into its radial and tangential components, V_r and V_t, as in Fig. 17.4. The vector V_t is reduced somewhat by "slip factor" in a real impeller, an effect that can be ignored in a qualitative discussion such as this. *The head output is proportional to the product of $U_2 V_t$.* For a given r/min, U_2 is constant; therefore, head is proportional to V_t.

Let us now look at what happens to the magnitude of the tangential component V_t as we vary the amount of flow passing through the impeller at constant r/min. As the flow is decreased, V_{rel} decreases. As V_{rel} decreases, angle alpha decreases markedly. This makes V_t increase, which increases head output. This head increase with decreasing flow is the *basic slope* of the stage characteristic.

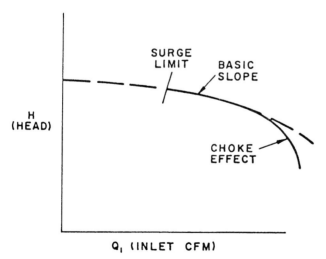

Figure 17.2 Characteristic shape of centrifugal compressor performance curve. (*Elliott Company, Jeannette, Pa.*)

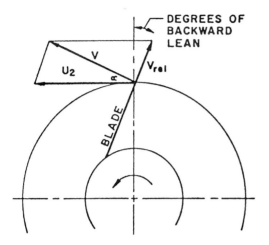

Figure 17.3 Blade angle and velocity relationships. (*Elliott Company, Jeannette, Pa.*)

17.2.2 Blade angle is a compromise

How does the degree of backward lean affect the steepness of the basic slope? Picture a radial blade (zero backward lean). V_{rel} is now the same as V_r in Fig. 17.4 and V_t is now equal to U_2. As we reduce the flow in this impeller, V_r and α decrease as before. V_t, however, remains constant. Head output therefore remains theoretically constant, regardless of flow. In a real impeller, of course, the head is reduced on increasing flow by a decrease in efficiency attributable to higher frictional losses. The

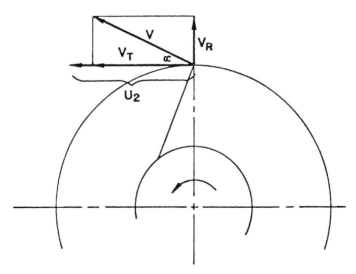

Figure 17.4 Resolution of vectors. Note that head output is proportional to product V_tU_2. (*Elliott Company, Jeannette, Pa.*)

resulting basic slope normally shows 2 or 3 percent head rise when going from design flow to minimum flow.

Now let us look at the opposite extreme, an impeller having a very high degree of backward lean—say 45° off radial at the tip. We can see that a change in flow, and therefore a change in the V_{rel} vector length, will cause very large changes in V_t, and therefore in head. Thus, such an impeller will typically produce a head rise of 20 percent or more, when moving from design flow to minimum flow.

It is evident from the foregoing that the effect of backward lean on head output is minimized at low flow; a high backward lean impeller will produce almost as much head at minimum flow as a low backward lean impeller running at the same tip speed. As we move out toward design flow, however, the head difference becomes quite dramatic, as seen in Fig. 17.5. The normal industry standard for conventional closed impellers is represented by the middle line, which is 25 to 35° of backward lean. This configuration is really a compromise between the high head obtainable at design flow with low backward lean blades and the steep basic slope obtainable with high backward lean blades.

One further point should be made concerning basic slope before we leave the subject. In the foregoing discussion, we used the term *flow* without elaboration, the implication being that impeller-tip-volume rate is dictated by inlet volume rate regardless of rotative speed and type of gas. This, of course, is not quite true—gases, unlike liquids, being compressible. It is well known that a heavy gas will be compressed to a greater extent in a given stage than a light gas, i.e., the heavy gas has a higher volume ratio.

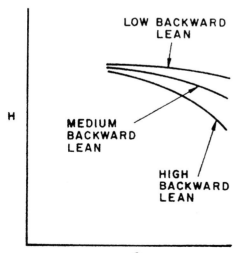

Figure 17.5   Effect of blade inclination on head rise. (*Elliott Company, Jeannette, Pa.*)

Therefore, for a given *inlet* cfm entering a given impeller at a given speed, the magnitude of V_{rel} is less for a heavy gas than for a light gas. If the impeller has backward lean, the magnitude of V_t will be greater for the heavy gas. Since head output is proportional to V_t, a given impeller running at a given speed will produce more head when compressing a heavy gas than when compressing a like *inlet* cfm of light gas. What is more, the magnitude of the difference increases at inlet flow increases, so the basic slope of a given backward lean impeller is actually less steep for a heavy gas than for a light gas. The higher the backward lean, the more pronounced this effect (see Fig. 17.6).

17.2.3   The fan law effect

The effect of volume ratio on what is known as *fan law* is worthy of mention. The fan law states that cfm potential of a stage is proportional to rotative speed and that the head produced is proportional to speed squared. Reexamination of Figs. 17.3 and 17.4 will demonstrate the logic of this law.

If V_{rel} were truly proportional to *inlet* cfm and we increased both inlet cfm and speed by 10 percent, then the head output would be 21 percent greater, because the tip-vector geometry would maintain exact similarity. Higher head produces higher volume ratio in a given gas; however, V_{rel} does not increase quite in proportion to speed and inlet cfm. By reasoning similar to that used in discussing heavy gas vs. light gas, then, the head output of a backward leaning stage handling 10 percent more

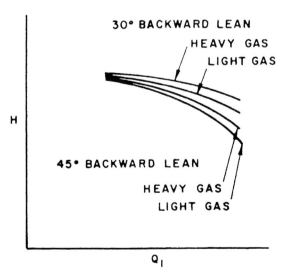

Figure 17.6   Effect of molecular weight on slope of performance curve. (*Elliott Company, Jeannette, Pa.*)

inlet cfm at 10 percent higher speed will increase somewhat *more* than 21 percent. By similar reasoning, if we reduce speed and inlet flow from 100 to 90 percent, the head produced will be slightly *less* than the 81 percent predicted by fan law, Table 17.1.

Fan law (affinity law) won its name from the fact that a fan is a low-head compressor normally handling air, a light gas. Since volume ratio effects are extremely small when imparting a small head to light gas, excellent accuracy can be obtained by fan law. As a general rule, the higher the head, the heavier the gas, and the greater the backward lean, the poorer the accuracy obtained by fan law will be. As a practical matter, speed changes up to 30 or 40 percent can be handled with sufficient accuracy for most purposes when dealing with typical single-stage air compressors. A little more discretion must be used on multistage compressors handling heavy gases, however, because fan law deviation can become quite significant for a speed change as small as 10 percent.

17.2.4 The choke effect

We have discussed at some length the basic slope of the head-flow curve and have avoided until now the choke or stonewall effect that

TABLE 17.1   Fan Laws (Affinity Laws)

1. With impeller diameter D held constant:

   A. $\dfrac{Q_1}{Q_2} = \dfrac{N_1}{N_2}$

   B. $\dfrac{H_1}{H_2} = \left(\dfrac{N_1}{N_2}\right)^2$

   C. $\dfrac{\text{Bhp}_1}{\text{Bhp}_2} = \left(\dfrac{N_1}{N_2}\right)^3$

   where Q = capacity, cfm
   H = total head, ft
   Bhp = brake horsepower
   N = compressor speed, r/min

2. With speed N held constant:

   A. $\dfrac{Q_1}{Q_2} = \dfrac{D_1}{D_2}$

   B. $\dfrac{H_1}{H_2} = \left(\dfrac{D_1}{D_2}\right)^2$

   C. $\dfrac{\text{Bhp}_1}{\text{Bhp}_2} = \left(\dfrac{D_1}{D_2}\right)^3$

When the performance (Q_1, H_1, & Bhp_1) is known at some particular speed (N_1) or diameter (D_1), the formulas can be used to estimate the performance (Q_2, H_2, & Bhp_2) at some other speed (N_2) or diameter (D_2). The efficiency remains nearly constant for speed changes and for small changes in impeller diameter.

occurs at flows higher than design and that must be superimposed on the basic slope, as in Fig. 17.2.

Just as basic slope is controlled by impeller-*tip*-vector geometry, the stonewall effect is normally controlled by impeller-*inlet*-vector geometry. In Fig. 17.7, we can draw vector U_1 to represent the tangential velocity of the leading edge of the blade (similar to U_2 at the tip). We can also draw vector V, representing absolute velocity of the inlet gas, which having made a 90° turn is now moving essentially radially—hence the term *radial inlet*. By vector analysis V_{rel}, which is gas velocity relative to the blade, is of the magnitude and direction shown, where $U_1 + V_{rel} = V$. At design flow, the direction of V_{rel} essentially lines up with the blade angle as shown.

17.2.5   Mach number considerations

The magnitude of V_{rel} compared to the speed of sound at the inlet is called the relative inlet Mach number: It is the magnitude of this ratio that dictates stonewall in a conventional stage. While true stonewall should theoretically not be reached until the relative inlet Mach number is unity, it is conventional practice not to exceed 0.85 or 0.90 at design flow.

It is evident from Fig. 17.7 that for a given r/min, the magnitude of V_{rel} will diminish with decreasing flow, since V is proportional to flow. If V_{rel} decreases, then relative inlet Mach number decreases, so the stonewall effect is normally not a factor at flows below design. It is also evident that at low flows, the direction of V_{rel} is such that the gas impinges on the leading side of the blade (positive incidence), a factor that is not very detrimental to performance until very high values of positive incidence are reached.

Let us now *increase* flow beyond design point. As V increases, so also do V_{rel} and relative inlet Mach number. In addition, V_{rel} now impinges

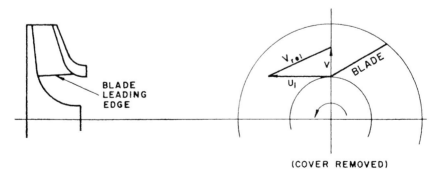

Figure 17.7   Velocity relationships at leading edge of impeller blade. (*Elliott Company, Jeannette, Pa.*)

on the *trailing* side of the blade, a condition known as *negative incidence*. It has been observed that high degrees of negative incidence
tend to contribute to the stonewall problem as "Mach One" is
approached, presumably because of boundary layer separation and
reduction of effective flow area in the blade pack.

17.2.6   Significance of gas weight

Since values of U_1 are typically in the 500 ft/s (152 m/s) range and values of V in the 250 ft/s (76 m/s) range, it is obvious that air at the speed
of sound at 80°F (27°C) = 1140 ft/s (347 m/s) and lighter gases suffer no
true impeller stonewall problems as described earlier, even at high
overloads. Some head loss below the basic slope will be observed, however, in even the lightest gases, in part because of increased frictional
losses throughout the entire stage and the extreme negative incidence
at high overloads.

The lightest common gas handled by conventional centrifugals for
which stonewall effect can be a definite factor is propylene with a speed
of sound of 740 ft/s (225 m/s) at –40°F (–40°C). In order of increasing
severity are propane at 718 ft/s (219 m/s) at –40°F (–40°C), butane,
chlorine at 630 ft/s (192 m/s) at –20°F (–29°C), and the various freons.
The traditional method of handling such gases is to use an impeller of
larger than normal flow area—to reduce V—and run it at lower than
normal r/min—to reduce U_1—thus keeping the value of V_{rel} abnormally
low. This procedure requires the use of more than the usual number of
stages for a given head requirement and sometimes even requires the
use of an abnormally large frame size for the flow handled.

17.2.7   Inducer impeller
increases head output

Much development work has been done in recent years toward the goal
of running impellers at normal speeds on heavy gases to reduce hardware costs to those incurred in the compression of light gases. One
approach has been to use inducer impellers, as in Fig. 17.8. The blades
on this impeller extend down around the hub radius so that the gas first
encounters the blade pack while flowing axially. Figure 17.8 shows the
vector analysis at the inducer outer radius. Assuming that the inducer
radius is the same as the leading edge radius of the conventional radial
inlet impeller, the vector geometries of the two are identical.

The advantage of the inducer lies in the fact that as we move radially
inward along the blade leading edge, the value of U_1 (and, therefore, of
V_{rel} and Mach number) decreases. As we move along the leading edge of
the *conventional* impeller, the vector geometry remains essentially constant. It can be seen, therefore, that while *maximum* Mach number for

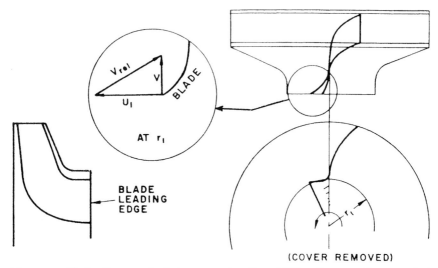

Figure 17.8   Vector diagrams for inducer-type impellers. (*Elliott Company, Jeannette, Pa.*)

the two styles is the same, the *average* Mach number for the inducer is *less,* for a given flow and speed. The inducer impeller can therefore be run somewhat faster, resulting in greater head output. The big disadvantage of the closed inducer impeller lies in the difficulty of fabrication. It is obviously more difficult to weld the longer and more curved blade path of an inducer impeller, than that of a conventional impeller. Other disadvantages are greater weight and the greater axial space requirement of the inducer impeller over that of the conventional impeller.

Another method of obtaining increased head output for a given Mach number is the reduction of backward lean. This expedient, however, has some disadvantages, not the least of which is the flatter curve that results.

17.2.8   Surge

Earlier in this text, surge was discussed from the point of view of compressor control. But, having discussed *basic slope* and *choke,* we are left with a vector-based discussion of minimum flow, or *surge.*

Surge flow has been defined by some as the flow at which the head-flow curve is perfectly flat and below which head actually decreases. This definition has a certain appeal, because straddling a surge flow so defined are myriad pairs of flow values producing identical heads—leading one to conjecture that the flow value is actually jumping back and forth between some such pair. However, since numerous centrifugal stages have been observed to run smoothly at flows below such a rate and others to surge at flows above such a rate, this definition must be considered imperfect at best. We must recall that unlike choke flow, which primarily hurts aerodynamic performance, surge can be quite damaging to a compressor and should be avoided. The higher the pressure level involved, the more important this statement becomes.

To understand what causes surge in a conventional stage, we must refer back to the tip vector geometry of Fig. 17.4. Because flow is reduced while speed is held constant, the magnitude of V_r decreases in proportion, and that of V_t remains constant for radial blades or increases for backward-lean blades. As flow decreases, therefore, the value of flow angle α decreases. In the normal parallel wall vaneless diffuser, this angle remains almost constant throughout the diffuser, so the path taken by a "particle" of gas is a log spiral in Fig. 17.9. The reason angle α remains constant in a parallel wall diffuser is that both V_r and V_t vary inversely with radius—V_r because radial flow area is proportional to radius and V_t because of the law of conservation of momentum.

It is evident from Fig. 17.9 that the smaller the angle α, the longer the flow path of a given gas particle between the impeller tip and the diffuser outer diameter. When angle α becomes small enough and the diffuser flow path long enough, the flow momentum at the walls is dissipated by friction to the point where pressure gained by diffusion causes a reversal of flow, and *surge results*. The angle α at which this occurs in a vaneless diffuser has been found to be quite predictable for various diffuser-impeller diameter ratios. The flow (and angle α) at which surge occurs can be lowered somewhat by reducing diffuser diameter but at the cost of some velocity pressure recovery.

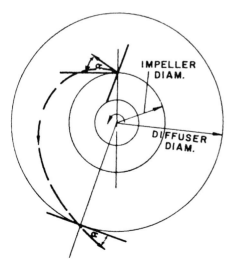

Figure 17.9 Spiral path taken by a molecule of gas in an impeller. (*Elliott Company, Jeannette, Pa.*)

17.2.9 Vaned diffusers

Before we discuss the foregoing in more detail, let us briefly discuss *vaned* diffusion, a device sometimes used in high-performance machines. Figure 17.10 shows the configuration of diffuser vanes. The vanes force the gas outward in a shorter path than unguided gas would take but not so short a path as to cause too rapid deceleration with consequent stream separation and inefficiency.

The leading edge of the diffuser vane is set for shockless entry of the gas at approximately design flow. It is evident that at flows lower than design, the gas impinges on the diffuser vanes with positive incidence. Conversely, at flows higher than design, negative incidence prevails. In a typical high-speed high-performance stage, positive incidence at the leading edge of the diffuser vane triggers surge on decreasing flow. On increasing flow, negative incidence at the inducer vanes can cause choking before impeller-inlet stonewall is reached. In spite of this disadvantage, vaned diffusion is sometimes used for air and certain other gases because stage efficiency is improved 2 to 3 percent. The short-flow range problem can, of course, be alleviated by making the diffuser vanes adjustable.

17.2.10 Vaneless diffusers

Having made our obeisance to vaned diffusion, let us return to the more common vaneless diffuser. We have seen that when V_r and α become too small, we will have surge. What can we do if our parameters are such that we are faced with a low value for α at design flow? We can artificially increase V_r and α by pulling our diffuser walls together, until α reaches the proper value at design flow. This brings to light an important

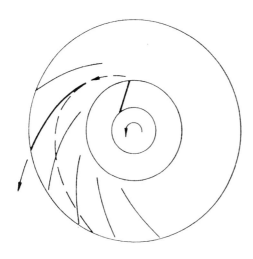

Figure 17.10 Vaned diffuser surrounding backward leaning impeller vanes. (*Elliott Company, Jeannette, Pa.*)

distinction: head output, as discussed earlier, is controlled by vector geometry in the *impeller tip* largely irrespective of what happens in the diffuser. Surge point is controlled by vector geometry in the *diffuser,* largely irrespective of what occurred in the impeller. In the common case where impeller tip width and diffuser width are the same (Fig. 17.1), the two sets of vector geometry are the same—ignoring impeller blade solidity. If such a stage has poor stability, it is frequently possible to lower the surge point by narrowing the diffuser without markedly changing basic slope or choke flow. This procedure can be carried only so far, however, because extreme positive incidence at the impeller inlet will eventually trigger surge regardless of diffuser geometry.

Just as we did when discussing choke flow, let us look at the effect of heavy gas compression on surge point. Since a heavy gas is compressed more at a given speed than is a light gas, it is evident that the critical value for α will be reached (on decreasing flow) at a higher *inlet* flow of heavy gas than that of a light gas. A given stage, therefore, has a higher surge flow, or a lower stable range, when compressing heavy gas than when compressing light gas at the same speed.

By similar reasoning, a given stage compressing a given gas at varying speed will surge at somewhat different inlet flows than those predicted by fan law. When speed is 10 percent above design speed, for instance, surge flow will be *more* than 10 percent higher than surge flow at design speed. When speed is 90 percent of design, the stage will surge at *less* than 90 percent of design-speed-surge flow.

We have discussed in some detail the three ingredients involved in a centrifugal compressor characteristic—basic slope, choke, and surge. We have considered how various physical design parameters such as backward lean, inlet blade angle, and diffuser flow angle affect these ingredients. We have also discussed the effect on characteristic shape of compressing different weight cases in a given stage. Now let us consolidate these bits and pieces into an overall look at the characteristic of a given stage used on various gases at various speeds.

In Fig. 17.11, we plotted the head-flow characteristic for a given conventional stage running at a given speed on various gases. There should be no surprises here, because we have discussed all the effects shown. Now if we divide the ordinate H by the speed squared, and the abscissa Q by speed, we have the same qualitative set of characteristics, except heavy gas becomes high speed, and light gas becomes low speed in Fig. 17.12. This figure illustrates departure from fan law.

17.2.11 Equivalent tip speeds

It is possible and quite appropriate to express Figs. 17.11 and 17.12 as a single plot. To do so, it is only necessary to use some nomenclature

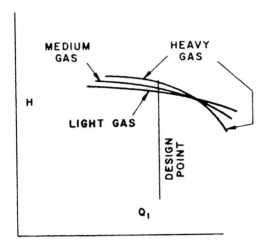

Figure 17.11 Head-flow characteristics of a given stage operating on different molecular weight gases. (*Elliott Company, Jeannette, Pa.*)

that includes both speed and type of gas. A convenient method of doing this is to multiply tip speed by the ratio of some reference acoustic velocity to actual gas acoustic velocity and call the resulting number *equivalent tip speed* in Fig. 17.13.

The reference acoustic velocity normally used is air at 80°F (27°C), since this is the gas on which components are usually tested. Using the equivalent tip speed concept permits the use of a single set of curves to describe the characteristic of a *given* stage when compressing any gas at any speed.

By way of example, we mentioned earlier that the sonic velocity of air at 80°F is 1140 ft/s and that of propylene at −40°F is 740 ft/s. If a stage is running at a tip speed of 780 ft/s on propylene, the *equivalent* tip speed is $780 \times 1140/740$ or 1200 ft/s (366 m/s). The stage characteristic

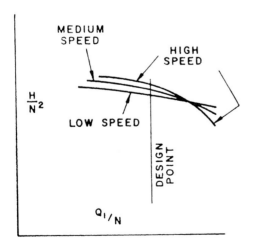

Figure 17.12 Speed-based performance curves illustrating departure from fan laws. (*Elliott Company, Jeannette, Pa.*)

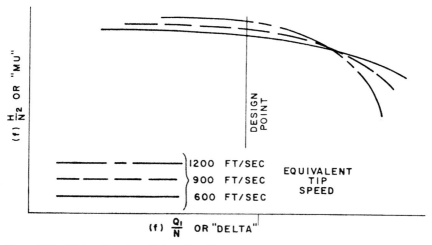

Figure 17.13 Plots of head coefficient ("mu") versus flow coefficient ("delta") allow prediction of stage performance with different gases. (*Elliott Company, Jeannette, Pa.*)

shape obtained at 780 ft/s tip speed on propylene is, therefore, the same as that obtained at 80°F air at 1200 ft/s (366 m/s) tip speed, but the head output of the latter is of course much higher.

The ordinate H/N^2 is commonly adjusted by some constants and called *head coefficient* or *mu*. The abscissa Q/N is likewise adjusted by constants and called *flow coefficient* or *delta*. Fig. 17.13 is in actuality, then, the common mu-delta curve used by some compressor investigators to predict stage performance. Others call the head coefficient "phi" (ϕ), and the flow coefficient "psi" (ψ); see Chap. 19.

17.3 Conclusions

Let us now review in practical terms just what we have learned. We now know that if conventional machinery is run at the usual tip speeds of 800 to 900 ft/s (244 m/s to 274 m/s) on propylene or heavier gases, the characteristic shape will be quite flat between design and surge, the stable range will be low, and the overload capacity almost nil. We can also see that even between 700 and 750 ft/s (213 m/s and 229 m/s) tip speed—the usual range selected for propylene, propane, and butane—we simply cannot expect the traditional 40 to 50 percent stable range and generous overload capacity obtainable on air machines. To obtain such a characteristic, it would be necessary to run between 500 and 600 ft/s (152 m/s and 183 m/s), which would almost double the number of stages required for a given amount of compression! The 700 to 750 ft/s range normally used is obviously a compromise between practical economics and desirable characteristic shape and range.

It is hoped that the foregoing has helped in a small degree to close the gap between the specialist and the generalist without too greatly offending the former or too greatly confusing the latter. Compressor performance is a difficult subject at best, and even today new insights are being gained through continuing development programs.

18

Applying Multiple Inlet Compressors*

The preceding segments of our text gave a thorough introduction into compressor performance prediction. In essence, we dealt with *single-inlet* machines. However, applying, analyzing, designing, and testing *multiple inlet* compressors may differ substantially from typical single-inlet centrifugals. Since multi-inlet, or sideload compressors are, nevertheless, quite common in the user industry, we will attempt to highlight certain aspects of this type of compressor that are mutually important to the user, contractor, and vendor. Further, an understanding of the sideload compressor is essential to provide a matched compressor and process installation.

18.1 Critical Selection Criteria

Careful consideration of all operating parameters is required to ensure a satisfactory compressor and process fit. Unfortunately, these parameters cannot be considered independently, but rather an overall operating analysis is required. This may result in certain operating parameters (or operating levels) being "desired" rather than being "critical" to successful operation of the overall process.

The common design parameter in API 617, the ASME code, and other codes may require some modification when applied to the sideload compressor. The usual guarantee of flow, discharge pressure, and

* Kenneth L. Peters (Elliott Company, Jeannette, Pa.), as published in *Hydrocarbon Processing,* May 1981. Adapted by permission of Gulf Publishing Company, Houston, Tex.

consumed horsepower likely will not ensure a proper compressor and process match.

The following will attempt to point out some parameters that will have a direct effect on compressor selection. This summary is not presented as absolute but rather to demonstrate overall analysis of the specified operating parameters. This discussion includes:

- Head rise to surge, surge margin, overload margin
- Head per compression section
- Compressor parasitic flows (i.e., balance piston leakage)
- Excess margins on other process equipment

18.1.1 Head rise to surge, surge margin, overload margin

Over the last few years, process engineers have asked for a characteristic curve shape guarantee including *head rise to surge* (HRTS), surge margin and/or overload margin. Other common terms referring to HRTS include pressure rise to surge, pressure ratio rise to surge, etc. Depending on other parameters specified, this addition to the guarantee may have no effect or may result in nonoptimum compressor selection. The typical compressor characteristic map shown in Fig. 18.1 illustrates this point.

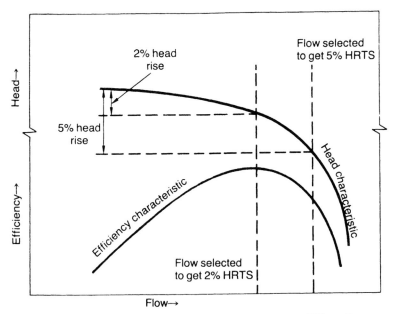

Figure 18.1 Effects of two and five percent head rise to surge. (*Elliott Company, Jeannette, Pa.*)

In this example, it is evident that a desirable level of 5 percent head rise to surge will result in nonoptimum efficiency and overload, while a 2 percent level will give the best efficiency and overload selection.

Another area of interplay on HRTS is with surge margin. The refrigeration process (the widest application of sideload compressors) requires operation at nearly constant discharge pressure (Fig. 18.2). Either driver speed or suction pressure must reduce as flow moves toward surge. Figure 18.3 shows an actual example where a 15 percent surge margin was specified with 5 percent HRTS. Further, the unit had a constant-speed drive with a trip-out if suction pressure dropped below atmospheric pressure. Design suction pressure was 15 psia.

With the required 5 percent HRTS, the unit would trip off stream at 97 percent design flow. Unfortunately, even 2 percent would lend the surge margin stipulation immaterial.

Because of various aerodynamic laws, the surge to stonewall flow range at the conditions of selection for refrigeration service is reduced, relative to that seen on various other less severe applications. While other applications may have overall flow ranges of 40 to 50 percent, range on a refrigeration selection is likely to be 20 to 30 percent. Hence, imposition of an excessive HRTS and/or surge margin criterion may result in only minimal overload capacity, as indicated in Fig. 18.3.

Conversely, an excessive overload margin stipulation may result in too low a HRTS or surge margin for safe, reliable, efficient operation.

18.1.2   Head per section

Another important operating parameter that must be evaluated is the required head per section split on the compressor.

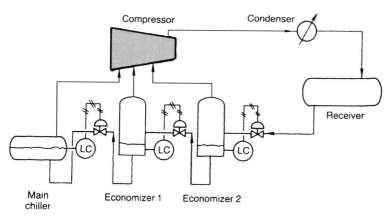

Figure 18.2 Simplified multilevel refrigeration process. (*Elliott Company, Jeannette, Pa.*)

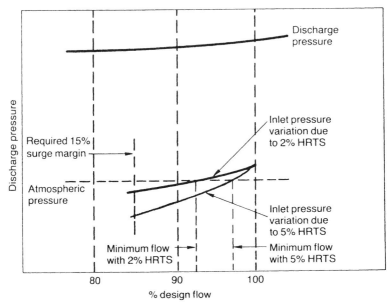

Figure 18.3 Effects on pressure for head rise to surge (HRTS) variations (constant-speed drives). (*Elliott Company, Jeannette, Pa.*)

Requests for smallest possible bearing spans may create or aggravate various aerodynamic considerations. The fewer impellers per section, the higher the head required per compressor stage. Higher head per stage requires increased rotative speed, resulting in operation at higher Mach numbers. Operation at higher Mach number levels tends to restrict the operating range, flattens the head rise characteristics, and reduces efficiency.

Operation at lower Mach levels is achieved by increasing the compressor stages per section and lowering rotative speed. Rotor dynamics criteria and hardware costs limit this approach.

While there is an optimum sidestream pressure level, as defined by the cycle, the pressure level may require only slight adjustment to allow better compressor selection.

18.1.3 Compressor parasitic flows

Reentry of seal equalizing line flow into the main gas stream can also influence compressor selection. If these flows are significant, reentry at a point other than the main inlet may be advantageous so that the sensitivity or tolerance effect on the operating range and efficiency may be reduced. If this option is taken, the manufacturer must take care that its effect on other aspects of the overall mechanical compressor design has been adequately considered. This effect can best be seen by reviewing Fig. 18.4.

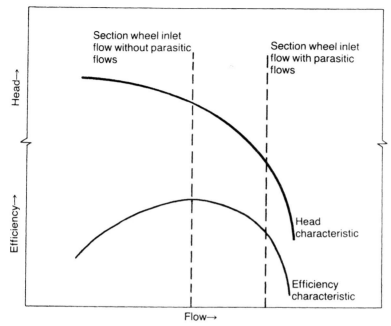

Figure 18.4   Possible performance shift due to parasitic flows in a multistage compressor. (*Elliott Company, Jeannette, Pa.*)

One can see that, for the same flow entering the main compressor inlet flange, the true operating point on the section characteristic may be significantly different. This problem is readily alleviated by, for instance, putting the parasitic flow into the first sideload. If the parasitic flow is, say, 5 percent of the first section, it may only be 1 to 2 percent of the overall flow (also consumed horsepower) entering the first wheel of the next section downstream of the sideload. Hence, variations due to balance piston seal leakage are negligible.

18.1.4   Excess margins on other process equipment

Design of excess margin into the various process components is also a very important consideration. For example, assume the process designer allows 5 percent excess in his specification to the contractor, the contractor takes in an additional 5 percent, and the compressor vendor elects to add 2 percent in flow. The process design flow now is approximately 88 percent of the compressor design flow. Operation at the process design flow may greatly reduce the stability margin of the compressor and in turn, the overall cycle.

Additionally, consider the situation where a constant-speed driver is used. The unit may be operating below atmospheric pressure at the inlet, as a result of the margins used in specification of the machine. (This is similar to the situation shown in Fig. 18.3 earlier.)

A cumulative accounting of all excess margins will aid in realizing successful compressor selection.

18.1.5 Representing compressor performance

In the past, typical performance for sideload units has been judged on a set of individual sectional curves based on constant inlet conditions to each section. These curves, however, have no direct relation to the way the unit will operate with a process.

Compressor performance should be presented as individual sectional curves in conjunction with a graph of expected performance based on the type of operation expected in the field. This *constant turndown* map easily can be generated from the individual sectional curves after making assumptions on mode of control, temperature, and other operating conditions.

This map is generated as follows from the sectional curves. Discharge or condenser pressure is assumed constant at design value. The sectional curves then are used to determine exact sideload pressures as a function of mass flow to each section. Normally, a constant turndown in mass flows is assumed. The net result is a characteristic curve for the compressor showing each sideload pressure (and inlet pressure) as a function of mass flow. Speed or inlet pressure (constant-speed units) are varied to meet design discharge pressure requirements. This provides a curve of the actual mode of operation of the unit.

Part load and overload operation data, at least initially, are generated by reducing or increasing all incoming mass flows by the same percentage. Also, initially, design temperatures are assumed constant. A minor modification can account for temperature variations at each inlet due to the flow or pressure change. Figures 18.5 and 18.6 are presented to show constant-speed and variable-speed applications, respectively.

The turndown map, if generated with input from the customer, can be used directly in a simulation. A good simulation analysis provides a much better evaluation of how a unit would operate in the field than only sectional curves. Review of the turndown map will be a great aid in analyzing interplay of requested levels of operating parameters.

Two important items observed directly from the map are pressure levels over the expected operation range of the system and potential stability of the variable-speed driver.

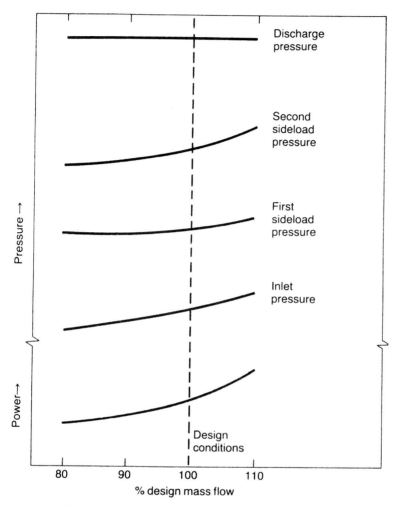

Figure 18.5 Typical turndown map with constant-speed driver. (*Elliott Company, Jeannette, Pa.*)

18.1.6 Practical levels of critical operating parameters

Because no code exists to establish recommended design parameters and limits for sideload units, the following list is recommended as a basis from which pertinent discussions can develop.

- Overall unit horsepower: This parameter can be made in accordance with the code as ± 4 percent, but because of limitations on test instrumentation and calculations, some assumptions may need to be included in the calculations.

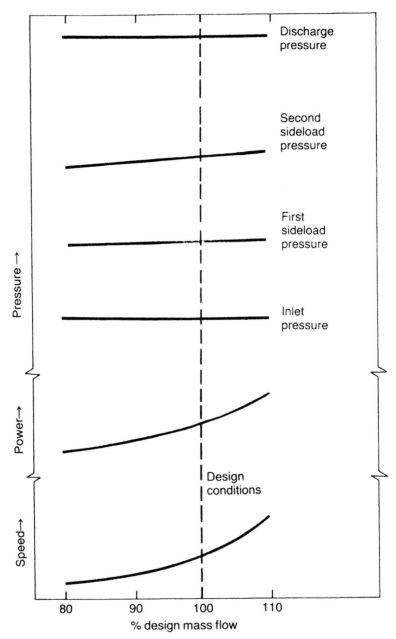

Figure 18.6 Typical turndown map with variable-speed driver. (*Elliott Company, Jeannette, Pa.*)

- Sectional head rise: This particular parameter should be set as low as possible both for constant-speed units and for variable-speed applications. Based on experience, a number as low as 2 to 3 percent overall can be adequate in most applications, assuming integration of the controls has been done properly.

- Overload and stability margins: These quantities are interrelated and collectively should be about 20 to 30 percent of the design flow value. Minimum margins should be 15 percent for surge and 5 percent for overload, or 20 percent combined range.

- Sideload pressure level: This particular parameter is usually the most unstated requirement on the sideload compressor. A reasonable tolerance is ±2 percent on the section head. Individual process designs may require these to be altered.

- Other desired tolerances or requirements need to be reviewed during the application phase of the compressor so that all parties realize their implications.

18.2   Design of the Sideload Compressor

Design of each sideload compressor application is dependent on the required parameters discussed previously. However, there are several design areas of a general nature that need to be highlighted.

18.2.1   Mixing area

One of the more complex design areas is in adequate calculation and physical occurrence of the mixing of two streams internally in the compressor. An analysis using conservation of mass, momentum, and energy provides the calculated properties of state following the interaction of the two streams. Any loss due to the mixing process is assumed to be a function of the amount of momentum lost by the main gas stream during the interaction. Also, the mixing process is assumed to occur at a constant static pressure (Fig. 18.7). That is to say, flowing conditions at station M are obtained by combining conditions of the main and sidestreams at station I and SL, using conservation of mass, momentum, and energy.

This analysis is a function of momentum. Thus velocities of the two streams at the point of mixing are critical. Close attention to the area ratio of both incoming streams is a must for optimum mixing.

A direct result of improper area control in the mixing chamber is that it may create different operating characteristics depending on location of the measuring equipment. With the static pressures of both streams equal at mixing, the velocities determine total pressure of

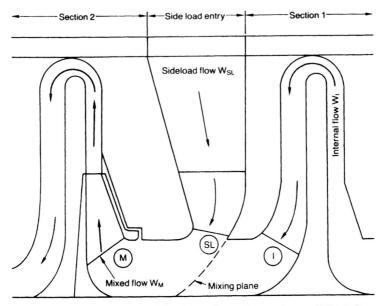

Figure 18.7 Cross section of sideload into the main gas stream. (*Elliott Company, Jeannette, Pa.*)

each stream. It is possible that total pressure calculated at station I will be higher than total pressure calculated at the sidestream flange. Hence, if process pressure controls are connected at the compressor flanges, it is possible the compressor will not be monitored accurately. Figure 18.8 depicts this problem graphically.

In Fig. 18.8, internal performance is measured from the mixing plane to the exit of the section (station I in Fig. 18.7). External flange-to-flange performance is measured at the respective compressor flanges. To match internal performance with external performance, control of the mixing areas is a must.

18.2.2   Aerodynamics

Sideload compressor applications are normally used in refrigeration systems that combine high molecular weight and low temperatures. This results in operation at higher Mach numbers. From the equation:

$$a_s = \sqrt{KgRT_s} \qquad (18.1)$$

where a_s = sonic velocity
   K = ratio of specific heats
   g = gravitational constant
   R = universal gas constant
   T_s = static temperature (°R)

As molecular weight goes up and T_s decreases, sonic velocity reduces. Hence, for the same relative velocity of the flowing gas, the relative inlet Mach number is higher than that for, say, air at standard conditions flowing at the same rate.

Notice that relative inlet Mach number is used—not simply Mach number. Care must be taken when reviewing Mach numbers that all values are based on the correct gas velocity. Very simply, the two Mach numbers usually referenced are *inlet Mach number* and *relative inlet Mach number*. Inlet Mach number is typically calculated as the Mach number as flow enters the impeller, measured either at the impeller eye or at the leading edge of the blade passage (Fig. 18.9).

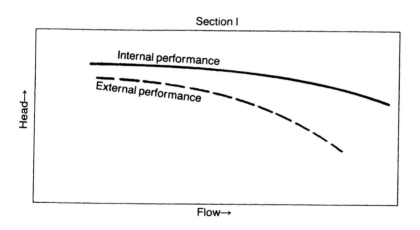

Section I

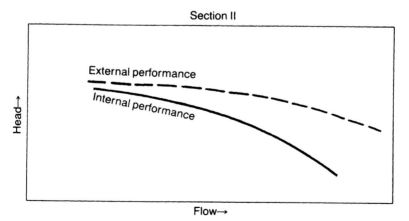

Section II

Figure 18.8   Possible effects of improper mixing area control as a function of measurement location. (*Elliott Company, Jeannette, Pa.*)

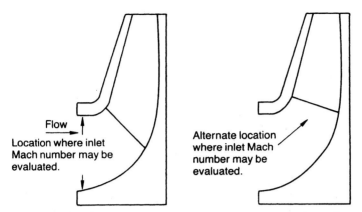

Figure 18.9 Locations sometimes used to measure inlet Mach number. (*Elliott Company, Jeannette, Pa.*)

In either location the area is unobstructed. Velocity is typically derived by dividing the inlet flow rate by the area, without considering the rotating disk or blade blockage.

Conversely, relative inlet Mach number denotes the Mach number of the flow stream perpendicular to the throat section of the impeller (Fig. 18.10).

Needless to say, relative inlet Mach numbers are higher than inlet Mach numbers because net throat area is less than area of the impeller eye or at the leading edge of the blade passage.

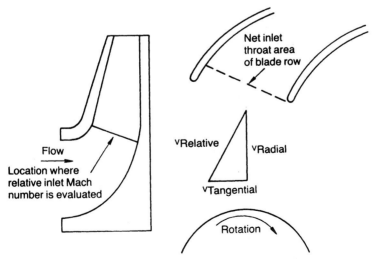

Figure 18.10 Impeller geometry for determining relative inlet Mach number. (*Elliott Company, Jeannette, Pa.*)

Proper application and basic knowledge of proven aerodynamic hardware have permitted successful operation of stages at relative inlet Mach numbers in the area of 0.95 and above.

As operating Mach numbers increase, the stable operating flow range decreases and the compressor characteristic curve flattens (Fig. 18.11). *Flattening* is the result of stage component efficiency curves displaying a sharper peak, with the peak moving toward higher flow coefficients with increased Mach numbers. The result is that head capability at part load is reduced relative to design, and the curve displays a lower slope.

With increasing Mach number, the maximum overload (choke or stonewall) flow coefficient moves steadily toward the design value. Similarly, on the low-flow end, the surge flow coefficient moves toward the design value. Very simply, the higher the Mach number, the smaller the stable operating flow range.

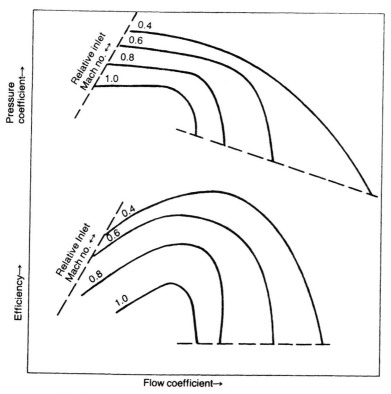

Figure 18.11   Effects of relative inlet Mach number. (*Elliott Company, Jeannette, Pa.*)

18.2.3   Temperature stratification

An area of concern is temperature stratification, both radial and circumferential. A good inlet and mixing area design will help minimize circumferential stratification; however, radial stratification will exist to some extent.

18.3   Testing

The acceptance performance test serves as a confirmation of all of the care taken in the application, design, and manufacture of the sideload unit. The ideal method needed to ensure this confirmation is to perform a controlled ASME-type test in actual service. However, for a number of reasons this is usually impractical. Thus a modified ASME equivalent performance test may be completed in the compressor vendor's shop prior to shipment.

The ASME equivalent performance test for sideload units, per se, is not clearly defined. In the absence of concise procedures, each section is generally regarded as an individual compressor and tested accordingly with the code applied as nearly as possible. However, some general guidelines should be highlighted to help clarify testing procedures.

18.3.1   Test setup

In actual service each sidestream typically originates at a distinctly different physical location, giving individual properties of state at each entry. To duplicate this condition in the vendor's shop would be extremely costly. In lieu of this, a simplified typical setup for a two sideload compressor is shown in Fig. 18.12.

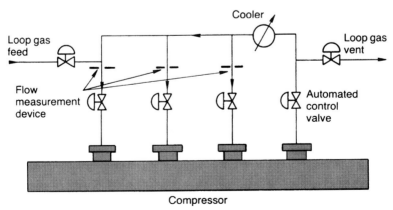

Figure 18.12   Typical test loop. (*Elliott Company, Jeannette, Pa.*)

Variations to the setup exist, but the basic closed loop concept is used throughout the industry, with control valves used to control pressures at the various inlets. Under normal conditions a bleed-feed system must be used to maintain loop test gas purity and pressure control. This system, or ones like it, have proven adequate over the past decades.

18.3.2   Instrumentation

Crucial to validity of a shop sideload performance test is the required instrumentation and its proper installation. The following general guidelines on instrumentation have been proven over the past several years and are recommended for general acceptance.

The ASME code provides guidelines for instrumenting the external flanges of the compressor and the flow measuring sections. This instrumentation will provide adequate readings at the compressor flanges and flow-measuring devices but will not suffice to obtain information on sectional performance or overall unit horsepower consumption. Actual performance depends on the properties of state of gas exiting each section and thus is not directly available from flange readings alone. Internal instrumentation is required, or assumptions must be made on temperature rise across each section. As methods of instrumenting sideload compressors improve, this guide may change.

18.3.3   Testing procedure

First, as nearly as possible, sidestream and main inlet flows should be held proportionately equal throughout the test and as close as possible to those anticipated in actual service. Failure to follow this approach may result in conflicting data. Because of varying splits between the sidestream and main gas stream, it is possible to achieve two different levels of performance for the same sectional inlet flow.

Just as important as the testing method is stabilization of individual testing points. Time is required at each testing point to permit all system components to become normalized because of heat transfer and other transients, before data are taken. Given constant inlet conditions, stabilization can be assumed when three consecutive readings of discharge pressures and temperatures taken at 3-min intervals, conform within 2 percent. A good practice to verify stabilization of the first point after each start-up of the test is by repeating it after a minimum 5-min interval. Subsequent readings can be taken when stabilization has been verified.

Because stabilization is a function of mass flow rate, equipment size, total heat capacity of the metal, curve shape, internal leakages, test loop configuration, and heat transfer, it cannot be assumed simply to be based on time. Properties of state of the flowing medium must be satisfied as constants.

18.3.4 Accuracy of test results

One question remains on testing, "How accurate are the test results?"

Testing of the entire unit at one speed may not give absolute results because of slight compromises required by code guidelines on volume ratio, Mach number, and Reynolds number. Some believe testing each section at a given speed or testing partially stacked rotors is a solution. Results of such testing, however, have not shown a sufficiently higher degree of accuracy to warrant the added expense and time.

18.3.5 Evaluation of results

Individual sectional curves do not depict a reasonable evaluation of the units performance during selection. Likewise, test results cannot be evaluated as individual sections. The sectional curves must be converted to a turndown map before any rational conclusion about overall unit acceptability may be discussed. The turndown map must be used to evaluate results of the sideload test.

Testing a sideload compressor is indeed complicated and involved and requires more expertise than the typical equivalent performance test. Testing requires a total understanding between the witness and the vendor.

The application, design, testing, and analysis of a sideload compressor all differ substantially from a typical centrifugal compressor. Failure to seriously consider the uniqueness and the complexity of the unit and process interaction—all the way from initial process design through to the actual field installation—may result in an unfavorable installation.

Predicting Compressor
Performance at New Conditions*

The performance of centrifugal compressors is generally documented during shop tests or during field tests at the installation site. Accurate test data will make it possible to predict compressor performance with different gas or operating characteristics at some future date. Moreover, since centrifugal compressor impeller dimensions are often custom-made for a specific operating condition, adjustments may be needed to meet a specific performance.

It is often necessary to arrive at a new speed for variable-speed driven machines or trim the impellers for motor-driven machines, if there is a considerable gap between proposed performance and test stand performance. Similarly, compressor owners may want to determine how a machine would perform with a different gas or with some new operating parameters. Dimensionless numbers provide a convenient way to predict off-design performance.

19.1   How Performance Tests Are Documented

Knowing the performance of an existing compressor is a prerequisite for determining performance at new conditions. However, not all compressor manufacturers document performance test results in the same format. Many of them now use computer programs to record or display the results. Still, in some places they resort to manual calculations that are then displayed in curve shape.

---

* Developed by Arvind Godse and originally published in *Hydrocarbon Processing,* June, 1989. Adapted by permission of Gulf Publishing Company, Houston, Tex.

Typical examples of performance curve presentation found after surveying a few performance reports show the combinations shown in Table 19.1.

19.2   Design Parameters: What Affects Performance

To judge the performance of a machine at off-design conditions, one must know which factors are of prime importance.

Examining a compressor is facilitated by understanding the governing laws of similarity, as well as through use of dimensionless numbers. This will provide insight as to how a machine can be adapted to varying demands such as load changes, unforeseen changes in thermodynamic properties, or trying to assess the feasibility of using the existing machine for a different application. When a machine is sold, the basic information on the impeller geometry often remains proprietary. However, using just three rules of thumb and understanding the dimensionless coefficients highlighted earlier in our text will allow us to closely approximate machine behavior:

- Head of a compressor varies as the square of tip speed.

- Flow handled by the compressor varies directly with tip speed and impeller diameter.

- The entire design concept of a centrifugal machine can be separated into two areas from a manufacturer's point of view: thermodynamic and mechanical.

19.2.1   Thermodynamic

For the proposed gas, with thermodynamic properties, flow rate and pressure ratios defined by the original user, the manufacturer selects the right impeller geometry. Here, the manufacturer will invariably employ dimensionless numbers versus pressure coefficient ψ, flow coef-

TABLE 19.1   Performance Curve Displays

Sequence	X axis	Y axis
1	Inlet volume (ICFM)	Discharge pressure
2	Inlet volume (ICFM)	Polytropic efficiency
3	Inlet volume (ICFM)	bhp
4	Inlet volume (ICFM)	Pressure ratio
5	Inlet volume (ICFM)	Power ratio
6	Inlet volume (ICFM)	Polytropic head
7	Mass flow lb/min	Pressure ratio
8	Mass flow lb/min	bhp

ficient ϕ, and their relationship with polytropic efficiency η_p. This will help in selecting impeller geometry, number, and configuration. While the manufacturer obviously uses these parameters for component selection, we, likewise, can use them to predict the performance of the machine at other conditions. These dimensionless numbers show the relationship between head, flow, tip speed, and efficiency for a selected blade geometry. They are rarely presented as standard information. However, one can calculate them and plot their graphs showing their relationship to performance test data. The simplicity of this exercise will be appreciated when the investigating engineer has many machines in his plant. And since the theory behind all this has been explained earlier, the rest of this write-up describes the associated formulas and provides an example.

19.2.2   Mechanical

After selecting the impellers, the manufacturer will consider the following:

- Material selection
- Type of split line orientation and casing configuration
- Rotor layout, depending on the kind of casing design and number of impellers. This will involve bearing span, bearing sizing, calculation of critical speeds, and making sure that they are away from the operating speed range. It will also require design decisions to finalize:

  - Seal design and seal system components
  - Lube system
  - Driver rating

From a *future* application point of view, one must have satisfactory answers to a number of questions before the decision can be made to operate an *existing* compressor under *new* conditions:

- Are the existing impellers able to meet flow and head requirements?
- Is the selected speed within an acceptable range for the existing machine?
- Is the casing pressure and type of split orientation satisfactory for the new application?
- Is the material selection of the installed machine acceptable?
- Is the seal system capable of handling the new application?
- Is the power requirement within the existing driver rating?

When one is trying to adapt the existing machine to changes in operating parameters within a narrow band, the answers will be satisfactory in most cases. The matter can be even simpler for a variable speed driver.

19.3   What to Seek from Vendors' Documents

There are many inspection records submitted by the vendor. It will be beneficial to reduce the data for all centrifugal machines bought through multiple sources to a common format. It is also important to know why reference is initially made to performance data obtained at the manufacturer's facility. Since the machine is new, passages are clean and clearances are per design recommendation. Therefore, these data would obviously help the user to track performance changes occurring with time.

19.3.1   Performance test data

As pointed out earlier, all manufacturers present the data in different ways. Therefore, arrange to segregate the data for five parameters that normally make up the performance test.

- Polytropic head
- Polytropic efficiency
- Inlet flow
- Mechanical losses
- Critical speed

The last two items are obtained from the mechanical running test. Mechanical run tests are mandatory for centrifugal compressors and are usually conducted prior to the optional shop performance test.
Other inspection records should be reviewed to ascertain:

- Number of impellers
- Individual impeller diameters

In addition, we need to know:

- Casing design pressure, test pressure—hydraulic or pneumatic and its split
- Type of seal and seal system pressure and flow
- Metallurgy of important parts
- Driver rating and speed range.

19.4   Illustrations and Example

Table 19.2, highlights our nomenclature and associated formulas. Next, an example will show the method employed for arriving at performance at new conditions. The results should always be compared with ratings and capabilities of the compressor and its driver and associated upstream/downstream equipment to verify feasibility.

Figure 19.1 represents a flowchart of the method and procedural sequence we apply to analyze performance of dynamic compressors at new conditions.

Example 19.1

Step 1: Purchase specification data or data known for existing compressor.

- Inlet conditions:

Pressure	540 psia (P_1)
Temperature	140°F (T_1)
Compressibility	0.901 (Z)
K	1.220
MW	22.23
Flowrate	2654 lb/min (G)

- Discharge conditions: Pressure   1330 psia (P_2)
- Rotor data: Impeller dia.   16.5 in (D_m and D_s)

  Number of impellers   5 (I)

TABLE 19.2   Nomenclature and Formulas

Description	Symbol	Formula
Molecular weight	MW	—
Adiabatic exponent	K	C_p / C_v
Polytropic efficiency	η_p	
Polytropic exponent	n	—
Polytropic compression exponent	x	$(n - 1)/n = (K - 1)/K\eta_p$
Gas constant	R	1544/MW
Compressibility	Z	—
Pressure ratio	r	P_2/P_1
Inlet temp., °F	T_1	—
Head, ft-lb/lb	H_p	$ZR(T_1 + 460)(r^x - 1)/x$
Speed, r/min	N	—
Mean impeller dia., in	D_m	—
Suction impeller dia., in	D_s	—
Number of impellers	I	—
Weight flow rate, lb/min	G	—
Suction volume, ft³/min	O_s	—
Flow coefficient	ϕ	$700\ Q_s/N(D_s)^3$
Pressure coefficient	ψ	$\dfrac{H_p(1300)^2}{IN^2D_m{}^2}$
Gas hp	W_G	$G\ H_p/33{,}000\ \eta_p$
Mechanical loss at speed N	W_M	
Mechanical loss at speed N_1	W_{M1}	$W_M(N_1/N)^2$
bhp at speed N	W	$W_G + W_M$

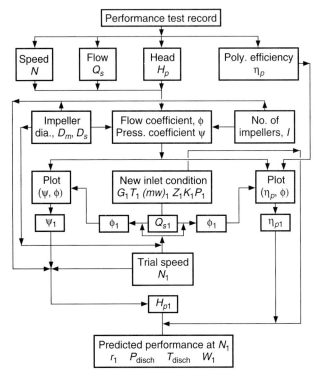

Figure 19.1 Flow diagram for reevaluating performance. (*Godse, Arvind, "Predicting Compressor Performance at New Conditions,"* Hydrocarbon Processing, *June 1989*)

Proposal speed	9600 (N)
Critical speeds	1st: 6,240
	2d: 15,280
Frictional hp at 9600 r/min	44

- Driver: Variable speed range 85 to 105%
       Rating: 4500 hp
- Performance test data at 9600 r/min

We now proceed to arrange available performance test data in the tabular format as follows. Alternatively, we could simply examine the compressor vendor's performance test curve and identify eight points on this curve. For each of these arbitrarily chosen points, we find corresponding head and flow values from the *X-Y* coordinates associated with this curve plot.

Let us suppose that we obtained the following:

Point	1	2	3	4	5	6	7	8
Head (H_p)	38,050	37,875	37,660	37,235	36,500	35,210	32,965	27,458
Flow (Q_s)	1,046	1,120	1,190	1,296	1,410	1,556	1,726	1,950
η_p	0.684	0.697	0.711	0.726	0.739	0.752	0.762	0.731

This now enables us to calculate corresponding values of ϕ and ψ for each of the eight points. We use the expressions listed in Table 19.2, or see reminder.*

Step 2:   Obtain the flow coefficient and head coefficient.

Point	1	2	3	4	5	6	7	8
ϕ	0.0169	0.0181	0.0194	0.0210	0.0228	0.02524	0.02800	0.03165
ψ	0.5125	0.5102	0.5073	0.5016	0.4916	0.4743	0.4440	0.3698

Step 3:   Plot the graph as shown in Fig. 19.2

Step 4:   New operating conditions:

- Inlet pressure: 560 psia (P_1)
- Inlet temperature: 130°F (T_1)
- Molecular weight: 24.45 (MW)
- Flowrate: 3000 lb/min (G)
- Other inlet conditions remain as in Step 1
- Required discharge pressure, $P_2 = 1330$ psia

To predict the performance of the compressor at the new operating conditions, take the following steps:

Step 5:   Find inlet volume [from $Q_s = ZGRT_1/\text{MW} (144) P_1$]

$$Q_s = \frac{0.901\ (3000)(1544)(130 + 460)}{24.45(144)(560)} = 1248.86 \text{ ft}^3/\text{min}$$

Step 6:   Since molecular weight is higher than the original case, a higher discharge pressure will be generated. Therefore, choose a lower speed.

Trial 1: Selected speed = 8900 r/min

Step 7:   Find flow coefficient ϕ at new inlet conditions and trial speed of 8900 r/min (from $\phi = (700)(1248.86)/(8900)(16.5)^3$.

$$\phi = 0.02186$$

---

* Please note again how, for instance, data for point #4 were calculated:

$\phi = 700Q_s/ND_s^{\ 3}$

$= (700)(1296)/(9600)(16.5)^3$

$= 0.0210$

$\psi = H_p (1300)^2/IN^2D_m^{\ 2}$

$= \dfrac{37{,}235 \times 1{,}690{,}000}{(5)(9216)(10^4)(272.25)}$

$= 0.5016$

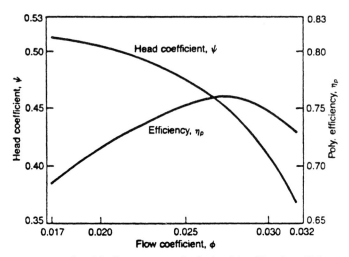

Figure 19.2   Graphically represented relationship of head coefficient, flow coefficient, and polytropic efficiency for a given compressor. (*Godse, Arvind, "Predicting Compressor Performance at New Conditions,"* Hydrocarbon Processing, *June 1989*)

From the graph in Fig. 19.2, find the corresponding:

- Polytropic efficiency, $\eta_p = 0.732$
- Head coefficient, $\psi = 0.497$

Step 8:   For the given impeller diameters and number of impellers, the head developed will be:

$$0.497 = \frac{H_p\,(1300)^2}{(5)(8900)^2(16.5)^2}$$

$$H_p = 31{,}709 \text{ ft·lb/lb}$$

Step 9:

$$\frac{(n-1)}{n} = \frac{(k-1)}{k\,\eta_p}$$

$$= \frac{0.22}{1.22(0.732)}$$

$$= 0.2463$$

Step 10:   Find pressure ratio r (from $H_p = ZRT(r^x - 1)/(x)(MW))$.

$$31{,}709 = \frac{(0.901)(1544)(590)(r^{0.2463} - 1)}{(0.2463)(24.45)}$$

$$\frac{P_2}{P_1} = 2.3378$$

Step 11: $P_2 = 560\,(2.3380) = 1309.17 \text{ psia}$

From the preceding it is clear that the speed needs to be increased, since the discharge pressure is less than 1330 psia. Let us choose 9000 r/min and find the results.

Step 12: Using the procedure as given in Steps 7, 8, 9, and 10, the following information is obtained:

$$\phi = 0.02162 \qquad \left[\text{from} \ \frac{(700)(1248.86)}{(9000)(16.5)^3} = \phi \right]$$

$$\eta_p = 0.731$$

$$\psi = 0.498$$

$$H_p = 32{,}491 \qquad \left[\text{from} \ \frac{(.498)(5)(9000)^2(16.5)^2}{(1300)^2} = H_p \right]$$

$$r = 2.3820$$

$$P_2 = 1334.0 \text{ psia}$$

The selected speed, 9000 r/min, is in order.

$$W_G = \frac{32{,}491(3000)}{33{,}000 \ (0.731)} = 4040.66 \qquad \text{from} \ \frac{(H_p)(G)}{33{,}000 \ \eta_p} = W_G$$

Step 13: Mechanical losses at 9000 r/min:

$$W_M = 44 \left(\frac{9000}{9600} \right)^2$$

$$= 38.67 \text{ hp}$$

Step 14:

$$\text{bhp} = 4040.66 + 38.67$$

$$W = 4079.33$$

Step 15: Discharge temperature from $T_{2,\text{actual}} = T_{2,\text{is}} = T_1 \ (P_2/P_1)^{(k-1)/k}$

$$T_{\text{DIS}} = 730.8°\text{R} \ (270.8°\text{F})$$

Properties of Common Gases

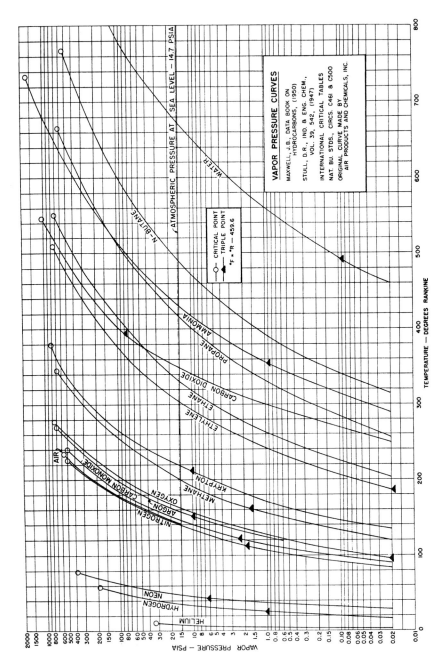

Figure A.1 Vapor pressure curves for common pure gases.

TABLE A.1   Properties of Hydrocarbon and Special Refrigerant Vapors

Gas	Chemical formula	Alternate designation	Molecular weight	Boiling point at 14.696 psia °F	Values at 14.696 psia & 60°F			Specific heat at constant pressure at 14.696 psia Btu/lb/°F (see notes)				Ratio of specific heats $K = C_p/C_v$ at 14.696 psia (see notes)				Molar heat capacity at 150°F and 14.696 psia Btu/°F/mole	Critical conditions	
					Specific gravity (air = 1.00)	Density lb/ft³	Specific volume ft³/lb	Minus 40°F	60°F	150°F	300°F	Minus 40°F	60°F	150°F	300°F		Temperature °Rankine	Pressure psia
Methane	CH₄	C₁	16.04	−259	0.555	0.0424	23.61	0.506	0.527	0.558	0.624	1.33	1.31	1.29	1.25	8.95	344	673
Acetylene	C₂H₂	—	26.04	−119	0.899p	0.0686p	14.58p	0.353	0.397	0.427	0.469	1.31	1.26	1.24	1.21	11.12	557	905
Ethylene	C₂H₄	Ethene	28.05	−155	0.969p	0.0739p	13.53p	0.312	0.362	0.406	0.478	1.29	1.24	1.21	1.17	11.39	510	742
Ethane	C₂H₆	C₂	30.07	−128	1.047	0.0799	12.52	0.365	0.410	0.458	0.543	1.22	1.19	1.17	1.14	13.77	550	708
Propylene	C₃H₆	Propene	42.08	−54	1.453p	0.1109p	9.021p	0.303	0.354	0.399	0.473	1.18	1.15	1.14	1.11	16.79	657	667
Propane	C₃H₈	C₃	44.09	−44	1.547	0.1180	8.471	0.333	0.389	0.443	0.534	1.16	1.13	1.11	1.09	19.53	666	617
Butadiene 1, 2	C₄H₆	—	54.09	+51	1.867p	0.1425p	7.018p	—	0.346	0.387	0.451	—	1.12	1.11	1.09	20.93	799	653
Butadiene 1, 3	C₄H₆	—	54.09	+24	1.867p	0.1425p	7.018p	—	0.341	0.392	0.468	—	1.12	1.10	1.09	21.26	766	628
Isobutylene	C₄H₈	—	56.10	+20	1.937p	0.1478p	6.766p	—	0.370	0.419	0.493	—	1.11	1.09	1.08	23.51	753	580
Butylene	C₄H₈	1-Butene	56.12	+21	1.937p	0.1478p	6.766p	—	0.355	0.406	0.484	—	1.11	1.10	1.08	22.78	756	583
Isobutane	C₄H₁₀	i-C₄	58.12	+11	2.068	0.1578	6.339	—	0.387	0.443	0.535	—	1.10	1.08	1.07	25.75	735	529
n-Butane	C₄H₁₀	n-C₄	58.12	+31	2.071	0.1581	6.327	—	0.391	0.444	0.532	—	1.10	1.08	1.07	25.81	766	551
Isopentane	C₅H₁₂	i-C₅	72.15	+82	2.491p	0.190p	5.262p	—	0.401b	0.439	0.529	—	1.07b	1.07	1.06	31.67	830	483
n-Pentane	C₅H₁₂	n-C₅	72.15	+97	2.491p	0.190p	5.262p	—	0.410b	0.441	0.528	—	1.07b	1.07	1.06	31.82	846	489
Benzene	C₆H₆	—	78.11	+176	2.697p	0.206p	4.860p	—	0.301b	—	0.360	—	1.09b	—	1.08	23.51	1012	714
n-Hexane	C₆H₁₄	n-C₆	86.17	+156	2.975p	0.227p	4.406p	—	0.443b	—	0.526	—	1.06b	—	1.05	38.17	915	440
n-Heptane	C₇H₁₆	n-C₇	100.20	+209	3.459p	0.264p	3.789p	—	0.474b	—	0.525	—	1.04b	—	1.04	47.49	973	397
n-Octane	C₈H₁₈	n-C₈	114.22	+258	3.943p	0.301p	3.324p	—	0.449b	—	0.524	—	1.04b	—	1.03	57.00	1025	362
Refrigerant 11 *	CCl₃F	—	137.38	+75	4.78b	0.365b	2.739b	—	0.134b	0.141	0.156	—	1.14b	1.13	1.10	19.37	848	635
Refrigerant 12 *	CCl₂F₂	—	120.92	−22	4.27	0.326	3.067	—	0.145g	—	0.526	—	1.14g	—	—	17.53g	694	597
Refrigerant 13 *	CClF₃	—	104.47	−115	3.62	0.276	3.624	0.133	0.150	0.164	0.183	1.17	1.15	1.13	1.12	17.13	544	561
Refrigerant 21 *	CHCl₂F	—	102.93	+48	3.63	0.277	3.608	—	0.136	0.148	0.169	—	1.18	1.16	1.13	15.23	813	750
Refrigerant 22 *	CHClF₂	—	86.48	−41	3.05	0.233	4.299	—	0.149	0.161	0.182	—	1.20	1.17	1.14	13.92	665	716
Refrigerant 113 *	C₂Cl₃F₃	—	187.39	+118	6.04b	0.461b	2.169b	—	0.159b	0.162	0.179	—	1.08b	1.08	1.07	30.36	877	495
Refrigerant 114 *	C₂Cl₂F₄	—	170.93	+38	6.08	0.464	2.155	—	0.157	0.168	0.188	—	1.09	1.08	1.07	28.72	754	474

For explanation of notes a, b, c, d, f, g, h, p, see Table A.2.

* This group of refrigerants is known by trade names such as Freon™, Genetron™, etc.

TABLE A.2 Properties of Miscellaneous Gases

Gas	Chemical formula	Alternate designation	Molecular weight	Boiling point at 14.696 psia °F	Specific gravity (air = 1.00)	Density lb/ft³	Specific volume ft³/lb	Cp −40°F	Cp 60°F	Cp 150°F	Cp 300°F	K −40°F	K 60°F	K 150°F	K 300°F	Molar heat at capacity at 150°F and 14.696 psia Btu/°F/mole	Critical Temperature °Rankine	Critical Pressure psia
Air (dry)*	—	—	28.97	−318	1.000	0.0763	13.106	0.240	0.240	0.241	0.243	1.40	1.40	1.40	1.39	6.98	239	547
Ammonia	NH₃	—	17.03	−28	0.594	0.0454	22.05	—	0.506	0.525	0.556	—	1.30	1.30	1.27	8.94	730	1639
Argon	A	—	39.94	−303	1.380	0.1053	9.497	0.125	0.125	0.125	0.124	1.67	1.67	1.67	1.67	4.99	272	705
Carbon dioxide	CO₂	—	44.01	−109	1.528	0.1166	8.576	0.189	0.201	0.213	0.254	1.34	1.30	1.28	1.25	9.37	548	1073
Carbon monoxide	CO	—	28.01	−312	0.967	0.0738	13.55	0.249	0.248	0.249	0.252	1.40	1.40	1.40	1.40	6.97	242	507
Chlorine	Cl₂	—	70.91	−30	2.48	0.1886	5.30	—	0.115	—	—	1.35	1.35	—	—	8.15d	751	1119
Ethylene oxide	CH₂CH₂O	—	44.05	+51	1.52	0.116	8.62	0.225h	0.264h	0.302h	0.355h	1.25h	1.21h	1.19h	1.15h	14.10	844	1043
Helium	He	—	4.003	−451	0.138	0.01054	94.91	—	—	1.248a	—	—	—	1.66a	—	5.00	†24	†151
Hydrogen	H₂	—	2.016	−423	0.0696	0.00531	188.32	3.324	3.409	3.442	3.462	1.42	1.41	1.40	1.40	6.94	†83	†327
Hydrogen chloride	HCl	—	36.47	−121	1.271	0.0970	10.31	—	0.194	—	—	—	1.41	1.41	—	7.08d	585	1200
Hydrogen sulphide	H₂S	—	34.08	−79	1.175	0.0897	11.15	0.233	0.238	0.243	0.251	1.34	1.33	1.32	1.30	8.28	673	1306
Methyl chloride	CH₃Cl	—	50.49	−11	1.777	0.1356	7.372	—	0.199f	—	—	—	1.29f	—	1.30	10.05f	749	969
Neon	Ne	—	20.19	−411	0.697	0.0532	18.81	0.246	0.246	0.246	0.239	1.66	1.66	1.66	—	4.97	80	385
Nitric oxide	NO	—	30.01	−240	1.038	0.0792	12.62	0.239	0.238	0.238	0.239	1.38	1.39	1.39	1.38	7.14	323	956
Nitrogen	N₂	—	28.02	−320	0.967	0.0738	13.55	0.249	0.249	0.249	0.250	1.40	1.40	1.40	1.40	6.98	227	492
Nitrous oxide	N₂O	—	44.02	−127	1.531	0.1168	8.56	—	0.21	—	—	—	1.30	—	—	9.24d	558	1054
Oxygen	O₂	—	32.00	−297	1.105	0.0843	11.86	0.218	0.219	0.221	0.226	1.40	1.40	1.39	1.38	7.07	278	732
Phosgene	COCl₂	—	98.92	+46	3.41	0.262	3.82	0.123	0.136	0.146	0.158	1.19	1.17	1.16	1.14	14.44	820	823
Sulphur dioxide	SO₂	—	64.06	+14	2.254	0.1720	5.814	—	0.147	—	—	—	1.25	—	—	9.42d	775	1142
Toluene	CH₃C₆H₅	—	92.13	+231	3.181p	0.243p	4.121p	—	0.346b	—	0.379	—	1.07b	—	1.06	31.87	1069	611
Water vapor	H₂O	Steam	18.02	+212	0.632b	0.0373b	26.80b	—	0.496b	—	0.55c	—	1.32b	—	1.31c	8.94	1165	3187

Values at 14.696 psia & 60°F (see notes). Specific heat at constant pressure at 14.696 psia Btu/lb/°F (see notes). Ratio of specific heats $K = C_p/C_v$ at 14.696 psia (see notes).

a—An average for 0–300°F.
b—At the boiling point.
c—Approximate average for 212–600°F and 14.7–200 psia.
d—At 60°F.
f—At 77°F.
g—At 86°F.
h—Within ±5%.
p—As a perfect gas.

* Normal atmospheric air contains some moisture. For convenience it is common to consider that, at 68°F and 14.696 psia, the air is at 36 percent relative humidity, weighs 0.075 lb/ft³, and has a k value of 1.395. (Based on ASME Test Code for Displacement Compressors.)
† These are effective values to be used only for generalized compressibility charts and gas mixtures. Actual values are

	T_c	p_c
Helium	9.7°R	33.2 psia
Hydrogen	59.7°R	188 psia

The generalized charts (Figs. A.2 through A.5) are redrawn by permission from those developed by Mr. L. C. Nelson and Prof. E. F. Obert and presented at the 1953 Annual ASME meeting. They were published in *Chemical Engineering* in July 1954, from which article Dresser-Rand (formerly Ingersoll-Rand) replotted these curves.

Four charts, based on a study of experimental data on 30 gases, have been prepared to cover a wide range of values.

Although steam (H_2O) and ammonia (NH_3) were considered, they do not coordinate well and since excellent tables and charts of their properties are available, their specific rather than generalized data should be used at all times.

Hydrogen and helium also cannot be correlated well with these charts, particularly below $T_r = 2.5$, unless *effective* or pseudo critical conditions are used in place of the *actual* critical conditions. Effective conditions are given below—for use *only* with generalized charts. These are as developed in 1960 by Dr. John M. Lenior, University of Southern California, Los Angeles, Cal. in the case of hydrogen and by Prof. Obert in his 1953 paper in the case of helium.

A.1   Effective Critical Conditions

Helium	$T_c = 24°R$	$p_c = 151$ psia
Hydrogen	$T_c = 83°R$	$p_c = 327$ psia

Note, however that three of these noncomformist gases have been included among the more accurate specific gas compressibility curves and one should always use the latter when suitable.

The four generalized charts cover the following ranges of reduced pressure and reduced temperature. The maximum indicated deviation from experimental data is also shown.

Chart no.	Range p_r	Range T_r	Max. error, %
1	0–0.65	0.7–5.0	1.0
2	0–6.5	1.0–15.0	2.5
3	6–12.5	1.0–15.0	2.5
4	10–42.5	1.0–15.0	5

A.2   Outline of Procedure

1. Calculate pseudo-critical temperature and pseudo-critical pressure for a given gas mixture using the method outlined on page 16. If working with a pure gas rather than a mixture (but do not have a specific compressibility curve for that gas) look up the critical temperature and pressure in App. A in Table 1 or 2.

2. If interested in compressibility at discharge conditions, estimate the discharge temperature T_2 from the following formula for adiabatic compression.

$$T_2 = T_1 \times r^{(k-1)/k}$$

3. Calculate the reduced temperature and pressure for the conditions in question using Eq. 1.19 and 1.20.

4. Read the compressibility factor Z from the applicable generalized chart on following pages.

5. Use this compressibility factor in proper formula to determine volume or horsepower.

Example A.1 Find the compressibility factors at inlet and discharge conditions for the following gas mixture when compressed from 315 psia and 100°F to 965 psia.

Gas component	H_2	N_2	CO_2	CO
Mol percent	61.4	19.7	17.5	1.4

Step 1: See Sec. 1.19, "Pseudo Critical Conditions and Compressibility."
Pseudo-critical temperature = 195°R
Pseudo-critical pressure = 493 psia

Step 2: Calculate theoretical discharge temperature.
$r = 965/315 = 3.06$
$k = 1.37$ (See sec. 1.18, "Ratio of Specific Heats")
Theoretical discharge temperature, $T_2 = 758°R$ (298°F)

Step 3:

	Inlet	Discharge
Pressure, psia	315	965
Temperature, °R	560	758
Reduced pressure	0.64	1.96
Reduced temperature	2.87	3.88
Compressibility (from Fig. A.3)	1.002	1.025

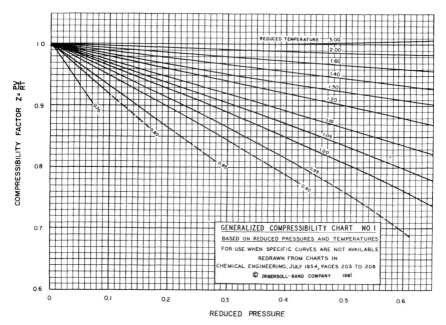

Figure A.2   Generalized compressibility chart for low values of reduced pressure.

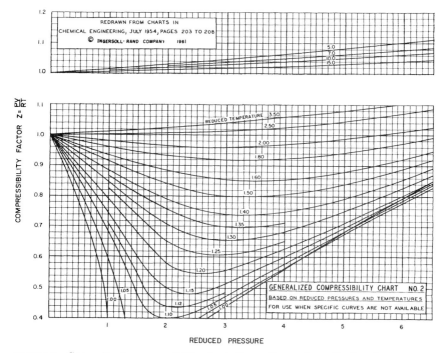

Figure A.3   Generalized compressibility charts for medium values of reduced pressure.

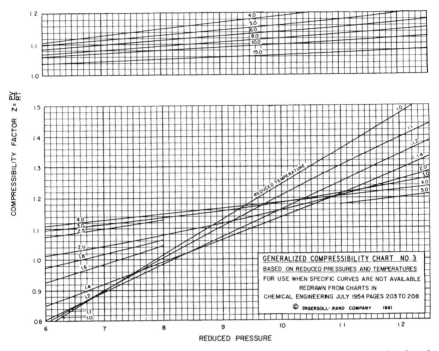

Figure A.4 Generalized compressibility chart for moderately high values of reduced pressure.

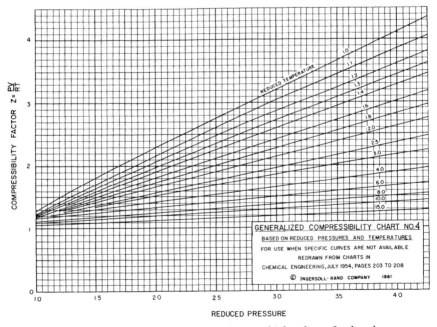

Figure A.5 Generalized compressibility chart for very high values of reduced pressure.

Shortcut Calculations and Graphical Compressor Selection Procedures

B.1 Selection Guide for Elliott Multistage Centrifugal Compressors*

---

* *Please Note:* This segment is reprinted from a 1994 Elliott Company Sales bulletin. The reference information contained herein is provided as an assist to developing your application. However, Elliott reserves the right to modify the design or construction of the equipment described and to furnish it, as altered, without further reference to the illustrations or information contained herein.

Thermodynamics

Compressor performance cannot be accurately predicted without detailed knowledge of the behavior of the gas or gases involved.

Mollier diagrams, of course, are readily available for most pure gases at "conventional" pressures and temperatures. However, in cryogenic areas or at very high pressure, some gases behave most peculiarly. Gas properties in these areas heretofore have been estimates arrived at through rather empirical methods.

The same is true of mixtures of gases, yet the preponderance of gas compression problems involve gas mixtures.

Through the knowledge and skill of Elliott·thermodynamicists, the behavior of a wide variety of gases— in any conceivable mixture—can now be accurately computed, plotted and offered to the process engineer. This knowledge has been computerized, and in minutes, made available as an actual Mollier diagram.

The only input required to obtain a plot of gas behavior is the identity and proportion of the gases involved (if a gas mix), and the limiting pressure and temperature values.

Performance calculations and selection of Elliott multistage compressors

Introduction

These are basic procedures that will help you to calculate compressor performance and estimate the right unit in your installation. The data herein cover most applications; unusual or special problems can be referred to your Elliott Representative.

Our computer, too, is ready and willing to assist you. From world-wide sales offices, we can access the main computer at the factory and thus eliminate many routine and time-consuming calculations. A good example of this would be the selection of an optimum compressor/driver arrangement, which requires analysis of many alternatives and especially so when high power and multiple-casing train setups are involved.

Another time-saver worthy of mention is the high degree of stan-dardization of Elliott compressor frames, impellers, seals, bearings and even mechanical-drive turbines. Many of these components are computerized to enable you to evaluate various alternatives in a minimum of time.

Calculation methods

The calculation procedures on the following pages apply to "straight" compression — the compression of a certain gas from a given suction pressure to a desired discharge pressure.

The methods outlined are:

1. The "N" method (so named because of the extensive use of the polytropic exponent "n"). It is used
a. when the fluid to be compressed closely approximates a "perfect" gas (air, nitrogen, oxygen, hydrogen).
b. when a chart of the properties of the gas or gas mixture is not available.

2. The "Mollier" method which involves use of a Mollier diagram and is used whenever a plot of the properties of the fluid being compressed is available.

Note that the final computerized selections use computerized data bases of actual impeller performance characteristics as well as sophisti-cated real-gas equations of state.

Note: For the convenience of the reader, a simplified Mollier diagram for ethylene can be found on p. 453. Actual Mollier diagrams are reproduced on p. 492 (English, Fig. B.8) and p. 493 (metric, Fig. B.9).

Thermodynamic equations

Fan Laws

Fan laws have been developed to estimate perform-ance of centrifugal compressors for operating con-ditions other than design. These are approximate calculations and as such, can be used to estimate off-design parameters.

The fan laws are:
1. Q α N
2. H α N^2
3. ln r_p α N^2
4. ΔT α N^2
5. HP or kW α N^3

where Q   = inlet volume flow
      H   = head
      N   = speed (r/min)
      r_p = absolute pressure ratio (P_2/P_1)
      ΔT = change in temperature
      HP or kW = power

Flow Calculations

Compressor flow conditions are often expressed in different forms, most common of which are:

1. Weight flow—lb/min, lb/h (kg/min, kg/h)
2. SCFM—60°F, 14.7 psia and dry
3. number of mols/h

None of these flows can be used directly in calculat-ing compressor performance. All must be con-verted to ACFM—actual cubic feet per minute. This is also commonly referred to as ICFM—inlet cubic feet per minute.

These conversions are:
$$ACFM = w \times v$$
$$ACFM = SCFM \times \frac{P_s}{P_1} \times \frac{T_1}{T_s} \times \frac{Z_1}{Z_s}$$

ACFM = no. of mols/ min $\times$ MW $\times$ v
w   = weight flow—lb/min (kg/min)
v   = inlet specific volume—ft^3/lb (m^3/kg)
P_s = standard pressure—usually 14.7 psi
       (1.013 bar) absolute
P_1 = inlet pressure—psi (bar) absolute
T_s = standard temperature—usually 520°R
T_1 = inlet temperature—°R
Z_1 = inlet compressibility
Z_s = standard compressibility—always 1.0
MW   = molecular mass

Gas Mixtures

Properties of a gas mixture necessary to select a compressor are:

1. Gas constant (dependent on molecular mass MW)
2. k (c_p and c_v)
3. P_1, T_1, v_1 and P_2
4. Compressibility, Z
5. Critical pressure, P_c
6. Critical temperature, T_c

Of the above properties of a gas mixture, MW, c_p, c_v, P_c, and T_c, are calculated by adding the products of the individual mol fractions of each constituent, times its specific property. The temperature of any

constituent is obviously the temperature of the mix-ture. The v (specific volume) of the mixture is obtained from Pv = ZRT. The compressibility of a mixture is obtained from Chart 1, using the calculated values of P_c and T_c. The k of a mixture is determined from

$$k = \frac{\Sigma Mcp}{\Sigma Mcp - 1.985}$$

The ΣMcp is the summation of the mol fraction times the molal ΣMcp of each constituent. The table below can be used to calculate the properties of a gas mixture.

Gas Mixture	(1) Mol% each gas	(2) Mols/h each gas	(3) Mol Mass (Table 1)	(4) (1) × (3)	(5) Mass %	(6) T_c (Table 1)	(7) P_c (Table 1)	(8) (1) × (6)	(9) (1) × (7)	(10) Mcp (Table 1)	(11) (1) × (10)
....	....	....	....	a	a/d × 100	....	....	....	....	....	....
....	....	....	....	b	b/d × 100	....	....	....	....	....	....
....	....	....	....	c	c/d × 100	....	....	....	....	....	....
				d				$T_{c\,(mix)}$	$P_{c\,(mix)}$		ΣMcp
Calculate $k_{(mixture)} = \dfrac{\Sigma Mcp_{(mix)}}{\Sigma Mcp - 1.985}$				Apparent Mol Mass of Mixture							

Determine the compressibility of the mixture Z_1 by finding the reduced temperature T_{R1} and the reduced pressure P_{R1} as follows:

$$T_{R1} = \frac{T_1}{T_{c\,(mix)}} \qquad P_{R1} = \frac{P_1}{P_{c\,(mix)}}$$

Then enter these values on Chart 1 to find Z.

Chart 1 Generalized compressibility chart

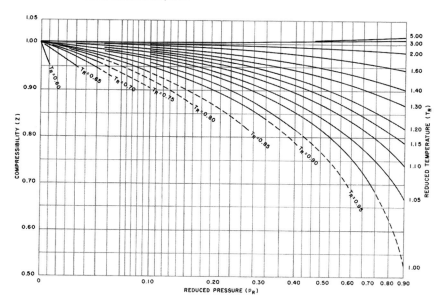

Chart 2 Polytropic to adiabatic efficiency conversion.

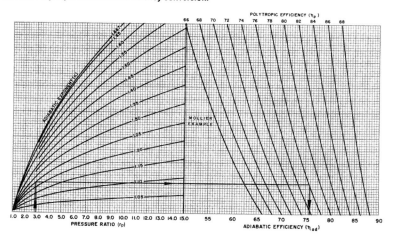

ENGLISH SECTION

Table 1 Gas Properties

(Most values taken from Natural Gas Processors Suppliers Association Engineering Data Book—1972, Ninth Edition)

Gas or Vapor	Hydrocarbon Reference Symbols	Chemical Formula	Molecular Mass	Specific Heat Ratio $k = c_p/c_v$ at 60°F	Critical Conditions Absolute Pressure p_c (psia)	Critical Conditions Absolute Temperature T_c (°R)	*Mcp at 50°F	*Mcp at 300°F
Acetylene	$C_2=$	C_2H_2	26.04	1.24	905	557	10.22	12.21
Air		N_2+O_2	28.97	1.40	547	239	6.95	7.04
Ammonia		NH_3	17.03	1.31	1636	731	8.36	9.45
Argon		A	39.94	1.66	705	272	4.97	4.97
Benzene		C_6H_6	78.11	1.12	714	1013	18.43	28.17
Iso-Butane	iC_4	C_4H_{10}	58.12	1.10	529	735	22.10	31.11
n-Butane	nC_4	C_4H_{10}	58.12	1.09	551	766	22.83	31.09
Iso-Butylene	iC_4-	C_4H_8	56.10	1.10	580	753	20.44	27.61
Butylene	nC_4-	C_4H_8	56.10	1.11	583	756	20.45	27.64
Carbon Dioxide		CO_2	44.01	1.30	1073	548	8.71	10.05
Carbon Monoxide		CO	28.01	1.40	510	242	6.96	7.03
Carbureted Water Gas (1)		-	19.48	1.35	454	235	7.60	8.33
Chlorine		Cl_2	70.91	1.36	1119	751	8.44	8.52
Coke Oven Gas (1)		-	10.71	1.35	407	197	7.69	8.44
n-Decane	nC_{10}	$C_{10}H_{22}$	142.28	1.03	320	1115	53.67	74.27
Ethane	C_2	C_2H_6	30.07	1.19	708	550	12.13	16.33
Ethyl Alcohol		C_2H_5OH	46.07	1.13	927	930	17	21
Ethyl Chloride		C_2H_4Cl	64.52	1.19	764	829	14.5	18
Ethylene	C_2-	C_2H_4	28.05	1.24	742	510	10.02	13.41
Flue Gas (1)			30.00	1.38	563	264	7.23	7.50
Helium		He	4.00	1.66	33	9	4.97	4.97
n-Heptane	nC_7	C_7H_{16}	100.20	1.05	397	973	39.52	53.31
n-Hexane	nC_6	C_6H_{14}	86.17	1.06	440	915	33.87	45.88
Hydrogen		H_2	2.02	1.41	188	60	6.86	6.98
Hydrogen Sulphide		H_2S	34.08	1.32	1306	673	8.09	8.54
Methane	C_1	CH_4	16.04	1.31	673	344	8.38	10.25
Methyl Alcohol		CH_3OH	32.04	1.20	1157	924	10.5	14.7
Methyl Chloride		CH_3Cl	50.49	1.20	968	750	11.0	12.4
Natural Gas (1)		-	18.82	1.27	675	379	8.40	10.02
Nitrogen		N_2	28.02	1.40	492	228	6.96	7.03
n-Nonane	nC_9	C_9H_{20}	128.25	1.04	345	1073	48.44	67.04
Iso-Pentane	iC_5	C_5H_{12}	72.15	1.08	483	830	27.59	38.70
n-Pentane	nC_5	C_5H_{12}	72.15	1.07	489	847	28.27	38.47
Pentylene	C_5-	C_5H_{10}	70.13	1.08	586	854	25.08	34.46
n-Octane	nC_8	C_8H_{18}	114.22	1.05	362	1025	43.3	59.90
Oxygen		O_2	32.00	1.40	730	278	6.99	7.24
Propane	C_3	C_3H_8	44.09	1.13	617	666	16.82	23.57
Propylene	C_3-	C_3H_6	42.08	1.15	668	658	14.75	19.91
Blast Furnace Gas (1)		-	29.6	1.39	—	—	7.18	7.40
Cat Cracker Gas (1)		-	28.83	1.20	674	515	11.3	15.00
Sulphur Dioxide		SO_2	64.06	1.24	1142	775	9.14	9.79
Water Vapor		H_2O	18.02	1.33	3208	1166	7.98	8.23

(1) Approximate values based on average composition.

*Use straight line interpolation or extrapolation to approximate Mcp (in btu/mol-° R) at actual inlet T. For greater accuracy, average T should be used.

Table 2 M-Line & MB-Line Frame Data

Frame	Nominal Flow Range (cfm)	Nominal Max No. of Casing Stages	Max Casing Pressure (psig)	Nominal Speed (r/min)	Nominal Polytropic Efficiency	Nominal H/N^2 (per stage)	Maximum Q/N
29M	750 – 9,500	10	750	11,500	0.78	7.5×10^{-5}	0.83
38M	6,000 – 22,000	9	625	7,725	0.79	1.52×10^{-4}	2.85
46M	16,000 – 34,000	9	625	6,300	0.80	2.28×10^{-4}	5.40
60M	25,000 – 58,000	8	325	4,700	0.81	3.85×10^{-4}	12.34
70M	50,000 – 84,000	8	325	4,200	0.81	5.67×10^{-4}	20.
88M	70,000 – 135,000	8	325	3,160	0.81	9.1×10^{-4}	42.7
103M	110,000 – 160,000	8	45	2,800	0.82	11.6×10^{-4}	57.1
110M	140,000 – 190,000	8	45	2,600	0.82	13.4×10^{-4}	73.1
10MB	90 – 1,600	12	10,000	18,900	0.77	2.6×10^{-5}	0.085
15MB	200 – 2,350	12	10,000	15,300	0.77	3.6×10^{-5}	0.153
20MB	325 – 3,600	12	10,000	12,400	0.77	6.2×10^{-5}	0.29
25MB	500 – 5,500	12	10,000	10,000	0.78	9.5×10^{-5}	0.55
32MB	2,000 – 8,000	10	10,000	8,300	0.78	1.39×10^{-4}	0.96
38MB	6,000 – 22,000	9	1,500	7,725	0.79	1.52×10^{-4}	2.85
46MB	16,000 – 34,000	9	1,200	6,300	0.79	2.28×10^{-4}	5.40
60MB	25,000 – 58,000	8	800	4,700	0.80	3.85×10^{-4}	12.34
70MB	50,000 – 84,000	8	800	4,200	0.80	5.67×10^{-4}	20.

(1) Number of casing stages is determined by critical speed margins. These numbers are a general guideline only.
(2) These values are typical. Flexibility in types of available staging can allow final computer selections to have significant variations in head and efficiency.

Selection Procedure

Step 1:
If MW, k, and Z are not given, determine gas mixture properties. By using the procedure and data on Pages 3 and 5, most gas compositions can be analyzed. For single gases or an analysis that has one gas consisting of up to 95% by volume, check to see if a Mollier Diagram is included, and use the Mollier method.

Step 2:
Calculate inlet volume flow (ACFM). Using the gas composition data from Step 1 and the relationships below or the Mollier charts, find the inlet volume entering the compressor. Note that for very large volumes and lower head requirements, compressors can have the flow divided in half having two inlets (double flow), one at each end of the machine. This gives the flexibility of having a smaller frame size handling larger volumes of flow. This can be important in a multi-body string such as a feed gas string in an ehtylene plant, or whenever a match in speed with other compressors or a particular driver is desired.

Step 3:
Select the compressor frame size. Using the inlet volume calculated in Step 2, enter Table 2 and select the proper frame size. Table 2 also contains other pertinent frame data to be used in the selection procedure.

Step 4:
Calculate the total head requirement. In order to determine the number of compression stages, it is necessary to know the total required head. It is important to remember that in a machine with more than one section, it is more accurate to total the heads from the various sections than to make an overall estimate.

Step 5:
Calculate the total number of casing stages. Reference the average H/N^2 values in Table 2. Multiply this by the speed squared (begin with nominal speed unless speed is fixed) to find an average amount of head developed by the impellers. Divide the total head requirement by this to determine the approximate number of casing stages.

Step 6:
Adjust the speed by using fan law relationships to agree with required discharge conditions.

Step 7:
The gas power (GHP) should be adjusted for balance piston or equalizing line leakage. For estimating purposes, we assume this to be a 2% increase. Mechanical losses can then be added to obtain shaft power (SHP).

Rough Out Example (N-method)

1) Given the following customer conditions

$w_1 = 1769$ lb/min	MW $= 29$
$P_1 = 80$ psia	k $= 1.4$
$T_1 = 90°$F (550° R)	Z $= 1.0$
$P_2 = 225$ psia	

2) Calculate inlet volume

$$v_1 = \frac{ZRT_1}{144\,P_1} = \frac{1.0\,(1545)\,(550)}{144\,(29)\,(80)} = 2.544 \text{ ft}^3/\text{lb}$$

$$Q = w_1 \times v_1 = 1769 \times 2.544 = 4500 \text{ ICFM}$$

3) Select compressor frame size
Based on an inlet volume of 4500 ICFM and knowing the required discharge pressure is 225 psia select a 29M frame size from Table 2.

4) Calculate the required head
Assume an efficiency of 0.78 from Table 2 and calculate the polytropic exponent.

$$\frac{n}{n-1} = \left(\frac{k}{k-1}\right)\eta_p = \left(\frac{1.4}{0.4}\right)0.78 = 2.73$$

Calculate the overall head

$$H = ZRT\frac{n}{n-1}\left[\frac{P_2}{P_1}^{\frac{n-1}{n}} - 1\right]$$

$$= 1.0\,\frac{(1545)}{29}\,(550)\,(2.73)\left[\frac{225}{80}^{0.3663} - 1\right]$$

$$H = 36837 \frac{\text{ft-lb}_f}{\text{lb}_m}$$

Check the discharge temperature for a need to inter-cool (Cool if $T_2 > 400°$F)

$$\frac{T_2}{T_1} = \left(\frac{P_2}{P_1}\right)^{\frac{n-1}{n}} = \left(\frac{225}{80}\right)^{0.3663} = 1.461$$

$$T_2 = 550\,(1.461) = 803°R = 343°F$$

No iso-cooling is therefore required.

5) Determine the number of casing stages.
From Table 2 the nominal speed for a 29M is 11500 r/min. Calculate the $Q_{/N}$

$$Q_{/N} = \frac{4500}{11500} = 0.391$$

From Table 2 $H_{/N^2} = 7.5 \times 10^{-5}$

$H_{/\text{stage}}$ would then be

$$H_{/N^2} \times N^2 = (7.5 \times 10^{-5})\,(11500)^2 = 9919 \frac{\text{ft-lb}_f}{\text{lb}_m}$$

Determine approximate number of casing stages.

$$\text{Number of stages} = \frac{36837}{9919} = 3.71 \cong 4 \text{ stages}$$

6) Adjust Speed
Adjust the nominal speed according to the casing stages.

$$4 \text{ stages must develop } 36837 \frac{\text{ft-lb}_f}{\text{lb}_m}$$

$$\text{or an average of } \frac{36837}{4} = 9209 \frac{\text{ft-lb}_f}{\text{lb}_m} \text{ per stage.}$$

Using Fan Law relationships adjust the speed.

$$H \alpha N^2$$

$$N = N_{NOM}\left[\frac{H_{REQ'D}}{H}\right]^{1/2} = 11500\left[\frac{9209}{9919}\right]^{1/2}$$

$$N = 11,081 \text{ r/min}$$

7) Calculate the approximate power

$$GHP = \frac{w_1 \times H}{33000 \times \eta_p} = \frac{1769 \times 36837}{33000 \times 0.78} = 2532 HP$$

Adjust for balance piston leakage
$$2532 \times 1.02 = 2583 HP$$

Add losses from Chart 4
$$SHP = 2583 + 78 = 2661 HP$$
(Assume Iso-Carbon Seal)

Rough Out Example (Mollier)
1) Given the following customer conditions
$w_1 = 1769$ lb/min
$P_1 = 80$ psia
$T_1 = 90°$ F ($550°$ R)
$P_2 = 225$ psia
Gas: ethylene

2) Calculate inlet volume
$v_1 = 2.6$ (from Mollier chart)
$Q = w_1 \times v_1 = 1769 \times 2.6 = 4600$ ICFM

3) Select compressor frame size
Based on an inlet volume of 4600 ICFM and knowing
the required discharge pressure is 225 psia select a
29M frame size from Table 2.

4) Calculate the required head
At given inlet conditions, determine inlet entropy (s)
and enthalpy (h) from Mollier chart:
$P_1 = 80$ $T_1 = 90$ $s_1 = 1.75$ $h_1 = 163$
At required discharge pressure and constant entropy
($s_1 = s_2$), determine h_2 from chart
$P_2 = 225$ $T_{2_i} = $ N/A $s_{2_i} = 1.75$ $h_{2_i} = 205$

Head required $= 778 (h_{2_i} - h_1)$

$H = 778 (205 - 163) = 32676 \dfrac{\text{ft-lb}_f}{\text{lb}_m}$ (adiabatic)

Check the discharge temperature for a need to inter-
cool. (Cool if $T_2 > 400°$ F)

Step 1 Determine adiabatic efficiency
$r_p = \dfrac{225}{80} = 2.81$ $k = 1.24$ $\eta_p = 0.78$
$\eta_{AD} = 0.76$ from Chart 2

Step 2 determine actual (not isentropic) Δh.
$\Delta h = \dfrac{h_{2_i} - h_1}{\eta_{AD}} = \dfrac{205 - 163}{0.76} = 55.3$

Step 3 Determine h_2 and read T_2 from Mollier
Chart.
$h_2 = h_1 + \Delta h = 163 + 55.3 = 218.3$
$T_2 = 232°$ F (from Mollier chart)
No iso-cooling is therefore required.

5) Determine the number of casing stages.
From Table 2 the nominal speed for a 29M is 11500
RPM. Convert adiabatic head to polytropic head by
the ratio of efficiencies.

$H = 32676 (0.78/0.76) = 33536$
From Table 2 $H_{/N^2} = 7.5 \times 10^{-5}$

$H_{/\text{stage}}$ would then be
$H_{/N^2} \times N^2 = (7.5 \times 10^{-5}) (11500)^2 = 9919 \dfrac{\text{ft-lb}_f}{\text{lb}_m}$

Determine approximate number of casing stages.
Number of stages $= \dfrac{33536}{9919} = 3.38 \cong 4$ stages

6) Adjust Speed
Adjust the nominal speed according to the casing
stages.

4 stages must develop $33536 \dfrac{\text{ft-lb}_f}{\text{lb}_m}$

or an average of $\dfrac{33536}{4} = 8384 \dfrac{\text{ft-lb}_f}{\text{lb}_m}$ per stage.

Using Fan Law relationships adjust the speed.
$H \alpha N^2$
$N = N_{NOM} \left[\dfrac{H_{REQ'D}}{H} \right]^{1/2} = 11500 \left[\dfrac{8384}{9919} \right]^{1/2} = 10573$

7) Calculate the approximate power
$GHP = \dfrac{w_1 \times H}{33000 \times \eta_p} = \dfrac{1769 \times 33536}{33000 \times 0.78} = 2305$ HP

Adjust for balance piston leakage
$2305 \times 1.02 = 2351$ HP

Add losses from Chart 4.
$SHP = 2351 + 70 = 2421$ HP
(Assume Iso-Carbon Seal)

Mechanical Losses

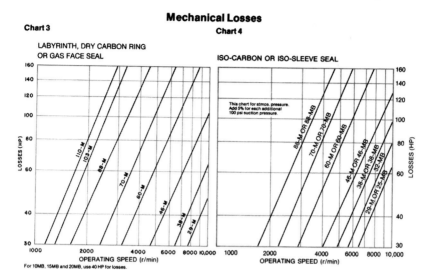

Chart 3

LABYRINTH, DRY CARBON RING
OR GAS FACE SEAL

Chart 4

ISO-CARBON OR ISO-SLEEVE SEAL

This chart for atmos. pressure.
Add 5% for each additional
100 psi suction pressure.

For 10MB, 15MB and 20MB, use 40 HP for losses.

Simplified Mollier Diagram for Ethylene

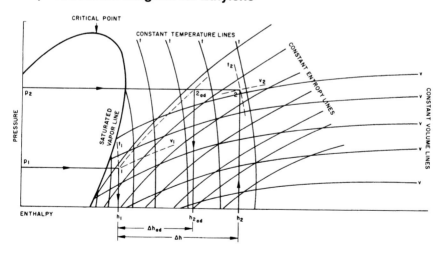

METRIC SECTION

Table 3 Gas Properties

(Most values taken from Natural Gas Processors Suppliers Association Engineering Data Book—1972, Ninth Edition)

Gas or Vapor	Hydrocarbon Reference Symbols	Chemical Formula	Molecular Mass	Specific Heat Ratio $k = c_p/c_v$ at 15.5°C	Critical Conditions Absolute Pressure P_c (bar)	Absolute Temperature T_c (K)	*Mcp at 0°C	at 100°C
Acetylene	C₂=	C_2H_2	26.04	1.24	62.4	309.4	42.16	48.16
Air		N_2+O_2	28.97	1.40	37.7	132.8	29.05	29.32
Ammonia		NH_3	17.03	1.31	112.8	406.1	34.65	37.93
Argon		A	39.94	1.66	48.6	151.1	20.79	20.79
Benzene		C_6H_6	78.11	1.12	49.2	562.8	74.18	103.52
Iso-Butane	iC₄	C_4H_{10}	58.12	1.10	36.5	408.3	89.75	116.89
n-Butane	nC₄	C_4H_{10}	58.12	1.09	38.0	425.6	93.03	117.92
Iso-Butylene	iC₄—	C_4H_8	56.10	1.10	40.0	418.3	83.36	104.96
Butylene	nC₄—	C_4H_8	56.10	1.11	40.2	420.0	83.40	105.06
Carbon Dioxide		CO_2	44.01	1.30	74.0	304.4	36.04	40.08
Carbon Monoxide		CO	28.01	1.40	35.2	134.4	29.10	29.31
Carbureted Water Gas (1)		-	19.48	1.35	31.3	130.6	31.58	33.78
Chlorine		Cl_2	70.91	1.36	77.2	417.2	35.29	35.53
Coke Oven Gas (1)		-	10.71	1.35	28.1	109.4	31.95	34.21
n-Decane	nC₁₀	$C_{10}H_{22}$	142.28	1.03	22.1	619.4	218.35	280.41
Ethane	C₂	C_2H_6	30.07	1.19	48.8	305.6	49.49	62.14
Ethyl Alcohol		C_2H_5OH	46.07	1.13	63.9	516.7	69.92	81.97
Ethyl Chloride		C_2H_4Cl	64.52	1.19	52.7	460.6	59.61	70.16
Ethylene	C₂—	C_2H_4	28.05	1.24	51.2	283.3	40.90	51.11
Flue Gas (1)			30.00	1.38	38.8	146.7	30.17	30.98
Helium		He	4.00	1.66	2.3	5.0	20.79	20.79
n-Heptane	nC₇	C_7H_{16}	100.20	1.05	27.4	540.6	161.20	202.74
n-Hexane	nC₆	C_6H_{14}	86.17	1.06	30.3	508.3	138.09	174.27
Hydrogen		H_2	2.02	1.41	13.0	33.3	28.67	29.03
Hydrogen Sulphide		H_2S	34.08	1.32	90.0	373.9	33.71	35.07
Methane	C₁	CH_4	16.04	1.31	46.4	191.1	34.50	40.13
Methyl Alcohol		CH_3OH	32.04	1.20	79.8	513.3	42.67	55.32
Methyl Chloride		CH_3Cl	50.49	1.20	66.7	416.7	45.60	49.82
Natural Gas (1)		-	18.82	1.27	46.5	210.6	34.66	39.54
Nitrogen		N_2	28.02	1.40	33.9	126.7	29.10	29.31
n-Nonane	nC₉	C_9H_{20}	128.25	1.04	23.8	596.1	197.07	253.10
Iso-Pentane	iC₅	C_5H_{12}	72.15	1.08	33.3	461.1	112.09	145.56
n-Pentane	nC₅	C_5H_{12}	72.15	1.07	33.7	470.6	115.21	145.94
Pentylene	C₅—	C_5H_{10}	70.13	1.08	40.4	474.4	102.11	130.37
n-Octane	nC₈	C_8H_{18}	114.22	1.05	25.0	569.4	176.17	226.17
Oxygen		O_2	32.00	1.40	50.3	154.4	29.17	29.92
Propane	C₃	C_3H_8	44.09	1.13	42.5	370.0	68.34	88.68
Propylene	C₃—	C_3H_6	42.08	1.15	46.1	365.6	60.16	75.70
Blast Furnace Gas (1)		-	29.6	1.39	—	—	29.97	30.64
Cat Cracker Gas (1)		-	28.83	1.20	46.5	286.1	46.16	57.31
Sulphur Dioxide		SO_2	64.06	1.24	78.7	430.6	38.05	40.00
Water Vapor		H_2O	18.02	1.33	221.2	647.8	33.31	34.07

(1) Approximate values based on average composition.

*Use straight line interpolation or extrapolation to approximate Mcp (in kJ/(kmol·K)) at actual inlet T. For greater accuracy, average T should be used

Table 4 M-Line & MB-Line Frame Data

Frame	Nominal Flow Range (m³/h)	Nominal Max No. of Casing Stages	Max Casing Pressure (bar)	Nominal Speed (r/min)	Nominal Polytropic Efficiency	Nominal H/N² (per stage)	Maximum Q/N
29M	1 275 – 16 140	10	52	11 500	0.78	2.25×10^{-4}	1.403
38M	10 200 – 37 380	9	43	7725	0.79	4.56×10^{-4}	4.84
46M	27 200 – 57 750	9	43	6300	0.80	6.84×10^{-4}	9.17
60M	42 500 – 98 550	8	23	4700	0.81	11.55×10^{-4}	20.97
70M	85 000 – 142 700	8	23	4200	0.81	17.01×10^{-4}	33.98
88M	119 000 – 229 400	8	23	3160	0.81	27.3×10^{-4}	72.6
103M	186 900 – 272 000	8	3	2800	0.82	34.8×10^{-4}	97.
110M	237 900 – 323 000	8	3	2600	0.82	40.2×10^{-4}	124.
10MB	150 – 2 700	12	690	18 900	0.77	8.0×10^{-5}	0.14
15MB	340 – 4 000	12	690	15 300	0.77	10.8×10^{-5}	0.26
20MB	550 – 6 120	12	690	12 400	0.77	18.6×10^{-5}	0.49
25MB	850 – 9 345	12	690	10 000	0.78	28.5×10^{-5}	0.94
32MB	3 400 – 13 600	10	690	8300	0.78	4.2×10^{-4}	1.64
38MB	10 200 – 37 380	9	103	7725	0.79	4.56×10^{-4}	4.84
46MB	27 200 – 57 750	9	83	6300	0.79	6.84×10^{-4}	9.17
60MB	42 500 – 98 550	8	55	4700	0.80	11.55×10^{-4}	20.97
70MB	85 000 – 142 700	8	55	4200	0.80	17.01×10^{-4}	33.98

(1) Number of casing stages is determined by critical speed margins. These numbers are a general guideline.
(2) These values are typical. Flexibility in types of available staging can allow final computer selections to have significant variations in head and efficiency

Selection Procedure

Step 1:
If MW, k, and Z are not given, determine gas mixture properties. By using the procedure and data on Pages 3 and 9, most gas compositions can be analyzed. For single gases or an analysis that has one gas consisting of up to 95% by volume, check to see if a Mollier Diagram is available, and use the Mollier method.

Step 2:
Calculate inlet volume flow (m³). Using the gas composition data from Step 1 and the relationships below or the Mollier charts, find the inlet volume entering the compressor. Note that for very large volumes and lower head requirements, compressors can have the flow divided in half having two inlets (double flow), one at each end of the machine. This gives the flexibility of having a smaller frame size handling larger volumes of flow. This can be important in a multi-body string such as a feed gas string in an ethylene plant, or whenever a match in speed with other compressors or a particular driver is desired.

Step 3:
Select the compressor frame size. Using the inlet volume calculated in Step 2, enter Table 4 and select the proper frame size. Table 4 also contains other pertinent frame data to be used in the selection procedure.

Step 4:
Calculate the total head requirement. In order to determine the number of compression stages, it is necessary to know the total required head. It is important to remember that in a machine with more than one section, it is more accurate to total the heads from the various sections than to make an overall estimate.

Step 5:
Calculate the total number of casing stages. Reference the average H/N² values in Table 4. Multiply this by the speed squared (begin with nominal speed unless speed is fixed) to find an average amount of head developed by the impellers. Divide the total head requirement by this to determine the approximate number of casing stages.

Step 6:
Adjust the speed by using fan law relationships to agree with required discharge conditions.

Step 7:
The gas power (GkW) should be adjusted for balance piston or equalizing line leakage. For estimating purposes, assume this to be a 2% increase. Mechanical losses can then be added to obtain shaft power (SkW).

Rough Out Example (N-method)

1) Given the following customer conditions

$w_1 = 802.4$ kg/min	MW $= 29$
$P_1 = 5.5$ bar	k $= 1.4$
$T_1 = 32°$ C (305 K)	Z $= 1.0$
$P_2 = 15.52$ bar	

2) Calculate inlet volume

$$v_1 = \frac{ZRT_1}{10^5 \, P_1} = \frac{1.0 \,(8314)\,(305)}{10^5\,(29)\,(5.5)} = 0.159 \text{ m}^3/\text{kg}$$

$$Q = w_1 \times v_1 = 802.4 \times 0.159 = 127.6 \text{ m}^3/\text{min}$$
$$127.6 \times 60 = 7656 \text{ m}^3/\text{h}$$

3) Select compressor frame size
Based on an inlet volume of 7656 m³/h, and knowing the required discharge pressure is 15.52 bar, select a 29M frame size from Table 4.

4) Calculate the required head
Assume an efficiency of 0.78 from Table 4 and calculate the polytropic exponent.

$$\frac{n}{n-1} = \left(\frac{k}{k-1}\right)\eta_p = \left(\frac{1.4}{0.4}\right)0.78 = 2.73$$

Calculate the overall head

$$H = ZRT\frac{n}{n-1}\left[\frac{P_2}{P_1}^{\frac{n-1}{n}} - 1\right]$$

$$= 1.0 \frac{(8314)}{29}(305)\,(2.73)\left[\frac{15.52}{5.5}^{0.3663} - 1\right]$$

$$H = 110\,350 \text{ } \frac{\text{Nm}}{\text{kg}}$$

Check the discharge temperature for a need to intercool (Cool if $T_2 > 205°$ C)

$$\frac{T_2}{T_1} = \left(\frac{P_2}{P_1}\right)^{\frac{n-1}{n}} = \left(\frac{15.52}{5.5}\right)^{0.3663} = 1.462$$

$$T_2 = 305\,(1.462) = 446 \text{K} = 173°\text{C}$$
No iso-cooling is therefore required.

5) Determine the number of casing stages.
From Table 4 the nominal speed for a 29M is 11 500 RPM. Calculate the $Q_{/N}$

$$Q_{/N} = \frac{7\,656}{11\,500} = 0.666$$

From Table 4 $H_{/N^2} = 2.25 \times 10^{-4}$
$H_{/stage}$ would then be

$$H_{/N^2} \times N^2 = (2.25 \times 10^{-4})\,(11\,500)^2 = 29\,756 \text{ } \frac{\text{Nm}}{\text{kg}}$$

Determine approximate number of casing stages.

$$\text{Number of stages} = \frac{110\,350}{29\,756} = 3.71 \cong 4 \text{ stages}$$

6) Adjust Speed
Adjust the nominal speed according to the casing stages.

4 stages must develop $110\,350 \frac{\text{Nm}}{\text{kg}}$

or an average of $\frac{110\,350}{4} = 27\,588 \frac{\text{Nm}}{\text{kg}}$ per stage.

Using Fan Law relationships adjust the speed.

$$H \alpha N^2$$

$$N = N_{NOM}\left[\frac{H_{REQ'D}}{H}\right]^{1/2} = 11\,500\left[\frac{27\,588}{29\,756}\right]^{1/2}$$

$$N = 11073 \text{r/min}$$

7) Calculate the approximate power

$$GkW = \frac{w_1 \times H}{60\,000 \times \eta p} = \frac{(802.4)\,(110\,350)}{(60\,000)\,(0.78)} = 1892 \text{kW}$$

Adjust for balance piston leakage
$$1892 \times 1.02 = 1930 \text{kW}$$

Add losses from Chart 6
$$SkW = 1930 + 58 = 1988 \text{kW}$$
(Assume Iso-Carbon Seal)

Rough Out Example (Mollier)
1) Given the following customer conditions
$w_1 = 802.4$ kg/min
$P_1 = 5.5$ bar
$T_1 = 32°C$ (305 K)
$P_2 = 15.52$ bar

Gas: ethylene

2) Calculate inlet volume
$v_1 = 0.163$ m³/kg (from Mollier chart)
$Q = w_1 \times v_1 = 802.4 \times 0.163 = 130.79$ m³/min
$130.79 \times 60 = 7847$ m³/h

3) Select compressor frame size
Based on an inlet volume of 7847 m³/h and knowing the required discharge pressure is 15.52 bar select a 29M frame size from Table 4.

4) Calculate the required head
At given inlet conditions, determine inlet entropy (s) and enthalpy (h) from Mollier chart:
$P_1 = 5.5$ bar $T_1 = 32°C$ $s_1 = s_2$ $h_1 = 379 \dfrac{kJ}{kg}$

At required discharge pressure and constant entropy $(s_1 = s_2)$, determine h_2 from chart
$P_2 = 15.52$ $T_{2_i} = N/A$ $s_2 = s_1$ $h_{2_i} = 477 \dfrac{kJ}{kg}$

Head required $= 1000 (h_{2_i} - h_1)$
$H = 1000 (477 - 379) = 98000$ (adiabatic) $\dfrac{Nm}{kg}$
Check the discharge temperature for a need to intercool. (Cool if $T_2 > 205°C$)

Step 1 Determine adiabatic efficiency
$r_p = \dfrac{15.52}{5.5} = 2.82$ $k = 1.24$ $\eta_p = 0.78$

$\eta_{AD} = 0.76$ from Chart 2

Step 2 Determine actual (not isentropic) Δh
$\Delta h = \dfrac{h_{2_i} - h_1}{\eta_{AD}} = \dfrac{(477 - 379)}{0.76} = 128.9$

Step 3 Determine h_2 and read T_2 from Mollier Chart.
$h_2 = h_1 + \Delta h = 379 + 128.9 = 507.9$
$T_2 = 109°C$ (from Mollier chart)
No iso-cooling is therefore required.

5) Determine the number of casing stages.
From Table 4 the nominal speed for a 29M is 11 500 r/min. Convert adiabatic head to polytropic head by the ratio of efficiencies.
$H = 98000 (0.78/0.76) = 100579 \dfrac{Nm}{kg}$
From Table 4 $H_{/N^2} = 2.25 \times 10^{-4}$
$H_{/stage}$ would then be
$H_{/N^2} \times N^2 = (2.25 \times 10^{-4})(11500)^2 = 29756 \dfrac{Nm}{kg}$

Determine approximate number of casing stages.
Number of stages $= \dfrac{100579}{29756} = 3.38 \cong 4$

6) Adjust Speed
Adjust the nominal speed according to the casing stages.
4 stages must develop $100579 \dfrac{Nm}{kg}$

or an average of $\dfrac{100579}{4}$ per stage. $= 25145$
Using Fan Law relationships adjust the speed.
$H \alpha N^2$
$N = N_{NOM} \left[\dfrac{H_{REQ'D}}{H} \right]^{1/2} = 11500 \left[\dfrac{25145}{29756} \right]^{1/2}$
$N = 10571$ r/min

7) Calculate the approximate power
$GkW = \dfrac{w_1 \times H}{60\,000 \times \eta_p} = \dfrac{(802.4)(100579)}{(60000)(0.78)} = 1724$ kW

Adjust for balance piston leakage
$1724 \times 1.02 = 1759$ kW
Add losses from Chart 6
$SkW = 1759 + 54 = 1813$ kW
(Assume Iso-Carbon Seal)

Metric Units

Mechanical Losses

Chart 5

Chart 6

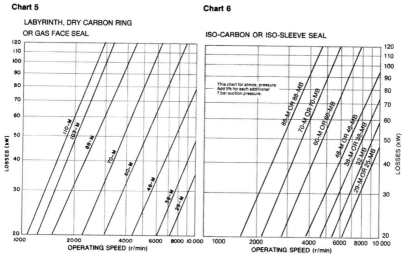

LABYRINTH, DRY CARBON RING OR GAS FACE SEAL

ISO-CARBON OR ISO-SLEEVE SEAL

For 10MB, 15MB and 20MB, use 30 kW for losses

Approximate dimensions and weights

Vertically Split

English units

 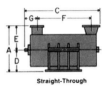

Straight-Through   Back-to-Back or Iso-Cooled

Technical Data

Elliott Compressor Frame	Material	(PSIG) Pressure Rating	* Total Weight lbs. Three Stages	Each Add'l Stage	Nozzle Size Inlet inches	Discharge inches	Rotation Facing inlet
15MB	Fgd. Steel	2000	5,035	350	6, 8	4, 6	CCW
15MBH	Fgd. Steel	4200	6,930	460	6, 8	4, 6	CCW
15MBHH	Fgd. Steel	7500	11,000	550	6, 8	4, 6	CCW
20MB	Plate	1500	9,560	660	8, 10	6, 8	CCW
20MBH	Fgd. Steel	4200	13,150	870	8, 10	6, 8	CCW
25MB	Fgd. Steel	1500 / 2000	18,140	1,250	14, 12, 10 or 8	10, 8 or 6	CW / CCW
25MBH	Fgd. Steel	3150 / 4200	25,000	1,655	12, 10 or 8	8 or 6	CW / CCW
25MBHH	Fgd. Steel	5150 / 10000	38,500 / 53,200	1,475 / 5,100	10, 8 or 6 / 8 or 6	6 or 4 / 6 or 4	CW / CCW
32MB	Fgd. Steel	1500 / 2000	23,800	2,490	16, 14 or 12	12 or 10	CW
32MBH	Fgd. Steel	3150 / 4200	36,500	3,650	12, 10 or 8	8 or 6	CW
32MBHH	Fgd. Steel	10000	56,600	7,250	10 or 8	8 or 6	CW
38MB	Fgd. Steel	700 / 1200 / 1500	30,045 / 36,300 / 51,300	3,440 / 4,130 / 5,250	24, 20 or 16 / 20 or 16 / 16 or 14	16 or 14 / 16 or 14 / 14 or 12	CW / CW
46MB	Fab. Steel	750 / 1200	40,700 / 50,700	4,000 / 4,800	30 or 24 / 24 or 20	20 or 16 / 26 or 14	CW
60MB	Fab. Steel	400 / 800	73,200 / 99,200	8,115 / 9,637	36 or 30 / 36 or 30	24 or 20 / 20 or 16	CW
70MB	Fab. Steel	800	152,300	18,800	42 or 36	30 or 24	CW
88MB	Fab. Steel	800	198,000	40,400	54 to 48	36 or 30	CW

NOTE: The drive end is normally the suction end.   *For back-to-back machines, add weight of two stages.

Approximate Dimensions (inches)

Elliott Compressor Frame	A	B	C Min. 3 Stages	CC Six Stages	Length Each Add'l Stage	D	E	F Min. 3 Stages	FF Six Stages	Length Each Add'l Stage	G
15MB	38	36	38.5	50	2.6	17	21	16	19	2.6	11
15MBH	39	38	40	52	2.6	17.5	21.5	16	21	2.6	13.75
15MBHH	45	41	48	61	3.3	22	23	17	23.5	3.3	19
20MB	47	44	48	62	3.2	21	26	19	23.5	3.2	13.75
20MBH	49	47	51	68	3.2	22	27	19	26	3.2	17
25MB	58	55	59	77	4	26	32	24	29	4	17
25MBH	60	58	63	84	4	27	33	24	32	4	21
25MBHH	69 / 83	63 / 70	73 / 76	93 / 98	5 / 6	34 / 41	35 / 42	26 / 29	36 / 48	5 / 6	29 / 31
32MB	72	71	68	83	5	33	39	29	44	5	18
32MBH	75	74	74	88	6	34	41	31	46	6	21
32MBHH	86	88	82	95	6	39	47	34	50	6	34
38MB	76 / 78 / 86	79 / 82 / 90	80 / 83 / 95	116 / 119 / 128	8 / 8 / 8	36 / 37 / 41	40 / 41 / 45	33 / 33 / 37	63 / 63 / 71	8 / 8 / 8	18 / 20 / 32
46MB	86 / 92	109 / 118	92 / 98	137 / 142	9 / 9	38 / 41	48 / 51	43 / 44	88 / 90	9 / 9	24 / 27
60MB	113 / 125	122 / 134	105 / 111	165 / 171	12 / 12	56 / 62	57 / 63	57 / 59	117 / 119	12 / 12	26 / 28
70MB	134	142	142	217	15	66	68	70	147	15	41
88MB	146	160	152	252	20	69	77	89	192	20	51

Metric units

Straight-Through   Back-to-Back
or Iso-Cooled

Technical Data

Elliott Compressor Frame	Material	(BAR) Pressure Rating	* Total Weight kg Three Stages	Each Add'l Stage	Nominal Nozzle Size Inlet inches	millimetres	Discharge inches	millimetres	Rotation Facing Inlet
15MB	Fgd. Steel	138	2290	160	6, 8	152, 203	4, 6	102, 152	CCW
15MBH	Fgd. Steel	290	3150	210		152, 203		102, 152	CCW
15MBHH	Fgd. Steel	520	5000	250	6, 8	152, 203	4, 6	102, 152	CCW
20MB	Plate	103	4350	300	8, 10	203, 254	6, 8	152, 203	CCW
20MBH	Fgd. Steel	290	6000	400	8, 10	203, 254	6, 8	152, 203	CCW
25MB	Fgd. Steel	103 / 138	8250	570	14, 12, 10, 8	356, 305, 254, 203	10, 8, 6	254, 203, 152	CW / CCW
25MBH	Fgd. Steel	217 / 290	11 315	750	12, 10, 8	305, 254, 203	8, 6	203, 152	CW / CCW
25MBHH	Fgd. Steel	355 / 690	17 464 / 24 131	670 / 2310	10, 8, 6 / 8, 6	254, 203, 152 / 203, 152	6, 4 / 6, 4	152, 102 / 152, 102	CW / CCW
32MB	Fgd. Steel	103 / 138	10 796	1130	16, 14, 12	406, 356, 305	12, 10	305, 254	CW
32MBH	Fgd. Steel	217 / 290	16 556	1655	12, 10, 8	305, 254, 203	8, 6	203, 152	CW
32MBHH	Fgd. Steel	690	25 674	3285	10, 8	254, 203	8, 6	203, 152	CW
38MB	Fgd. Steel	48 / 83 / 103	13 628 / 16 467 / 23 270	1558 / 1870 / 2378	24, 20, 16 / 20, 16 / 16, 14	610, 508, 406 / 610, 508, 406 / 406, 356	16, 14 / 16, 14 / 14, 12	406, 356 / 406, 356 / 356, 305	CW
46MB	Fab. Steel	52 / 83	18 458 / 23 014	1815 / 2175	30, 24 / 24, 20	762, 610 / 610, 508	20, 16 / 26, 14	508, 406 / 660, 356	CW
60MB	Fab. Steel	27 / 55	33 180 / 44 998	3676 / 4365	36, 30 / 36, 30	914, 762 / 914, 762	24, 20 / 20, 16	610, 508 / 508, 406	CW
70MB	Fab. Steel	55	69 083	8515	42, 36	1067, 914	30, 24	762, 610	CW
88MB	Fab. Steel	55	89 813	18 300	54, 48	1372, 1219	36, 30	914, 762	CW

NOTE: The drive end is normally the suction end.   *For back-to-back machines, add weight of two stages.

Approximate Dimensions (millimetres)

Elliott Compressor Frame	A	B	C Min. 3 Stages	CC Six Stages	Length Each Add'l Stage	D	E	F Min. 3 Stages	FF Six Stages	Length Each Add'l Stage	G
15MB	965	914	978	1270	66	432	533	406	483	66	279
15MBH	990	965	1016	1320	66	445	546	406	533	66	350
15MBHH	1143	1041	1219	1550	84	559	584	432	597	84	483
20MB	1194	1118	1219	1575	81	533	660	483	597	81	350
20MBH	1245	1194	1295	1725	81	559	686	483	660	81	432
25MB	1470	1400	1500	1960	100	660	810	610	740	100	430
25MBH	1520	1470	1600	2130	100	690	840	580	810	100	530
25MBHH	1750 / 2110	1600 / 1780	1850 / 1930	2360 / 2490	130 / 150	860 / 1040	890 / 1070	660 / 740	910 / 1220	130 / 150	740 / 790
32MB	1830	1800	1730	2110	130	840	990	740	1120	130	460
32MBH	1900	1880	1880	2240	150	860	1040	790	1170	150	530
32MBHH	2180	2240	2080	2410	150	990	1190	860	1270	150	860
38MB	1930 / 1980 / 2180	2010 / 2080 / 2290	2030 / 2110 / 2410	2950 / 3020 / 3250	200 / 200 / 200	910 / 940 / 1040	1020 / 1040 / 1140	840 / 840 / 940	1600 / 1600 / 1800	200 / 200 / 200	460 / 510 / 810
46MB	2180 / 2340	2770 / 3000	2340 / 2490	3480 / 3610	230 / 230	970 / 1040	1220 / 1300	1090 / 1120	2240 / 2290	230 / 230	610 / 690
60MB	2870 / 3180	3100 / 3400	2670 / 2820	4190 / 4340	300 / 300	1420 / 1570	1450 / 1600	1450 / 1500	2970 / 3020	300 / 300	660 / 710
70MB	3400	3610	3610	5510	380	1680	1730	1780	3730	380	1040
88MB	3710	4060	3860	6400	510	1750	1960	2260	4880	510	1300

All dimensions and weights are approximate and to be used only for preliminary planning. See your Elliott Representative for more accurate data.

Approximate dimensions and weights

Horizontally Split

English units

Straight-through compression

Iso-Cooled with external volutes

Iso-Cooled with internal volutes

Technical Data

Elliott Compressor Frame	Material	Min. Casing Length (stages)	Total Weight lbs. Min.	Total Weight lbs. Add'l Stage	Weight Heaviest Part, lbs. Min.	Weight Heaviest Part, lbs. Add'l Stage	Nozzle Size, inches Inlet	Nozzle Size, inches Discharge	Rotation* Facing Inlet
29M	C.I.	3	8,405	886	3,855	400	16	6, 8 or 10	CW
	C.S.	3	8,052	935	3,855	400	16	6, 8 or 10	CW
	C.S.	3	9,025	1,034	4,915	500	16	6, 8 or 10	CW
	F.S.	3	9,025	1,034	4,915	500	12,8	8	CW
38M	C.I.	3	15,124	2,462	8,624	950	20	16	CW
	F.S.	3	15,597	2,276	7,953	850	20 or 24	16	CW
	F.S.	3	18,905	2,400	9,965	1,000	20 or 24	16	CW
46M	C.I.	2	23,534	2,992	10,350	1,800	24	20	CW
	F.S.	3	25,888	3,950	12,359	2,000	30	20	CW
	F.S.	3	29,954	4,189	15,072	2,300	30	20	CW
60M	C.I.	3	46,904	6,688	22,373	2,200	36	24	CW
	F.S.	3	41,861	6,688	20,409	2,500	36	24	CW
70M	C.I.	2	54,412	10,876	27,293	3,100	30	30	CW
	F.S.	2	59,616	11,952	30,021	3,400	42 or 48	30	CW
88M	C.I.	2	98,716	21,860	48,904	8,000	54 or 48	36 or 30	CW
	F.S.	2	105,305	24,290	52,531	8,200	54 or 48	36 or 30	CW
103M	C.I.	2	88,000	26,000	40,000	13,000	66 or 60	42	CCW
	F.S.	2	95,000	28,000	44,000	13,800	66 or 60	42	CCW
110M	C.I.	2	115,715	29,872	52,545	15,000	72	48	CCW
	F.S.	2	124,364	31,740	56,746	16,000	72	48	CCW

Approximate Dimensions (inches)

Elliott Compressor Frame	A	B	Overall Length C Min. Stages	Overall Length CC Six Stages	Overall Length CCC Four Stages	Overall Length Each Add'l Stage	Nozzle Distance F Min. Stages	Nozzle Distance FF Six Stages	Nozzle Distance FFF Four Stages	Nozzle Distance Each Add'l Stage	G	E	EE	D
29M	61	58	52	74	—	4.5	24	38	—	4.5	17½	32	32	27
	61	58	52	74	—	4.5	24	38	—	4.5	17½	32	32	27
	61	58	52	74	—	4.5	24	38	—	4.5	18½	32	32	27
	61	58	52	74	—	4.5	24	38	—	4.5	18½	29	29	27
38M	68	83	65	86	—	7	31	52	—	7	20	35	35	27
	68	83	65	86	87	7	31	52	57	7	20	35	39	27
	68	83	65	86	87	7	31	52	57	7	21	35	39	27
46M	84	97	73	100	—	9	39	66	—	9	21	42	42	28
	71	79	87	114	119	9	39	66	69	9	22	44	52	22
	71	79	87	114	119	9	39	66	69	9	23	44	52	22
60M	124	119	105	141	—	12	51	86	—	12	22	68	68	24
	92	103	105	141	148	12	51	86	93	12	25	57	64	24
70M	146	131	103	148	—	15	50	95	—	15	30	80	84	22
	120	128	103	148	157	15	53	98	106	15	23	68	77	24
88M	125	131	115	175	—	20	65	123	—	20	24	72	75	24
	142	131	115	171	161	20	65	123	127	20	24	84	96	24
103M	141	144	131	194	—	21	71	132	—	21	23	78	84	24
	156	148	133	194	198	21	71	132	139	21	27	82	102	24
110M	158	176	128	210	—	24	83	155	—	24	25	92	98	24
	177	176	130	210	222	24	83	155	162	24	29	94	114	24

*The normal drive end is the discharge end. For units requiring opposite rotation, the drive end is the suction end.

Metric units

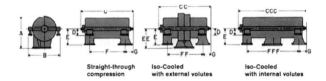

Straight-through compression | Iso-Cooled with external volutes | Iso-Cooled with internal volutes

Technical Data

Elliott Compressor Frame	Material	Min. Casing Length (stages)	Total Weight kg Min.	Total Weight kg Add'l Stage	Weight, Heaviest Part kg Min.	Weight, Heaviest Part kg Add'l Stage	Nominal Nozzle Size Inlet inches	Nominal Nozzle Size Inlet millimetres	Nominal Nozzle Size Discharge inches	Nominal Nozzle Size Discharge millimetres	Rotation Facing Inlet*
29M	C.I.	3	3813	402	1749	180	16	406	6, 8, 10	152, 203, 254	CW
	C.S.	3	3652	424	1749	180	16	406	6, 8, 10	152, 203, 254	CW
	C.S.	3	4093	469	2229	230	16	406	6, 8, 10	152, 203, 254	CW
	F.S.	3	4093	469	2229	230	12,8	305, 203	8	203,	CW
38M	C.I.	3	6860	1117	3912	430	20	508	16	406	CW
	F.S.	3	7075	1032	3607	390	20, 24	508, 610	16	406	CW
	F.S.	3	8575	1089	4520	450	20, 24	508, 610	16	406	CW
46M	C.I.	2	10 675	1357	4695	820	24	610	20	508	CW
	F.S.	3	11 743	1792	5606	910	30	762	20	508	CW
	F.S.	3	13 587	1900	6837	1040	30	762	20	508	CW
60M	C.I.	3	21 276	3034	10 148	1000	36	914	24	610	CW
	F.S.	3	18 988	3034	9258	1130	36	914	24	610	CW
70M	C.I.	2	24 681	4933	12 380	1410	42	1067	30	762	CW
	F.S.	2	27 042	5421	13 618	1540	42, 48	1067, 1219	30	762	CW
88M	C.I.	2	44 778	9916	22 183	3630	54, 48	1372, 1219	36, 30	914, 762	CW
	F.S.	2	47 766	11 018	23 828	3720	54, 48	1372, 1219	36, 30	914, 762	CW
103M	C.I.	2	39 917	11 794	18 144	5900	66, 60	1676, 1524	42	1067	CCW
	F.S.	2	43 092	12 701	19 958	6260	66, 60	1676, 1524	42	1067	CCW
110M	C.I.	2	52 488	13 550	23 834	6800	72	1829	48	1219	CCW
	F.S.	2	56 412	14 397	25 740	7250	72	1829	48	1219	CCW

Approximate Dimensions (millimetres)

Elliott Compressor Frame	A	B	Overall Length C Min. Stages	Overall Length CC Six Stages	Overall Length CCC Four Stages	Overall Length Each Add'l Stage	D	E	EE	Nozzle Distance F Min. Stages	Nozzle Distance FF Six Stages	Nozzle Distance FFF Four Stages	Nozzle Distance Each Add'l Stage	G
29M	1550	1470	1320	1880	—	110	690	810	810	610	970	—	110	440
	1550	1470	1320	1880	—	110	690	810	810	610	970	—	110	440
	1550	1470	1320	1880	—	110	690	810	810	610	970	—	110	470
	1550	1470	1320	1880	—	110	690	810	810	610	970	—	110	470
38M	1730	2110	1650	2180	—	180	690	890	890	790	1320	—	180	510
	1730	2110	1650	2180	2210	180	690	890	990	790	1320	1450	180	510
	1730	2110	1650	2180	2210	180	690	890	990	790	1320	1450	180	530
46M	2130	2460	1850	2540	—	230	710	1070	1070	990	1680	—	230	530
	1800	2010	2210	2900	3020	230	560	1120	1320	990	1680	1750	230	560
	1800	2010	2210	2900	3020	230	560	1120	1320	990	1680	1750	230	580
60M	3150	3020	2670	3580	—	300	610	1730	1730	1300	2180	—	300	560
	2340	2620	2570	3580	3760	300	610	1730	1730	1300	2180	2360	300	640
70M	3710	3330	2620	3760	—	380	560	2030	2130	1270	2410	—	380	760
	3050	3250	2620	3760	3990	380	610	1730	1960	1350	2490	2690	380	580
88M	3175	3330	2920	4440	—	510	610	1830	1900	1650	3120	—	510	610
	3610	3330	2920	4340	4090	510	610	2130	2440	1650	3120	3230	510	610
103M	3580	3660	3330	4930	—	530	610	1980	2130	1800	3350	—	530	580
	3960	3760	3380	4930	5030	530	610	2080	2130	1800	3350	3530	530	690
110M	4010	4470	3250	5330	—	610	610	2340	2490	2110	3940	—	610	640
	4500	4470	3300	5330	5640	610	610	2390	2900	2110	3940	4110	610	740

*The normal driving end is the discharge end. For units requiring opposite rotation, the drive end is the suction end.

B.2   Quick Selection Methods for Multistage Compressors*

Among the many purely graphical methods of rapidly selecting multi-stage compressors is one developed around 1965 by Don Hallock of the Elliott Company, Jeannette, Pa. To use these charts, the following quantities must be known:

1. W—weight flow in lb/min, or scfm—standard ft^3/min.

2. P_1—inlet pressure in psia

3. R_p—pressure ratio (discharge psia/inlet psia)

4. t_1—inlet temp., °F

5. M—molecular weight

6. K—ratio of specific heats

Determine inlet cfm, Q_1.   If W is known, use Fig. B.1, proceeding through P_1, t_1, and M to find Q_1.

If scfm is known, use Fig. B.2, proceeding through P_1, t_1, and "temperature standard" to find Q_1.

Determine head H.   On Fig. B.3, enter R_p and proceed through K, t_1, and M as shown. If head H exceeds 80,000 to 90,000, more than one compressor body will be required.

Determine number of stages required.   On Fig. B.4, enter head H and proceed through M to read the number of stages required. Round this off to the next higher even number.

Determine speed and size of machine.   On Fig. B.5, enter Q_1 and read maximum width in inches. Proceed to the stepped lines and read r/min and flange sizes. Proceed through number of stages and read length of machine in inches. In the example shown, the ICFM is 45,000 and the gas is between propane and chlorine in mole weight. The speed is shown to be 4000 r/min and the flanges are 36 and 24 in. A slightly higher flow requires 3500 r/min and 42- and 30-in flanges.

---

* Segment developed and contributed by Don Hallock, Elliott Company, Jeannette, Pa. Adapted by permission of *HP* and the Elliott Company. Originally published in the October 1965 issue of *Hydrocarbon Processing.*

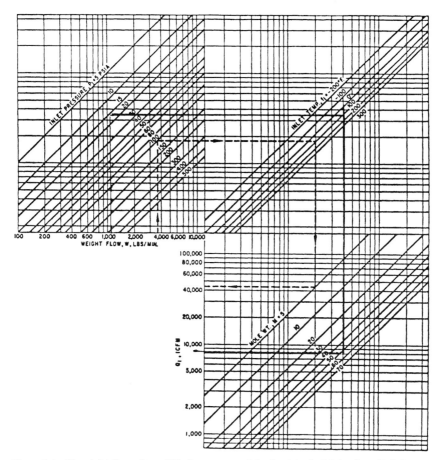

Figure B.1   If weight flow of gas W is known, use this chart to find inlet flow Q_1 icfm.

Determine horsepower requirement. On Fig. B.6, enter W, proceed through Q_1 and H, and read HP. If W is not known, work backward from Q_1 on Fig. B.1 to find W before using Fig. B.6.

For uncooled, constant weight flow compression, such as alkylation, wet gas, recycle, or air under 50 psig, the foregoing is sufficient to determine price, size, and driver requirement. For cooled or variable weight flow compression, proceed as follows:

Cooled compression. Assume one cool and two compression sections, each section handling a pressure ratio equal to the square root of the overall pressure ratio.

- Determine discharge temperature t_2 from Fig. B.7, proceeding through R_p, Q_1, K, and t_1.

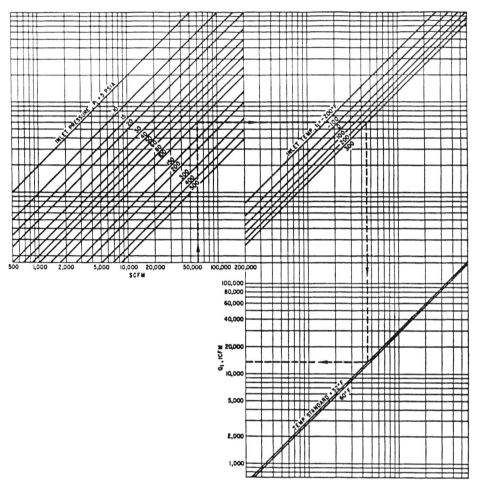

Figure B.2 If scfm is known, use this chart to find inlet flow Q_1 icfm.

- Assuming this t_2 is satisfactory, proceed through all the figures for each of the separate sections. Speed and width of the compressor will be dictated by the first sections. Total HP is the sum of the sections.

- If one cool does not depress t_2 sufficiently, or if still more horsepower saving is desired, try two cools or more. R_p per section for a two-cool, three-section arrangement is the cube root of the overall R_p; for a three-cool, four-section arrangement, it is the fourth root. Bear in mind that more than one set of cooler openings is seldom available on a single compressor body. When more than one cooler is chosen, therefore, more than one compressor body is likely to be required.

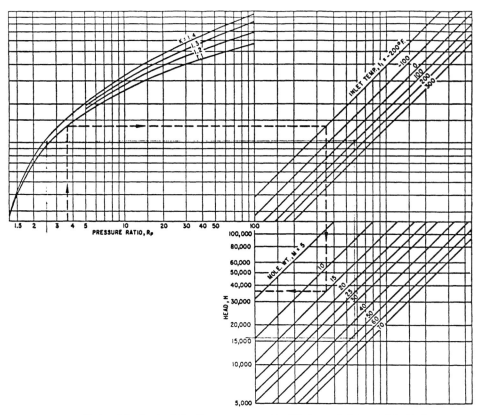

Figure B.3 Enter this chart at R_p, the pressure ratio (discharge/inlet, psia) to find Head H.

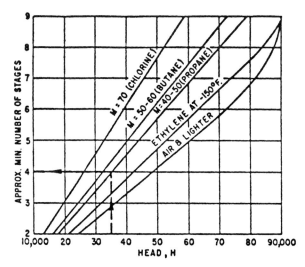

Figure B.4 Enter this chart with H found on Fig. B.3 to find number of stages required.

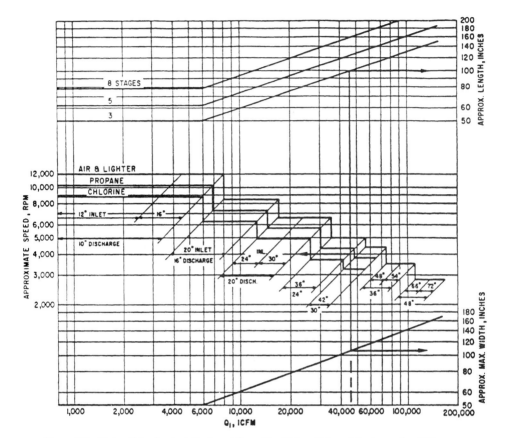

Figure B.5   Enter this chart at Q_1 found from Fig. B.1 or B.2 and find speed, width, length, and flange sizes.

Considerable judgment is required in choosing the number of coolers to use. Once temperature limits are satisfied, the use of additional coolers becomes a matter of economics between compressor and cooler cost, and horsepower evaluation.

Variable weight flow.   For applications having side flows either in or out, it is necessary to consider each constant flow compression section separately. Mixture temperature to the second section after the first "inward" side flow must be calculated by finding the discharge temperature of the first section from Fig. B.7, multiplying by the first section weight flow, adding in the product of the side stream temperature and weight flow, and dividing by the sum of the weight flows. With mixture t_1, P_1, W, M, and K known, the figures can now be used for the second section, and so on through the machine.

M and K of the side stream will generally be the same or quite close to those of the inlet, so mixture calculations for these quantities will

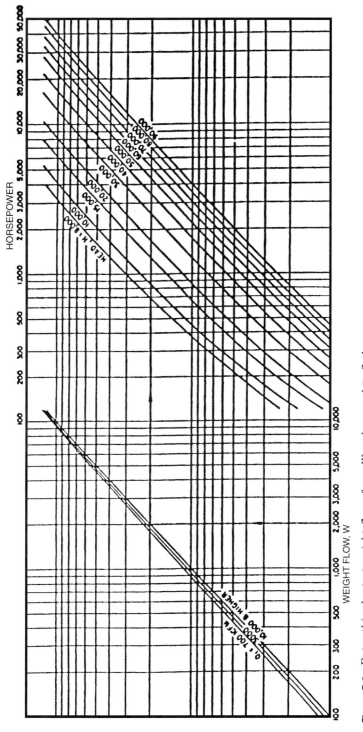

Figure B.6 Enter this chart at weight flow of gas W and proceed to find compressor horsepower required.

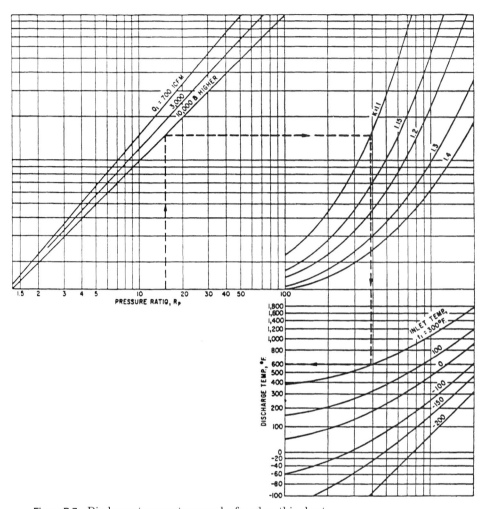

Figure B.7   Discharge temperature can be found on this chart

normally be unnecessary. For extraction side flows, the second section inlet conditions are the same as the first section discharge conditions, except for W.

Normally, the first section will "see" the largest Q_1, in which case the first section Q_1 will dictate the size and speed of the machine. An occasional refrigeration process, however, will show the second section Q_1 to be the largest. In this case, *that* Q_1 will dictate machine size and speed.

To determine the number of stages required, add the stages for each compression section and add in a blank stage for each large side load. It is impossible to give criteria for exactly what constitutes a "large" side load, but experience has shown that a typical propylene unit will require a blank stage for the first side load only, whereas a typical ethylene machine may require two blank stages. If the total number of stages, including blanks, exceeds nine, a second machine will probably be required.

B.3   DeLaval Engineering Guide to Compressor Selection*

---

* Reprinted by permission of IMO DeLaval, Inc., DeLaval Turbine Division, Trenton, N.J.

Delaval Engineering Guide to Compressor Selection

TABLE OF CONTENTS
I.   Definition of Symbols
II.   The Gas Compression Theory
III. Determining Z,k,MW
IV.   Selection Procedure
      A. Qualifying the selection
         procedure
      B. Input data required
      C. Method of calculation
      D. Fabricated multi-stage
         selection procedure
      E. Cast casing compressor
         selection
      F. Design considerations
      G. Compressor Model numbers
V.   Sample Calculation
VI.   Data Section
      A. Fabricated compressors
         (specifications)
      B. Cast casing compressors
         (weights and dimensions)

I. Symbols

The following are the symbols (and their definitions) and the units of measurement used throughout this section.

C	Specific heat of mixture	N	Number of compression stages
C_p	Specific heat at constant pressure	n	Polytropic exponent
		P_R	Reduced pressure
C_v	Specific heat at constant volume	P	Pressure (psia) (bar)
D	Impeller diameter (inches) (mm)	P_c	Critical pressure (psia) (bar)
G	Weight of mixture	P_e	Brakepower (kW)
g	Gravitational constant (32.2 ft/sec²)	P_i	Gaspower (kW)
		Q	Capacity (cfm) (m³/sec)
H	Head (ft lbf/lbm) (Nm/kg)	R	Gas constant $\left(\frac{1544}{MW}\right)\left(\frac{8314}{MW}\right)$
k	Ratio of specific heats (C_p/C_v)		
lbf	Pound force	r	Pressure ratio (P_2/P_1)
lbm	Pound mass	T	Absolute temperature (°R = °F + 460) (°K = C + 273)
ṁ	Mass flow (lbm/min) (kg/sec)		
M_{cp}	Molar specific heat at constant pressure	T_c	Critical temperature (°R) (°K)
MW	Molecular weight	T_R	Reduced temperature
MWP	Maximum working pressure (psi) (bar)	ΔT	Change in temperature (°)
		U	Tip speed (ft/sec) (m/sec)
		V	Total volume (ft³) (m³)

v	Specific volume (ft³/bm) (m³/kg)
W	Weight of gas (lbm)
Z	Compressibility factor
BHP	Brake horsepower (hp)
GHP	Gas horsepower (hp)
η	Efficiency
Φ	Flow coefficient
Ψ	Head coefficient
Subscripts	
ad	Adiabatic process
p	Polytropic process
s	Standard conditions (14.7 psia, 60°F, dry Q₁) (1.0135 bar and 0 °C)
1	Inlet to section
2	Discharge from section
x	At a specific point (inlet, discharge, etc.)

II. The Gas Compression Theory

The relationship between the volume, absolute pressure and absolute temperature of a perfect gas, based on Charles' and Boyle's Laws, is: PV = WRT; or, on a mass basis, Pv = RT.

An important characteristic of gases is specific heat.

Specific heat is defined as the amount of heat (BTU) (kJ) required to raise the temperature of one pound (kilogram) of gas one degree Fahrenheit (Kelvin). The amount varies depending on whether the gas volume or pressure is kept constant during the heating process. This is defined by: R ∝ C_p-C_v. The ratio (k) of specific heat of a gas at constant pressure to that at constant volume (C_p/C_v) is used in gas calculations.

If heat is neither added nor removed from the gas during compression, the process is defined as isentropic or adiabatic. The relationship of pressure and volume for a perfect gas undergoing isentropic compression is defined as PV^k, a constant.

Because many gases do not perfectly obey the theoretical laws, the deviation must be accounted for. The deviation, termed compressibility (Z), is defined as the ratio of actual gas volume at a given temperature and pressure to the volume calculated by the theoretical law (Pv = RT).

The general equation for adiabatic work is:

$$H = ZRT\left[\frac{\left(\frac{P_2}{P_1}\right)^{\frac{k-1}{k}} - 1}{\frac{k-1}{k}}\right] \text{ ft lbf/lbm (Nm/kg)}$$

The actual compression path seldom follows the adiabatic process but is generally in the form PV^n, a constant. This is called a polytropic process and is defined as reversible with heat transfer.

n is the exponent of polytropic compression and is found from:

$$\frac{n-1}{n} = \frac{k-1}{k}\left[\frac{1}{\eta_p}\right]$$

where η_p is the polytropic compression efficiency. Figure 1 shows the relationship between polytropic and adiabatic efficiency.

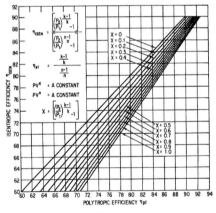

Figure 1

1

III. Determining Z, k, and MW

Before a compressor cycle can be calculated, it is necessary to know the specific heat ratio, k; molecular weight, MW; and compressibility, Z, of the gas. For pure gases or air, these values can be taken from Figure 2. For a mixture of gases, the values must be calculated. Mixtures are generally specified in volumetric or mole percentages.

The properties of the mixture are determined by the composite properties of the constituent gases.

The values for the compressibility (Z) of gas mixtures can be calculated if the gas analysis is known.

Z can be derived from the rule of corresponding states using reduced temperature and pressure. To calculate reduced temperature (T_R) and reduced pressure (P_R), see the following information. The critical constants T_c and P_c for various gases are given in Table 1.

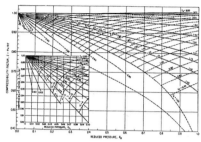

Figure 2

English & Metric

To facilitate the broadest use of the information contained in this brochure, terms have been stated in both English and Metric.

Use of parentheses, blue type or light blue shading over areas of type indicate Metric values.

Physical Constants of Gases

Compound	Formula	Mol. wt. MW	C_p and C_p/C_v at 14.7 psia and 60°F		C_p and k at 1.01325 bar and 0°C		Critical constants		Critical constants		MC_p at 60°F	MC_p at 100°F	MC_p at 200°F	MC_p at 0°C	MC_p at 25°C	MC_p at 100°C
			C_p	C_p/C_v	C_p kJ/(kg·k)	k	psia P_c	°R T_c	bar P_c	K T_c						
Acetylene	C_2H_2	26.036	0.3966	1.238	1.6345	1.243	905.0	557.4	61.4	308.3	10.33	10.69	11.53	42.56	43.72	47.62
Air	$N+O_2$	28.966	0.2470	1.395	1.0048	1.400	547	238.7	37.7	132.4	6.96	6.96	6.99	29.11	29.11	29.11
Ammonia	NH_3	17.032	0.5232	1.310	2.0323	1.317	1,657	731.4	112.8	405.5	8.91	8.57	9.02	34.61	35.08	37.25
Benzene	C_6H_6	78.108	0.2404	1.118	0.9429	1.128	714	1,013.0	49.0	562.2	18.78	20.47	24.46	73.65	82.09	105.05
1,2-Butadiene	C_4H_6	54.088	0.3458	1.12	1.3934	1.124	653	799.0	45.0	443.7	18.70			75.37	80.17	93.71
1,3-Butadiene	C_4H_6	54.088	0.3412	1.12	1.3615	1.126	628	766.0	43.3	425.4	18.45			73.65	79.72	95.75
N-Butane	C_4H_{10}	58.120	0.3970	1.094	1.5625	1.101	550.7	765.6	38.0	425.2	23.07	24.51	26.16	90.84	97.83	117.83
Isobutane	C_4H_{10}	58.120	0.3872	1.097	1.5433	1.102	529.1	734.9	36.5	408.1	22.50	23.96	27.62	89.70	97.10	118.08
N-Butene	C_4H_8	56.104	0.3703	1.105	1.4160	1.117	583	755.6	40.2	419.6	20.77	22.09	25.18	79.45	85.74	103.31
Isobutene	C_4H_8	56.104	0.3701	1.106	1.4872	1.111	579.8	752.5	40.0	417.9	20.76			83.44	89.03	105.66
Butylene	C_4H_8	56.104	0.3703	1.105	1.4687	1.112	583	755.6	41.0	428.6	20.78	21.94	24.86	82.41	87.86	103.88
Carbon dioxide	CO_2	44.010	0.1991	1.300	0.8223	1.299	1,073	548.0	73.8	304.2	8.76	9.00	9.35	36.19	37.04	39.80
Carbon monoxide	CO	28.010	0.2484	1.403	1.0467	1.397	510	242.0	35.0	132.9	6.96	6.96	6.98	29.32	28.97	28.85
Chlorine	Cl_2	70.914	0.1149	1.366	0.4731	1.330	1,120	751	77.2	417.2	8.15			33.55	33.85	35.03
Ethane	C_2H_6	30.068	0.4097	1.193	1.6462	1.202	708.3	550.1	48.8	305.4	12.32	12.96	14.68	49.50	52.88	62.72
Ethyl alcohol	C_2H_5OH	46.069	0.3070	1.130	1.5240	1.135	927.0	929.6	63.8	516.3	14.14			70.21	73.49	84.10
Ethylene	C_2H_4	28.052	0.3622	1.243	1.4562	1.256	742.1	509.8	50.3	282.4	10.16	10.68	12.08	40.85	43.58	51.42
N-Hexane	C_6H_{14}	86.172	0.3984	1.062	1.5416	1.067	439.7	914.5	30.1	507.4	34.33	36.23	41.08	132.85	143.24	172.90
Helium	He	4.003	1.2480	1.659	5.2000	1.667	480	510	2.3	5.2	5.00			20.82	20.82	20.82
Hydrogen	H_2	2.016	3.408	1.408	14.3649	1.404	188.0	60.2	13.0	33.33	6.87	6.90	6.95	28.96	28.66	28.41
Hydrogen sulfide	H_2S	34.076	0.254	1.323	0.9797	1.333	1,306	672.7	90.1	373.5	8.66	8.18	8.36	33.38	33.53	34.58
Methane	CH_4	16.042	0.5271	1.311	2.1637	1.316	673.1	343.5	46.1	190.6	8.46	8.65	9.30	34.71	35.90	39.62
Methyl alcohol	CH_3OH	32.042	0.2700	1.203	1.3398	1.241	1,157.0	924.0	80.9	512.6	8.65			42.93	45.08	51.11
Nitrogen	N_2	28.016	0.2482	1.402	1.0467	1.396	492.0	227.2	126.2	126.2	6.95	6.96	6.963	29.32	28.97	28.74
N-Octane	C_8H_{18}	114.224	0.3998	1.046	1.5349	1.050	362.1	1,025.2	24.9	568.8	45.67			175.33	189.39	228.04
Oxygen	O_2	32.00	0.2188	1.401	0.9169	1.396	730	278.2	50.8	154.8	7.00	7.03	7.120	29.34	29.21	29.61
N-Pentane	C_5H_{12}	72.146	0.3972	1.074	1.5541	1.080	489.5	845.9	33.7	469.7	28.66	30.30	34.41	112.12	120.83	145.36
Isopentane	C_5H_{12}	72.146	0.3880	1.075	1.5248	1.082	483.0	830.0	33.8	460.4	27.99	29.90	34.44	110.02	119.02	144.76
Propane	C_3H_8	44.094	0.3885	1.136	1.5516	1.139	617.4	666.2	42.5	369.8	17.13	18.21	20.90	68.42	73.85	89.58
Propylene	C_3H_6	42.078	0.3541	1.154	1.4202	1.162	667	657.4	46.1	364.8	14.90	15.77	17.88	59.76	63.96	76.15
Sulfur dioxide	SO_2	64.060	0.1470	1.246	0.6029	1.275	1,142	775.0	78.9	430.7	9.42			38.62	39.43	42.11
Toluene	C_7H_8	92.134	0.2599	1.091	1.0224	1.097	611	1,069.5	41.1	591.8	23.95			94.21	104.16	131.24
Water	H_2O	18.016	0.4446	1.335	1.8715	1.328	3,206	1,165.4	221.2	647.4	8.01	8.03	8.12	33.72	33.42	33.49
Natural gas		19.27	0.488	1.269	1.799	1.316	670	380	46.2	211.1	8.47	8.72	9.37	34.66	35.90	37.60

Table 1

For example, for a gas mixture with a composition (by volume) of 14% ethane, 85% methane and 1% nitrogen, T_C and P_C would be calculated as follows:

Gases	V (%/100)	T_C	T_C	VT_C	VT_C	P_C	P_C	VP_C	VP_C
C_2H_6	0.14	550.1	305.4	77.01	42.76	708.3	48.8	99.16	6.83
CH_4	0.85	343.5	190.6	292.00	162.01	673.1	46.1	572.19	39.14
N_2	0.01	227.2	126.2	2.27	1.26	492.0	33.9	4.92	0.34

For mixture T_C = 371.28°R (206.03°C) P_C = 676.27psia (46.31 bar)

Using the above values, and assuming gas conditions of 90°F (30°C) and 124.5 psia (8.5 bar):

$$T_R = \frac{T}{T_C} = \frac{90 + 460}{371.3} \quad \frac{30 + 273.15}{206.03} = 1.48$$

$$P_R = \frac{P}{P_C} = \frac{124.5}{676.3} \quad \frac{8.5}{46.31} = 0.18$$

Using the calculated values of reduced temperature and pressure, the value of Z (.98) can be read from Figure 2, a generalized curve that can be used for any gas mixture. Figure 3 is a curve directly showing compressibility factors of natural gas at various pressures and temperatures.

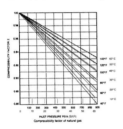

Figure 3

The molecular weight of a gas mixture is equal to the sum of the products of the proportional volume of each constituent and its molecular weight.

$$MW = m_1 v_1 + m_2 v_2 \dots + m_n v_n$$

A simplified method for finding the ratio of specific heats (k) makes use of the molal specific heat M_{cp} expressed as

$$k = \frac{M_{cp}}{M_{cp} - 1.99} \qquad k = \frac{M_{cp}}{M_{cp} - 8.33}$$

Calculation of the properties of a gas mixture can best be done in tabular form. The following example determines the properties of a typical natural gas.

Gas	V (%/100)	MW	V(MW)	M_{cp} at 100°F	VM_{cp}	M_{cp} at 25°C	VM_{cp}
C_2H_6	0.14	30.07	4.21	12.96	1.814	52.88	7.40
CH_4	0.85	16.04	13.63	8.65	7.353	35.80	30.43
N_2	0.01	28.02	0.28	6.96	0.069	28.97	0.29
Total	1.00		MW = 18.12		M_{cp} = 9.237		M_{cp} = 38.12

$$k = \frac{9.237}{9.237 - 1.99} = 1.275$$

$$k = \frac{38.12}{38.12 - 8.33} = 1.280$$

IV. Selection Procedure

A. QUALIFYING THE SELECTION PROCEDURE

This procedure is intended to aid the user in making rapid preliminary compressor selections and estimating compressor performance. Only Delaval engineering will issue formal selections.

The method is to be used on a sectional basis. It examines a gas before it enters and after it leaves the compressor or compressor section (Figure 4). In the case of intercooled or side loaded compressors, the sections must be dealt with separately; the section with the largest inlet flow (Q) governs the frame size.

Figure 4

B.   INPUT DATA REQUIRED

Selection of a compressor frame size and calculation of performance requires the following data: k, Z, MW, P_1, P_2, T_1, Q_1, (or $\dot{m}$). If the gas analysis is provided, values for k, Z, and MW can be calculated.

C. METHOD OF CALCULATION

The Delaval process compressor line was designed around the concept of component and performance similarity throughout the various frame sizes. Using the non-dimensional impeller flow coefficient (Φ) as a basis for determining aerodynamic performance of an impeller of any size, a common link between frame sizes results. In this way, theoretical and test data have been combined to define compressor characteristics for any size unit. The following procedure utilizes this approach for compressor selection.

D. FABRICATED MULTI-STAGE SELECTION PROCEDURE

Steps:

1. Calculate volmetric inlet flow (ACFM) from either of the following methods:

a. From mass flow rate ($\dot{m}$),
$$ACFM_X = v_X (\dot{m}) \text{ where } v_X = \frac{Z_X RT_X}{144 P_X}$$

b. From moles/hour,
$$ACFM_X = \frac{(Moles/hour)(MW)(v_X)}{60}$$

c. From SCFM,
$$ACFM_X = SCFM \frac{(P_S)(T_X)(Z_X)}{(P_X)(T_S)(Z_S)}$$

1. Calculate volumetric inlet capacity from either of the following methods:

a. From mass flow rate
$$Q_X = \dot{m}(v_X) \text{ where } v_X = \frac{Z_X(R)T_X}{P_X}$$

b. From moles/hour
$$Q_X = \frac{(Moles/hour)(MW) v_X}{3600}$$

c. From standard volumetric inlet flow,
$$Q_X = \frac{Q_S(R_S)(T_X)(Z_X)}{(P_X)(T_S)(Z_S)}$$

2. Calculate adiabatic head based on inlet conditions to section,
$$H_{ad} = Z_1 RT_1 \frac{\left[r^{\frac{k-1}{k}} - 1 \right]}{\frac{k-1}{k}}$$

3. Estimate discharge temperature (T_2) due to compression cycle[1]
$$\Delta T = T_1 \frac{\left[r^{\frac{k-1}{k}} - 1 \right]}{\eta_{ad}} \text{ (assume } \eta_{ad} = .75)$$
$$T_2 = T_1 + \Delta T$$

4. Determine minimum frame size from Figure 5.

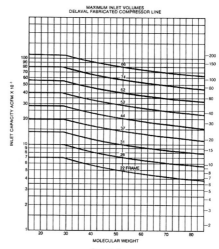

Figure 5

Footnote:
[1]Nominal temperature limitations are 450°F (250°C) for labyrith seals and 375°F (190°C) for oil face or bushing seals.

5. Find impeller wheel diameters from following table

Frame	D	Frame	D (mm)
22	13.65"	22	347
26	16.25"	26	413
31	19.25"	31	489
37	22.875"	37	581
44	27.25"	44	692
52	32.5"	52	826
62	38.5"	62	978
74	45.6"	74	1158
88	54.25"	88	1378

6. Determine maximum impeller head per stage from Figure 6. Minimum number of compression stages required from:

$$\text{No. of stages} = \frac{H_{ad}}{\text{Head per stage}}$$

Round off quantity to the next higher integer.

7. Calculate tip speed[1]

$$U = \sqrt{\frac{H_{ad}(g)}{N\psi}} \qquad U = \sqrt{\frac{H_{ad}}{N\psi}}$$

Select nominal ψ

MW	ψ
6	.45
18	.46
29	.48
44	.50
71	.51

8. Calculate inlet and discharge flow coefficients[2]

$$\Phi_x = \frac{3.056\, Q_x}{UD^2} \qquad \Phi_x = \frac{4Q_x}{\pi U D^2}$$

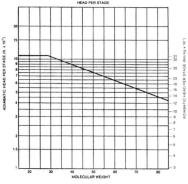

Figure 6

Footnotes:
[1]For initial sizing, limit tip speed to 900 ft/sec (275 m/sec); or 800 ft/sec (245 m/sec) if low-yield material is required.
[2]Discharge flow coefficient is calculated from discharge conditions in this procedure. It is normally determined from conditions prior to the last stage of compression.

9. Use Figure 7 to determine first and last stage efficiency and average to get overall efficiency., If Φ falls to the right of the efficiency curve, select a larger frame size. If Φ falls to the far left, select a smaller frame size.

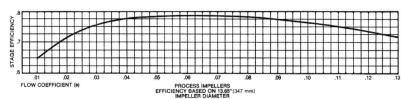

Figure 7

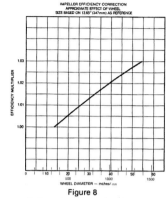

Figure 8

10. Correct efficiency for wheel size using Figure 8.

11. Calculate compressor running speed

$$RPM = \frac{229U}{D}$$ $$RPM = \frac{60U}{D}$$

12. Calculate horsepower

a. $$GHP = \frac{H_{ad}\,[\dot{m}]}{33,000[\,\eta_{ad}\,]}$$ $$P_i = \frac{H_{ad}\,[\dot{m}]}{1000\,[\,\eta_{ad}\,]}$$

b. Determine mechanical losses from Figure 9. (Divide total by 2 if labyrinth end seals are used.)

c. Calculate balance drum leakage (2% of GHP or P_i)[1]

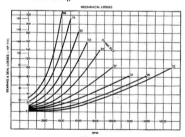

Figure 9

d. BHP = GHP + Mech. losses + balance drum leakage

$$P_e = P_i + \text{Mech. losses} + \text{balance drum leakage}$$

13. Determine casing split

The density of the gas and the maximum working pressure of the compressor will determine the casing split. The following chart is provided as a general guide:

Frame Size	22	26	31	37	44	52	62	74	88
MWP for horizontally split casing (psi)	800	600	600	600	450	300	300	300	300
(bar)	56	42	42	42	32	22	22	22	22

MWP = 1.10 (max. discharge pressure)

If the gas contains over 70% hydrogen, the casing will be vertically split between 200 to 285 psi (14 to 20 bar) MWP and above.

E. CAST CASING COMPRESSOR SELECTION

Although the calculation method presented in this section is based on the Delaval fabricated line, some performance data for cast case units can be calculated from the previous procedure. Once head and inlet flow are determined, the figures presented on pages 11 and 13 should be used to select the proper frame size and number of stages. Impeller diameter and efficiency corresponding to case size is presented below.

MULTISTAGE:

Case	12/12	16/16	18/18	20/20	24/24	30/30	36/36
Nominal Impeller Dia. (Inches)	14	14	23	23	30	38	45
(mm	355	355	584	584	762	965	1143
Avg. Adiabatic Efficiency	.78	.78	.80	.80	.81	.82	.82

SINGLESTAGE (opposed nozzles):

Case	20/20	24/24	30/30	36/36
Nominal Impeller Dia. (inches)	18	32	32	38
(mm)	457	813	813	965
Avg. Adiabatic Efficiency	.80	.81	.82	.84

Note: All single-stage units are available in either axial inlet or opposed nozzle configurations. Refer to factory for axial inlet efficiencies.

Substituting this information for steps 8 through 10 allows a quick estimation of pipeline compressor performance.

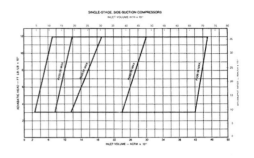

FRAME SIZE SELECTION FOR SINGLE-STAGE, OVERHUNG COMPRESSORS

Determine actual inlet flow into the compressor as well as the total head requirement to find frame size

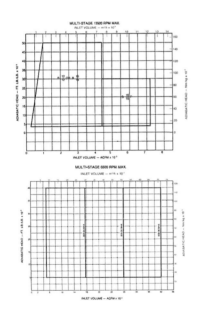

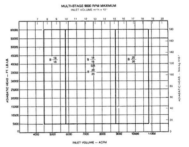

FRAME SIZE SELECTION FOR MULTI-STAGE CAST CASING COMPRESSORS

Calculate the actual inlet flow into the compressor as well as the total head requirements using the driver speed to determine the correct sizing graph. Assume a maximum of 10,000-11,000 ft. (30,000-33,000 Nm/kg) of head per stage to pinpoint the number of compression stages required.

F. DESIGN CONSIDERATIONS

In many cases, a centrifugal compressor must be designed to match special process or driver requirements. By physical arrangement of inner components or the casing structure, specific requirements can be met while still delivering maximum performance. Variations include:

Double-flow arrangement which permits the unit to be smaller in frame size and higher in rotational speed. The inlet flow is split in half and undergoes parallel compression (see Figure 10).

Figure 10   Figure 11

Back-to-back arrangement of two sections in an intercooled machine, which keeps hot discharge temperatures away from end seals and reduces or eliminates aerodynamic thrust forces (see Figure 11).

Overframing the casing and diaphragms, which is sometimes used to increase compressor efficiency. The diffuser plate diameter is increased while impeller diameter is held constant.

Uprating flow capacity of the compressor, which may only require an increase in speed for small changes in flow or an inner bundle change-out for large variations. Nozzle sizes and internal dimensions of the casing will determine the maximum flow capability of the compressor. Consult factory for specific information.

Rotational speed, which can be varied by two methods while the section still produces constant head.

(a) Addition of one impeller permits speed reduction as shown in the equation

$$\text{Revised RPM} = \sqrt{\frac{N}{N+1}}\ \left[\text{RPM}\right]$$

(b) Wheel triming by reducing the outside diameter of the impeller can allow for up to a 10% increase in rotational speed.

G. COMPRESSOR MODEL NUMBERS

Every Delaval centrifugal compressor is designated by a model number that describes that particular unit. Typical model numbers (and their meanings) for process and pipeline units are shown below.

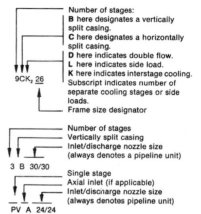

Number of stages:
B here designates a vertically split casing.
C here designates a horizontal split casing.
D here indicates double flow.
L here indicates side load.
K here indicates interstage cooling.
Subscript indicates number of separate cooling stages or side loads.
Frame size designator

$9CK_2\ 26$

Number of stages
Vertically split casing
Inlet/discharge nozzle size
(always denotes a pipeline unit)

3 B 30/30

Single stage
Axial inlet (if applicable)
Inlet/discharge nozzle size
(always denotes pipeline unit)

PV A 24/24

V. Sample Calculations (English)

Given: $k = 1.275$, MW = 18.12, Z = .98
$P_1 = 124.5$ psia, $P_2 = 500$ psia,
$\dot{m} = 5470$ lbm/min
$T_1 = 90°F = 550°R$

Steps:

1. $ACFM_1 = \dot{m} \, v_1; \; v_1 = \dfrac{.98 \left[\dfrac{1544}{18.12}\right] 550}{124.5 \, (144)} = 2.56 \, \dfrac{ft^3}{lbm}$

$ACFM_1 = 5470 \, (2.56) = 14000 \, ft^3/min$

2. $H_{ad} = .98 \left[\dfrac{1544}{18.12}\right] (550) \left[\dfrac{\left[\dfrac{500}{124.5}\right]^{\frac{1.275-1}{1.275}} - 1}{\dfrac{1.275 - 1}{1.275}}\right] = 74,460 \, ft.$

3. $\Delta T = 550 \dfrac{\left[\dfrac{500}{124.5}\right]^{\frac{1.275-1}{1.275}} - 1}{.75} = \begin{array}{r} 256° \\ +90° \\ \hline 346° \end{array}$

$T_2 = 346°F$ (no intercooling required)

4. From Figure 5, inlet flow is close to maximum of 31 frame and well within the range of 37 frame.

5. Wheel diameters   31 ⟶ 19.25"
                       37 ⟶ 22.875"

6. From Figure 6, maximum head per stage = 11,000 ft.

Minimum number of stages = 74,460/11,000 = 6.77 or 7 stages.

7. $U = \sqrt{\dfrac{74,450 \, (32.2)}{(7) \, (.46)}} = 863$ ft/sec

Conversion Table

TO OBTAIN	MULTIPLY	BY
inches	mm	0.0394
ft³	m³	35.31
ft/sec	m/sec	3.281
cfm	m³/h	0.5883
head (ft)	Nm/kg	0.335
lbm/min	kg/sec	132
psi	bar	14.22
hp	kW	1.341

8. Φ_1 for 31 frame $= \dfrac{3.056 \, (14000)}{863 \, (19.25)^2} = .134$

According to Figure 7, a 31 frame is marginal.

Φ_1 for 37 frame $= \dfrac{3.056 \, (14000)}{863 \, (22.875)^2} = .095$

Φ_2 is calculated from Q_2

$Q_2 = \dot{m} \, v_2$

$v_2 = \dfrac{Z_2 R T_2}{144 \, P_2}$; $Z_2 = .99$ (from example on page 27, Z_2 is found on Figure 2 from T_R and P_R)

$v_2 = \dfrac{.99 \left[\dfrac{1544}{18.12}\right] (806)}{144 \, (500)} = .944 \, \dfrac{ft^3}{lbm}$

$Q_2 = 5470 \, (.944) = 5166 \, ft^3/min$

$\Phi_2 = .035$

9. From Figure 7:

$\eta \, \Phi_1 = .775$; $\eta \, \Phi_2 = .775$; η avg. = .775

10. Determine impeller efficiency correction from Figure 8:

$1.0075 \, (.775) = .781$

11. $RPM = \dfrac{229 \, (863)}{22.875} = 8640$ RPM

12. $GHP = \dfrac{74460 \, (5470.)}{33,000 \, (.781)} = 15,803$ hp

Mechanical losses = 81 hp.

$BHP = 1.02 \, (15,803) + 81 = 16,200$ hp

13. A discharge pressure of 500 psia corresponds to a 550 psi MWP casing. Therefore, casing is horizontally split. Model selected is a seven-stage, 37-frame horizontally split: 7C37.

V. Sample Calculations (Metric)

Given k = 1.280, M = 18.29, z = 0.98
P_1 = 8.5 bar, P_2 = 34.5 bar
$\dot{m}$ = 42 kg/sec, T_1 = 30°C = 303.15 K

Steps.

1. $Q_1 = \dot{m}\,(v_1) = \dot{m}\,\dfrac{(Z_1)\,(R)\,(T_1)}{P_1}$

$= 42\,\dfrac{(0.98)\,(8314.34)\,(303.15)}{(18.129)\,(8.5 \times 10^5)}$

$= 6.732$ m³/sec $= 24235$ m³/hr

2. Had $=$

$= .98\left[\dfrac{8314.34}{18.129}\right](303.15)\dfrac{\left[\dfrac{34.5}{8.5}^{\frac{1.280-1}{1.280}} - 1\right]}{\frac{1.280-1}{1.275}}$

$=223350$ Nm/kg

3. $\Delta T = \dfrac{303.15}{0.75}\left[\left(\dfrac{34.5}{8.5}\right)^{\frac{1.280-1}{1.290}} - 1\right] = 145$ K

$T_2 = 303.15 + 145 = 448.15$ K $= 175$°C
(no intercooling required)

4. $Q_1 = 24235$ m³/hr. From Figure 5, inlet flow is close to maximum of 31 frame and well within the range of 37 frame.

5. Wheel diameters   31 —frame ──▶ 489 mm
37 —frame ──▶ 581 mm

6. From Figure 6, maximum head per stage = 33000 Nm/kg.

Therefore N $= \dfrac{223350}{33000} = 6.77$ or 7 stages

7. U $= \sqrt{\dfrac{223350}{7\ \ 0.46}} = 263$ m/sec.

Conversion Table

TO OBTAIN	MULTIPLY	BY
mm	inches	25.40
m³	ft³	0.0283
m/sec	ft/sec	0.305
m³/h	cfm	1.6992
Nm/kg	ft (head)	2.989
kg/sec	lbm/min	7.58 x 10⁻³
bar	psi	0.0703
kW	hp	0.746

8. Φ_1, 31-frame $= \dfrac{4\,(6.732)}{263\,(\pi)\,(0.489^2)} = 0.136$

According to Figure 7, a 31-frame is marginal.

Φ_1, 37-frame $= \dfrac{4\,(6.732)}{262\,(\pi)\,(0.581^2)} = 0.097$

Φ_2 is calculated from Q².

$Q_2 = \dot{m}v_2 = \dot{m}\,\dfrac{(Z_2)\,(R)\,T_2}{P_2}$ Find Z_2 from reduced temperature and pressure (from example shown on page 27)

$T_R = \dfrac{T_2}{T_c} = \dfrac{448.15}{206.3} = 2.18$

$P_R = \dfrac{P_2}{P_c} = \dfrac{34.5}{46.31} = 0.74$

From Figure 2 : $Z_2 = 0.99$

Therefore $Q_2 =$

$= 42\,\dfrac{(0.99)\,(8314.34)\,(448.15)}{(18.129)\,(34.5 \times 10^5)} = 2.477$ m³/sec

$\Phi_2 = \dfrac{(4)\,2.477}{(263)\,(\pi)\,(0.581^2)} = 0.036$

9. From Figure 7: $\eta_{\Phi_1} = 0.775$

$\eta_{\Phi_2} = 0.775$ average = 0.775

10. Determine impeller efficiency correction from Figure 8.

$(1.0075)\,(0.775) = 0.781$

11. N $= \dfrac{60\,(263)}{\pi\,(0.581)} = 8645$ RPM

12. $P_i = \dfrac{(42)\,223350}{0.781\,(1000)} = 12011$ kW

Mechanical losses = 63 kW (from Figure 9)
P_e = (1.02) (12011) + 63 = 12314 kW

13. A discharge pressure of 34.5 bar corresponds to a 42 bar MWP casing. Therefore, casing is horizontally split. Model selected is a seven-stage, 37-frame horizontally split : 7C37.

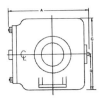
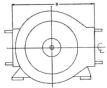

Model B 12/12
Model B 16/16

Weights and dimensions

Cast Casing Compressors

FRAME SIZE	MAX. NO. STAGES	A		B		C		D		NOZZLE SIZE SUCTION		DISCHARGE		TOTAL WEIGHT		MAX. MAINT. WT.	
		in.	mm	in.	mm	in.	mm	in.	mm	in.	mm	in.	mm	lb.	kg	lb.	kg
B 12/12	5	53	1346	48	1219	38	965	27	685	12	305	12	305	13,000	5900	3,050	950
B 16/16	5	57	1448	48	1289	49	1295	29	737	16	406	16	406	16,200	7300	3,100	1400
B 18/18	5	103	2616	96	2348	69	1753	36	915	18	457	18	457	34,300	19600	10,300	5200
B 20/20	5	105	2667	96	2438	72	1829	30	760	20	508	20	508	45,000	20900	11,500	5300
B 24/24	3	103	2616	102	2591	84	2134	45	1143	24	610	24	610	48,000	21800	12,900	5900
B 30/30	3	125	3175	144	3658	90	2286	54	1372	30	762	30	762	85,000	38500	15,500	7000
B 36/36	3	126	3200	160	4064	117	2972	67	1702	36	914	36	914	120,000	54500	18,600	8500
PV 20/20	1	87	2210	81	2057	64	1626	33	838	20	508	20	508	30,000	13600	3,400	1550
PV 24/24	1	96	2438	120	3048	77	1956	43	1092	24	610	24	610	40,500	18300	4,960	2300
PV 30/30	1	102	2591	134	3404	81	2057	45	1143	30	762	30	762	51,000	23100	6,250	2850
PV 36/36	1	104	2642	144	3658	104	2642	53	1346	36	914	36	914	66,000	29000	8,300	3800

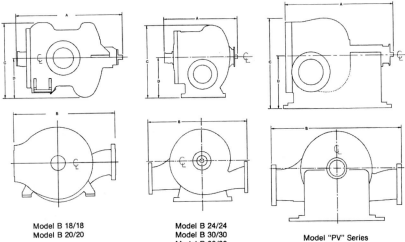

Model B 18/18
Model B 20/20

Model B 24/24
Model B 30/30
Model B 36/36

Model "PV" Series

Note: Axial inlet (PVA type compressor) is located in end cover at ₵ of shaft.

11

Weights and dimensions

Fabricated Centrifugal Compressors

Summary of Specifications

NOTES:
1. Speed and flow ratings are based on 275m/sec (900ft/sec) tip speed and largest standard impeller using standard materials. These ratings may be affected by gas and process conditions or special material requirements.
2. Weights given apply for listed pressure rating and nozzle sizes only.
3. Max. maint. weight is the greatest weight necessary to handle in normal maintenance.
4. All dimensions are approximate.

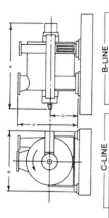

FRAME SIZE	RATED SPEED rpm	RATED INLET FLOW (f) dfm	m³/hr	MAX. INLET NOZZLE DIA. in	mm	MAX. DISCH. NOZZLE DIA. in	mm	NO OF STAGES	A in	mm	B in	mm	C in	mm	D in	mm	C-LINE TOTAL WEIGHT lb	kg	C-LINE MAX. MAINT. lb	kg	B-LINE TOTAL WEIGHT lb	kg	B-LINE MAX. MAINT. lb	kg	ROTOR WEIGHT lb	kg
22	15,100	7,000	12,000	16	406	12	305	5	64	1625	56	1420	55	1400	25	635	13000	5900	5000	2300	15500	7100	3500	1600	425	200
								6	69	1750	56	1420	55	1400	25	635	14000	6400	5500	2500	16500	7500	4000	1800	475	215
								7	73	1850	56	1420	55	1400	25	635	15000	6800	6000	2700	17500	8000	4500	2000	525	240
								8	78	1980	56	1420	55	1400	25	635	16000	7250	6500	3000	18500	8400	5000	2300	600	275
26	12,700	10,000	17,000	18	457	14	356	5	68	1725	64	1625	62	1580	28	715	18000	8200	6500	3000	22500	10200	5000	2300	675	300
								6	73	1850	64	1625	62	1580	28	715	19500	8900	7000	3200	24500	11100	6000	2700	775	350
								7	78	1980	64	1625	62	1580	28	715	21300	9700	7500	3400	26500	12000	7000	3200	875	400
								8	84	2130	64	1625	62	1580	28	715	23000	10500	8000	3700	28500	13000	8000	3700	975	450
31	10,700	14,000	24,000	20	508	16	406	5	75	1905	69	1750	69	1750	31	790	26000	11800	8500	3900	27000	12300	8000	3700	1100	500
								6	81	2060	69	1750	69	1750	31	790	28500	12900	9000	4100	29500	13400	9500	4300	1250	550
								7	88	2235	69	1750	69	1750	31	790	31000	14000	9200	4200	32000	14500	11000	5000	1400	650
								8	94	2390	69	1750	69	1750	31	790	33500	15200	10000	4600	34500	15700	12500	5700	1550	700
37	9,000	20,000	34,000	24	610	20	508	5	82	2080	77	1955	77	1960	35	890	37000	16800	13000	5900	36000	16400	13000	5900	1700	800
								6	89	2260	77	1955	77	1960	35	890	41000	18600	14000	6400	39500	18000	15250	6800	1950	900
								7	96	2440	77	1955	77	1960	35	890	45000	20400	15000	6800	43000	19500	17500	8000	2200	1000
								8	105	2670	77	1955	77	1960	35	890	49000	22250	16000	7300	46500	21000	19000	8600	2450	1100
44	7,550	28,000	48,000	30	762	24	610	5	92	2340	84	2130	84	2130	38	970	50000	22700	16000	7300	45000	20400	20000	9100	2850	1300
								6	104	2640	84	2130	84	2130	38	970	56000	25400	17000	7700	50000	22700	24000	11000	3300	1500
								7	112	2845	84	2130	84	2130	38	970	62000	28100	18000	8200	55000	25000	28000	12700	3750	1700
								8	124	3150	84	2130	84	2130	38	970	68000	30850	19000	8600	60000	27200	32000	14500	4200	1900
52	6,390	40,000	68,000	36	914	30	762	5	107	2720	98	2490	98	2490	45	1145	70000	31750	18000	8200	64000	29000	34000	15500	4200	1900
								6	117	2970	98	2490	98	2490	45	1145	78000	35400	19000	8600	71000	32700	41000	18600	4850	2200
								7	127	3225	98	2490	98	2490	45	1145	87000	39500	21000	9500	80000	36300	48000	21800	5500	2500
								8	140	3555	98	2490	98	2490	45	1145	96000	43500	24000	11000	88000	39900	55000	25000	6150	2800
62	5,350	56,000	95,000	42	1067	36	914	5	118	3000	112	2845	114	2900	53	1350	108000	49000	28000	12700	100000	45400	53000	22000	6500	3000
								6	130	3300	112	2845	114	2900	53	1350	122000	55500	30000	13600	112000	50800	64000	29000	7550	3400
								7	145	3680	112	2845	114	2900	53	1350	136000	62000	34000	15500	125000	56700	75000	34000	8600	3900
								8	157	3990	112	2845	114	2900	53	1350	150000	68000	38000	17300	137000	62100	86000	39000	9650	4400
74	4,500	79,000	135,000	48	1219	42	1067	5	134	3400	130	3300	135	3430	62	1575	161000	73000	40000	18000	150000	68000	80000	36300	10000	4500
								6	154	3910	130	3300	135	3430	62	1575	182000	82500	43000	19500	169000	77000	96000	43500	11650	5300
								7	168	4270	130	3300	135	3430	62	1575	203000	91500	50000	22700	187000	85000	112000	50800	13300	6000
								8	181	4600	130	3300	135	3430	62	1575	223000	101000	56000	25400	206000	93350	127000	57600	14900	6800
88	3,800	110,000	190,000	60	1524	48	1219	5	156	3960	150	3810	156	3965	71	1810	252000	114500	63000	28600	236000	107000	127000	57800	16100	7300
								6	179	4550	150	3810	156	3965	71	1810	285000	129500	67000	30400	266000	121000	152000	69000	18700	8500
								7	195	4950	150	3810	156	3965	71	1810	318000	144500	78000	35400	296000	134000	177000	80300	21400	9700
								8	211	5360	150	3810	156	3965	71	1810	351000	159000	89000	40400	326000	148000	202000	91500	24000	11000

B.4   Shortcut (Graphical) Method of Determining Approximate Performance of Sulzer Centrifugal Compressors*

The calculation procedures given in the following pages permit

To determine   Compressor size and type

- Nominal diameter D (m)
- Number of stages z

Power input P (kW)

Speed n (r/min)

Absolute discharge temperature T_2 (K)

Using   Mass flow m (kg/s)

Suction pressure p_1 (bar abs)

Absolute suction temperature T_1 (K)

Relative humidity ϕ_1 (%)

Discharge pressure p_2 (bar abs)

Molecular mass M (kg/kmol)

Isentropic exponent k

Compressibility factor Z

The following factors, symbols and indices are also used:

Actual suction volume flow V_1 (m³/s)

Absolute humidity x

Peripheral speed u (m/s)

Head (polytropic) h_p (kJ/kg)

Temperature difference

$(\Delta T = T_c - T_1)$ ΔT (K)

Intercooling power factor f

Indices   Suction conditions   1

Discharge conditions   2

Dry t

Wet f

Polytropic p

per casing G

per group of stages

(between two coolings) S

Uncooled $*$

After cooling c

Total T

Number of casings i

Number of intercoolings j

How to use   A guide to the selection diagrams and two examples are given
the diagrams   in Table B.1, one with air in one casing, the other with gas in
two casings.

---

* *Please note:*   These graphical methods are intended for screening studies only. Contact the manufacturer for more definitive layout and performance prediction.

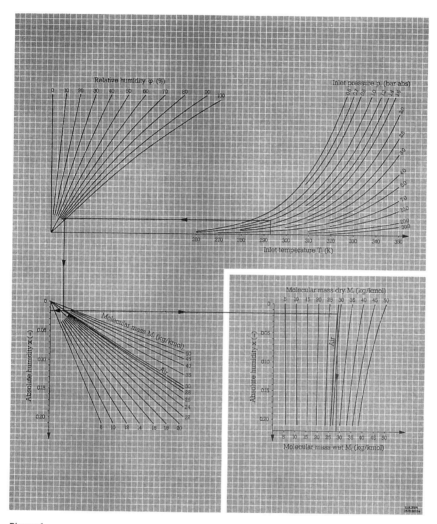

Diagram 1
Determination of the absolute humidity x
$(T_1 \rightarrow p_1 \rightarrow \varphi_1 \rightarrow M_t \rightarrow x)$

Diagram 2
Determination of the molecular mass M_f
of the wet gas $(x \rightarrow M_t \rightarrow M_f)$

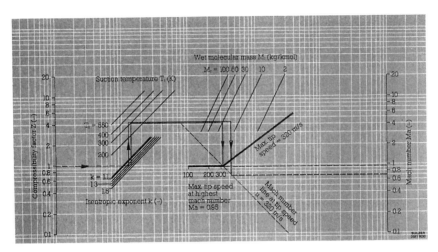

Diagram 3
Determination of the max. permissible
peripheral speed u_{max} ($Z \to k \to T_1 \to M_f \to u_{max}$)

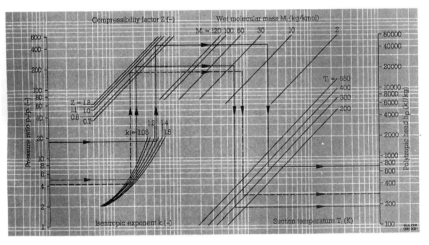

Diagram 4
Determination of the polytropic head h_p
($k \to p_2/p_1 \to Z \to M_f \to T_1 \to h_p$)

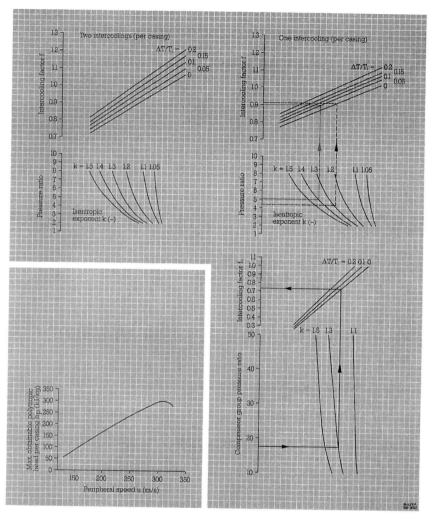

Diagram 5
Determination of the obtainable polytropic head per casing $h_{pG\,max}$ ($u_{max} \to h_{pG\,max}$)

Diagram 6
Determination of the influence of intercooling on the required shaft power

($p_2/p_{1G} \to K \to \Delta T \to T_1 \to$ estimated number of intercoolings per casing $i \to f$)

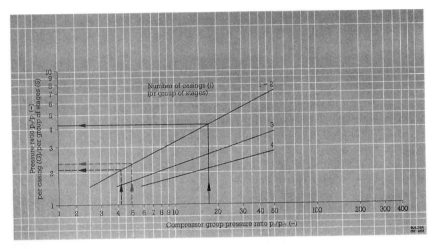

Diagram 7
Determination of the pressure ratio per
casing p_2/p_G^*

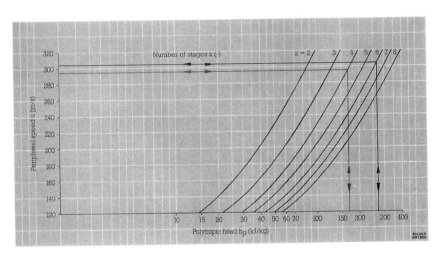

Diagram 8
Determination of the number of stages Z
of the compressor ($h_p \to z \leftrightarrow u$).

From u_{max} determined with Diagram 3,
round off Z to the whole number and

correct peripheral speed accordingly

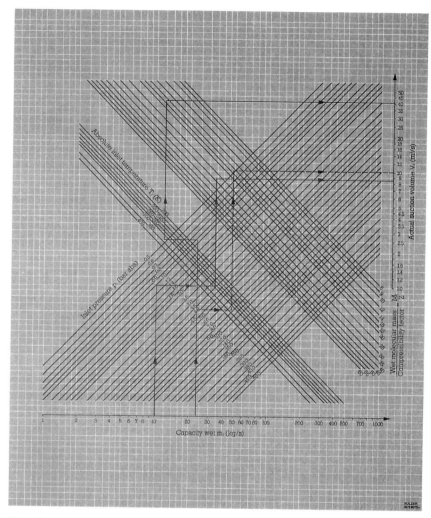

Diagram 9
Determination of the actual suction
volume V_1 ($\dot{m}_f \rightarrow p_1 \rightarrow T_1 \rightarrow M_f \rightarrow Z \rightarrow \dot{V}_1$)

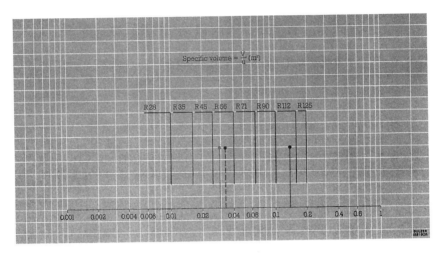

Diagram 10
Selection of the compressor size: nominal
diameter D (cm) as a function of $\dfrac{\dot{V}_1}{u}$,

where $\dot{V}_1$ = suction volume (m³/s) and
u = peripheral speed (m/s)

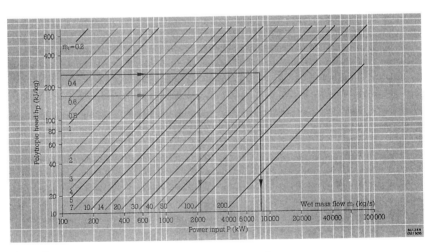

Diagram 11
Determination of the power input P
$(h_{pG} \rightarrow \dot{m}_f \rightarrow P)$

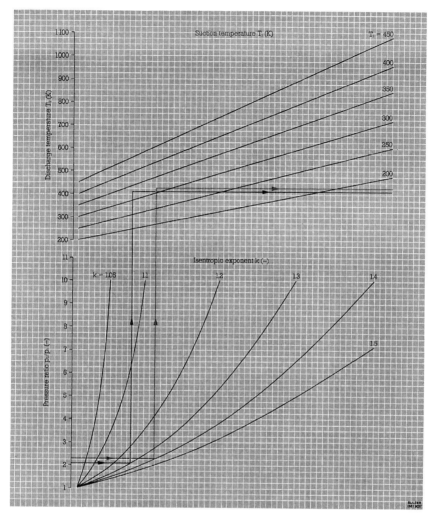

Diagram 12
Determination of the discharge
temperature T_2 ($p_2/p_1 \rightarrow K \rightarrow T_1 \rightarrow T_2$)

k = isentropic exponent (–)

TABLE B.1 Selection and Performance Calculation of a Centrifugal Compressor Train

	Calculation example 1: air compressor, one casing	Calculation example 2: gas compressor, two casings
Given		
Capacity	$\dot{m}_t = 10$ kg/s	$\dot{m}_t = 23.66$ kg/s
Suction pressure	$p_1 = 1$ bar abs	$p_1 = 0.92$ bar abs
Suction temperature	$T_1 = 293$ K	$T_1 = 333$ K
Relative humidity	$\varphi_1 = 90\%$	$\varphi_1 = 0\%$
Discharge pressure	$p_2 = 5$ bar abs	$p_2 = 16.1$ bar abs
Dry molecular mass	$M_t = 28.95$ kg/kmol	$M_t = 17.03$ kg/kmol
Isentropic exponent c_p/c_v	$k = 1.4$	$k = 1.29$
Compressibility factor	$Z = 1$	$Z = 1$
Calculation instructions		
1. Determination of the absolute humidity x (from T_1, p_1, φ_1, M_t) with Diagram 1	$x = 0.016$	$x = 0$
2. Determination of the wet molecular mass M_f (from x, M_t) with Diagram 2	$M_f = 28.7$ kg/kmol	$M_f = M_t = 17.03$ kg/kmol
3. Calculation of the wet mass flow $\dot{m}_f = \dot{m}_t \, (1 + x)$	$\dot{m}_f = 10(1 + 0.016)_f$ $= 10.16$ kg/s	$\dot{m}_f = \dot{m}_t$ $= 23.66$ kg/s
4. Determination of the max. permissible peripheral speed u_{max} (from Z, k, T_1, M_f) with Diagram 3	Electric motor $u_{max} = 320$ m/s Turbine $u_{max} = 290$ m/s	Electric motor $u_{max} = 320$ m/s Turbine $u_{max} = 290$ m/s
For further calculation motor drive has been selected		
5. Determination of the total polytropic head h^*_{pT} (from k, p_2, p_1, Z, M_f, T_1) with Diagram 4	$h^*_{pT} = 186$ kJ/kg	$h^*_{pT} = 722.8$ kJ/kg
6. Determination of the max. polytropic head obtainable per casing $h_{pG\,max}$ (from u_{max}) with Diagram 5	$h_{pG\,max} = 300$ kJ/kg	$h_{pG\,max} = 300$ kJ/kg
7. Calculation of number of casings $i = h_{pT}/h_{pG\,max}$, with $h_{pT} = h^*_{pT} \cdot f_T$, whereby f_T has to be estimated with Diagram 6	$i = 1$	$i = 2$ with $f_T = 0.73$
8. Determination of the pressure ratio per casing p_2/p_{1G} with Diagram 7	$p_2/p_{1T} = p_2/p_{1G} = 5$	$p_2/p_{1G} = 4.27$
9. Determination of the polytropic head per casing h^*_{pG} (from k, p_2/p_{1G}, Z, M_f, T_1) with Diagram 4	$h^*_{pG} = h^*_{pT} = 186$ kJ/kg	$h^*_{pG} = 293$ kJ/kg

From now on if two or more casings are necessary, the calculation has to be made for each casing separately (one after the other)

	First casing	Second casing
10. Determination of the influence of intercooling on the required shaft power (from p_2/p_{1G}, K, ΔT, T_1, and estimated number of intercoolings per casing j) with Diagram 6	$f = 0.9$ with $\Delta T = 20$ and $j = 1$	$f = 0.91$ with $\Delta T = 0$ and $j = 1$
11. Calculation of the fictitious polytropic head $h_{pG} = h^*_{pG} \cdot f$	$h_{pG} = 186 \cdot 0.9$ $= 167.4$ kJ/kg	$H_{pG} = 293 \cdot 0.91$ $= 266.6 \cong 267$
12. Determination of the number of stages z per casing and the definite peripheral speed u (from h_{pG}; $z \rightarrow u$) with Diagram 8 (round off z to whole number and correct peripheral speed correspondingly)	$z = 4$ $u = 295$ m/s	$z = 6$ $u = 304$ m/s
13. Determination of the actual suction volume V_1 (from $\dot{m}_f$, p_1, T_1, M_f, Z) with Diagram 9	$\dot{V}_1 = 8.59$ m³/s	$\dot{V}_1 = 10.2$ m³/s
14. Selection of the compressor size (nominal diameter D) as a function of $\dot{V}_1$ with Diagram 10	$D = 56$ cm	$D = 56$ cm
15. Type designation (from steps 10, 12, 14)	RZ 56-4	RZ 56-6
16. Calculation of the speed n $\quad n = \dfrac{60 \cdot u}{\pi \cdot D} = $ (D in meters)	$n = \dfrac{60 \cdot 295}{\pi \cdot 0.56} = 10060$ r/min	$n = \dfrac{60 \cdot 304}{\pi \cdot 0.56} = 10368$ r/min
17. Determination of the power input P (from h_{pG}, $\dot{m}_f$) with Diagram 11	$P = 2173$ kW	$P = 8100$ kW
		Total train 16200 kW
18. Determination of the discharge temperature T_2 (from p_2/p_1, between intercooling, k, T_1) with Diagram 12 whereby T_1 is the suction temperature after preceding intercooling and pressure ratio p_2/p_1 between intercooling has to be determined with Diagram 7	$T_2 = 424$ K with $T_1 = 313$ K and $p_2/p_1 = 2.3$	$T_2 = 413$ K with $T_1 = 333$ K and $p_2/p_1 = 2.1$

Second casing additional (right of steps 15–17):

RZ 112-6	
$n = \dfrac{60 \cdot 304}{\pi \cdot 1.12} = 5184$ r/min	
$P = 8100$ kW	

For second casing:

First casing	Second casing
$D = 112$ cm	$D = 56$ cm
RZ 112-6	RZ 56-6
$n = \dfrac{60 \cdot 304}{\pi \cdot 1.12} = 5184$ r/min	$n = \dfrac{60 \cdot 304}{\pi \cdot 0.56} = 10368$ r/min
$P = 8100$ kW	$P = 8100$ kW
$T_2 = 413$ K with $T_1 = 333$ K and $p_2/p_1 = 2.1$	$T_2 = 413$ K with $T_1 = 333$ K and $p_2/p_1 = 2.1$

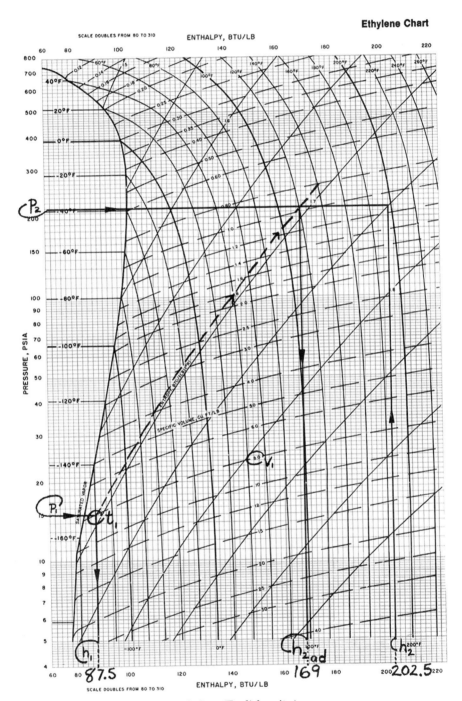

Figure B.8 Mollier diagram for ethylene (English units.)

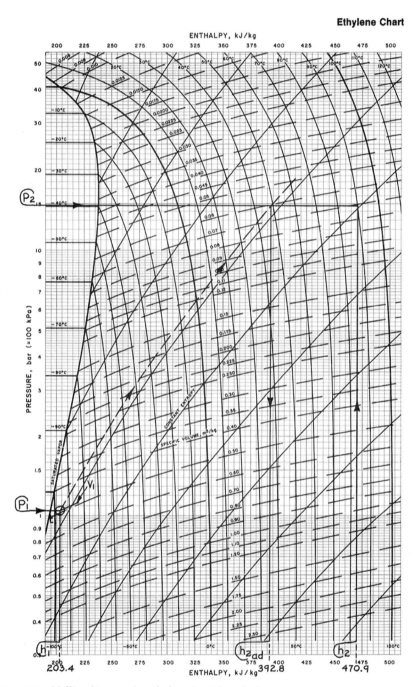

Figure B.9 Mollier diagram for ethylene (metric units.)

Bibliography
and List of Contributors

A-C Compressor Corporation, Appleton, Wisconsin

"Ro-Flo Sliding Vane Compressor," Form 101.

Aerzen USA, Coatesville, Pennsylvania

"Aerzen Environmental Technology," Form G1-003/01/EN.
"The Compact II," Form 92-318.
"Aerzen Process Gas Screw Compressors," Form 800/4.91.
"Aerzen VMTS-type Air Screw Compressors," Form 1000/6.91.
"Aerzen Rotary Screw Compressors, VMY-series," Form 300/1.94.
"Aerzen Screw Compressor Package Series VML," Form 2000/8.88.
"Aerzen VM 37-series Screw Compressors," Form 300/1.92.
"Aerzen Screw Compressors," Form 1000/9/93.

Anglo Compression, Inc., Mount Vernon, Ohio

"Modern High-Efficiency Compressor Valves," Bulletin 201.
"High-Reliability Reciprocating Compressor Pistons," Form 401.
"Customized Compressor Cylinder Design and Manufacture," Bulletin 321.

Ausdel Pty, Ltd., Cheltenham, Victoria, Australia

"Hydroscav Gas Stripping Technology Bulletin," Form 234.

Bently-Nevada Corporation, Minden, Nevada

"TorXimitor® Torque Meters," Form AN-058.

BHS-Voith Getriebewerk G.m.b.H., Sonthofen, Germany

"Gear Units," Form P 1/8-91.

Cooper Industries, Mount Vernon, Ohio

"Compressor Systems for the Process, Petrochemical, and Production Industries," Form 9-206A.
"Damped-Plate Compressor Valves," Form DPVB-681(A)-5MH.

"EPV-750 High-Efficiency Poppet Valves," Form 1-205.
"ESV-500 Compressor Valves," Form CCVB-979-10 MHG.
"Penn Horizontal Balanced-Opposed Process Compressors," Form 9-201 A.
"Reciprocating Gas Compression Systems for Enhanced Oil Recovery," Form EOR-0384-5MHOL.
"Tandem/Truncated Cylinder Configurations," Form 9-204 A.

Compressor Controls Corporation, Des Moines, Iowa

"Compressor Control and Surge Abatement Instruction Manual," Form IM30.
"Compressor Control Strategies," installation and teaching aids.
Batson, Brett, "Control System Objectives and Implementation," (special manuscript prepared for inclusion in this text).

Coupling Corporation of America, York, Pennsylvania

"Flexxor Coupling," Form 5M281.

Demag Delaval Turbomachinery, Trenton, New Jersey

"MH/MV Centrifugal Compressors," Mannesmann Demag Verdichter Form MA 25.40 E/10.93.
"Process Compressors," Mannesmann Demag Verdichter und Drucklufttechnik Form MA 10.15 en/4.89.
"Centrifugal Compressors," Mannesmann Demag Verdichter und Drucklufttechnik Form MA 25.40 en/10.82.
"Centrifugal Compressors for the Oil and Gas Industry," Mannesmann Demag Verdichter und Drucklufttechnik Form 25.70/02.90 E.
Salisbury, Roy J., "Lube, Seal and Control Oil Systems for Turbomachinery," customer training course and technical manuscript.

Dresser Industries, Inc., Roots Division, Connersville, Indiana

"Universal RAI Rotary Positive Displacement Blowers," Form B-5125.
"Whispair Blowers," Form B-5219.

Dresser Rand Company, Engine Process Compressor Division, Painted Post, New York

"BDC Balanced-Opposed Compressors for Process Applications," Bulletin 3650.
Beyer, R. W., "Reciprocating Compressor Performance and Sizing Fundamentals."
"Gas Properties and Compressor Data," Form 3519-D.
"HHE-FA/FB Balanced-Opposed Process Compressors," Form 85077.
"HHE Heavy Duty Process Compressors," Form 85084.
"HHE Heavy Duty Reciprocating Process Compressors," Form 3596.
Lentek, G. A., "Reciprocating Compressors."
"PHE Balanced-Opposed Process Compressors," Form 85068.
Schaad, R. G., "Reciprocating Compressor Drive Systems."

———, "Reciprocating Process Compressor Designs and Applications."
"TCV Engine Compressors," Form 85083.
Woollatt, D., H. Wertheimer, and R. Beyer, "Design and Application of Compressor Valves for Reliability and Efficiency."

Elliott Company, Jeannette, Pennsylvania

"Elliott PAP-Plus Plant Air Package Compressor," Bulletin P-29.
Hallock, Donald C., "Centrifugal Compressors and the Cause of the Curve," Elliott Reprint 93-476-MOY.
"Quick Selection Methods for Elliott Multistage Compressors," Bulletin P25A.
"Elliott Multistage Centrifugal Compressors," Bulletin P-25C.
Peter, Kenneth L., "Applying Multiple Inlet Compressors," (also published in *Hydrocarbon Processing*, May 1981).

Flexibox, Inc., Houston, Texas

"Flexible Couplings," Engineering Catalog 100CC.

Indikon Division of Metravib Instruments, Cambridge, Massachusetts

"Indikon Torquemeters and On-Stream Alignment Monitors," Form 484.

KMC Inc., West Greenwich, Rhode Island

"Flexure Pivot Tilt Pad Bearings," Bulletin 101.
"Thrust-Type Flexure Pivot Bearings," Bulletin 201.

Lincoln Division of McNeil Corporation, St. Louis, Missouri

"Modular Lube Centralized Lubricating Systems," Form 440503.

Lubrication Systems Company, Houston, Texas

"Customized Lube and Seal Oil Skids for Turbomachinery," Form 101.

Lubriquip, Inc., Cleveland, Ohio

"Pump-To-Point Lubricators," Bulletins 10102, 51020, and 51040.

Lucas Aerospace Corporation, Bendix Fluid Power Division, Utica, New York

"Bendix Contoured Diaphragm Coupling."
"Bendix Flexible Diaphragm Couplings," Pub. No. 67U-6-717A/B.
"Contoured Flexible Diaphragm Couplings," Pub. No. 67-U-9-7811-B.

'Lucas Contoured Diaphragm Couplings," Pub. No. 67-U-6-919A.
"Lucas Contoured Diaphragm Couplings," Pub. No. 67-U-6-8811A.

Nash Engineering Company, Norwalk, Connecticut

"Liquid Ring Compressors For The Process Industries," Bulletins 819-A, 836-A, and 455-A.

Nuovo Pignone, Florence, Italy

Beni, P., and A. Traversari, "Approaches to the Design of a Safe Secondary Compressor for High Pressure Polyethylene Plants."
Agostini, M., and E. Giacomelli, "Safety, Operation and Maintenance of LDPE Secondary Compressors," Quaderni Pignone 52.
Tosi, G., A. Timori, and M. Stangarone, "Rotordynamics in Centrifugal Compressors," Quaderni Pignone 48.

PPI Division, The Duriron Company, Inc., Warminster, Pennsylvania

"Metal Diaphragm Compressors," Bulletins PD-400C, HP-100, HP-400, and HP-410.

Rotordynamics—Seal Research, North Highlands, California

"Honeycomb Labyrinth Seal Technology for Turbomachinery," Form 401.

Sulzer-Burckhardt Engineering Works, Ltd., Winterthur, Switzerland

Klaey, "The Laby Compresses Gases Economically," *Sulzer Technical Review,* 3/1990.
"Labyrinth Compressors," Bulletin 21.05.14.40.
Matile, C., "Industrial Reciprocating Compressors for Very High Pressures," *Sulzer Technical Review,* 2/1971.
"Reciprocating Process Compressors," Bulletin 22.00.14.40.

Sulzer Turbosystems International, Houston, Texas; New York, New York; and Winterthur, Switzerland

"Turbocompressors," Form STI 892/3M.
"Sulzer Axial Compressors," Form 26.13.10.40—Bhi 30.
"Sulzer Isotherm Turbocompressors," Form 20.14.10.40—Bid 50.
"Sulzer Turbocompressors," Form 25.01.10.40—Aid 15.
Buchel, A., "Basic Design Features of Sulzer Turbocompressors—Applications in the Hydrocarbon Processing Industries," Form e/27.13.10-Cgh.
Wachter, M., "Some Special Design Aspects of Turbocompressors for the Oil, Gas and Petrochemical Industry."
"Sulzer Centrifugal Compressors," Form 27.20.10.40—Chi 30.
"Sulzer Centrifugal Barrel Compressors," Form 27.24.10.40—Bhi 50.

Marriott, A., J. Ryrie, and D. Gilon, "Mopico Compressor for Gas Pipeline Stations," *Sulzer Technical Review*, 1/1991.

Marriott, A., and D. Gilon, "Initial Experience With A New High-Speed, High-Pressure, Oil-Free Motor Compressor Unit (HOFIM)." *Instit. of Mechanical Engineers*, C449/038/93 (1995).

Torquetronics, Inc., Allegany, New York

"Torquetronics Shaft Torque Monitoring Instrumentation," Form 345.

Revolve Technologies, Inc., Calgary, Alberta, Canada

Brallean, G. E., W. M. Grasdal, V. Kulle, C. P. Oleksuk, and R. A. Peterson, "The Application of Active Magnetic Bearings to a Power Turbine," ASME Paper 90-GT-199, 1990.

Eakins, P. S., and T. J. Al-Himyary, "Dry Gas Seal and Magnetic Bearing Systems for Natural Gas Pipeline Compressors," Nova Corporation, Calgary, Alberta, 1991.

Eakins, P. S., C. R. Feldmeyer, and A. G. St. Onge, "Operating Experience with Active Magnetic Bearings in a Power Turbine," *Proceedings, CGA Symposium on Industrial Applications of Gas Turbines*, 1991.

Foster, E. G., V. Kulle, and R. A. Peterson, "The Application of Active Magnetic Bearings to a Natural Gas Pipeline Compressor," ASME Paper 86-GT-61, 1986.

Hesje, R. C., and R. A. Peterson, "Mechanical Dry Seal Applied to Pipeline (Natural Gas) Centrifugal Compressors," ASME Paper 84-GT-3, 1984.

Sears, J. E., and S. O. Uptigrove, "Development and Operation of a Dual Hard Face Dry Gas Seal." *Proceedings, Revolve '89*, 1989.

Zurn Industries, Inc., Erie, Pennsylvania

"Ameriflex and Amerigear High Performance Couplings," Form 673ADV, 10/88.

"Amerigear: Introduction, Design and Manufacturing," Form 462ADV, 8/81.

"Ameriflex Flexible Diaphragm Coupling," Form No. 271-ADV, Rev. 4/74.

Individual Contributors

Bloch, H. P., and P. W. Noack, "The Expanded Capabilities of Screw Compressors," *Chemical Engineering*, Feb., 1992, pp. 100–108.

Bloch, H. P., and P. W. Noack, "Recent Experience With Large Liquid-Injected Rotary Screw Process Gas Compressors," presented and published in the *Proceedings of the 20th International Turbomachinery Symposium*, Texas A&M University, Dallas, Texas, 1991.

Calistrat, M., "Flexible Couplings," ISBN 0-9643099-0-4, Caroline Publishing, P.O. Box 451611, Houston, Texas 77245-1611.

Chow, R., B. McMordie, and R. Wiegand, "Coatings Limit Compressor Fouling," *Turbomachinery International*, Jan./Feb. 1995.

Godse, A., "Predict Compressor Performance at New Conditions," *Hydrocarbon Processing*, June 1989, pp. 77–79.

Index

Index

Abrasion cleaning, 357
Abrasive wear, 320
Access covers, 45
Accumulators, 371
Acoustic velocity, 389, 408
Acoustical vibration, 90
Acetylene compression, 149
Actual volume flow, in screw compressors, 157
Actuators, for valve unloading, 71–75
Adiabatic efficiency, of screw compressors, 171
Adiabatic (isentropic) compression, 22, 158
    in screw compressors, 158
Adiabatic process, 127
Adjustable clearance pockets, 76–78
Adjustable rollers, 284, 286
Adjustable-speed drivers, 77, 78
Aeration services, 141, 154, 155
Aerodynamic effects, 316
Aerodynamic loss, 240
Aerodynamics, of sideload compressors, 420–423
Affinity laws (fan laws), 401
    deviations from, 401, 407
Air, composition of, 13
Air purge, 335
Air stripping principle, for oil purification, 372
Alarm horn, 95
Aluminum bearings, 40
Amagat's law, 11
Amonton's law, 10
Ampere meter, 95
Analog studies, 125
Annunciator, 95
Antichoke controller, 389
Anti-instability (honeycomb) seals, 259–261, 262, 263

Antisurge control algorithm, 391, 392
Antisurge control loops, 393
Antisurge valve, 389, 390, 393
API (American Petroleum Institute)
    Standard 617, 227, 277, 278, 287, 300, 309
    modification recommended for multiple inlet compressors, 411
API Standard 618, 38, 43, 51, 53, 54, 57, 64, 65
API Standard 619, 159
API Standard 670, 312
Application range:
    for different types of compressors, 2, 216
    for labyrinth piston compressors, 108, 109
    for poppet valves, 58
Approach temperature, 198
Armored reflex gauge, 91
ASME criteria, 276, 277, 279
ASME-type testing:
    of sidestream compressors, 424
    typical instrumentation requirements, 425
Automatic trap, 91, 92
Auxiliary landing system:
    development efforts, 337
    for magnetic bearings, 334, 335
Availability, of screw compressors, 171
Avogadro's law, 11
Axial blading, 258
Axial compressors, 212–216, 258
    application range, 216
    driver availability, 216
    performance map, 214
Axial inlet retrofit, 338, 339
Axial load, variable, 340
Axial thrust, 239, 256–258, 270
Axial-radial flow compressor, 232

Babbitt, 40, 55
Backflow control, 141
Back-to-back design, 281, 282
Back-to-back flow, 233, 235, 236, 270
Backup rings, 86
Backup seal, 327
Backward-lean impeller, 217, 299, 301,
    398–400
Backward-leaning vanes, 217, 299, 301
Balance line, 341
Balance piston, 235, 256, 257, 266, 270, 340
Balance piston leakage, 341
Balanced-opposed compressor(s), 34–36
Balancing drum (balance piston), 235,
    256, 257, 266, 270, 340
Balancing operation, 257
Barrier water sealing, schematic diagram,
    169
Basic elements, 4
Battery backup system, for magnetic
    bearings, 334
Beam-type construction, 335, 340
Bearing damping, 315
Bearing span (may create performance
    problems), 414
Bearing stiffness, 314, 315
Bearing support, 45
Bearings, 40, 55, 268–274
    aluminum alloy, 40
    for centrifugal compressors, 268–274
    radial, 268–270, 273
    for screw compressors, 165
    thrust, 270–274
    trimetal, 40
Blade angle, 398–400
Block-and-bleed valves, 89
Boiling point, 12
Bolting materials, 278
Boyle's law (isothermal law), 10
Breakaway friction, 69
BRITE program, 345
Buffer gas, 185
Buffer gas injection:
    in screw compressors, 165
    used with dry seals, 330
Buffer gas systems, 86, 87
Built-in compression ratio, 155, 156
Built-in volume ratio, 155
Bundle rollers, 284
Bundles, 244, 277, 285
Bypass control, 163
Bypass (recycle), 70

Campbell (interference) diagram, 318
Capacitance switch, 91
Capacity:
    of diaphragm compressors, 136
    of hyper compressors, 111
    of liquid ring compressor, 145, 146
    of reciprocating compressors, 34, 180,
        181
    of sliding vane compressors, 143
Capacity control:
    for dynamic compressors, 217–220,
        373–394
    for hyper compressors, 124
    for reciprocating compressors, 69–78
    for screw compressors, 160–164, 175–177
Capacity display, 96
Capacity loss:
    as function of clearance, 181
    as function of compression ratio, 182
    due to throttling, 183–185
    due to valve preload, 183
    of reciprocating compressors, 181–190
Capacity tolerance, for screw compres-
    sors, 159
Capillary temperature switch, 89, 90
Carbon labyrinth seal, 166
Carbon ring seals, 165
Cartridge seal, 329
Casing split:
    horizontal (axial), 209, 211, 227–244
    selection based on pressure, 277
    vertical (barrel-type), 209, 211, 227–244
Casing welding, 289–292
Casing(s), configurations for liquid ring
        compressors, 147
    deflection, 278, 279
    design, 229–244, 276–285
    manufacturing techniques, 285–295
    materials, 243, 277–280
    size selection, 230, 280, 445–491
Cast-iron:
    cylinder, 47
    frame, 45
Cavity, 127
Cavity plate, 128, 131–134
Central valves, 119–121
Centrifugal compressor:
    application range, 209–213
    dimensions and weights 458–461, 480,
        481
    multistage, 209, 210, 232–244
    single-stage, 209, 210
Ceramic coating, 57

Certification, of metallurgy, 307, 308, 310
Channel valves, 56
Characteristic curves, 376
Charles' law, 10
Check valves, 129, 131, 132, 135
Choke limit, 220, 389, 396, 397, 401, 402, 405, 406
    for multiple inlet compressors, 413, 414
Circumferential sealing rings, 121
Clearance (clearance space), 7, 180, 181
    effect at high ratios, 30, 31
    in hyper compressors, 112
Clearance pockets, 75–78
    ineffective range, 76
Clearance volume, 54, 76, 128, 129
Closed cooling system, 51, 53, 64, 80
Closed loop control, 383, 392
Coalescer, 372
Coalescing filter, used with dry gas seals, 328, 330
Coast-down bearings, 334
    future developments, 337
    life expectancy, 335
Coating case study, 321, 322
Coatings, 57
Code thickness requirement, 277
Coefficient of reaction, of hyper (LDPE) compressors, 111
Coke-oven gas compression, 167
Combination axial-centrifugal compressor, 232
Combination of compressor types, 2
Combined lube and seal oil system, 367, 368
    materials of construction, 371
Compound compressors, 233, 235
Compressibility charts, 443, 444, 449, 471
Compressibility correction, 15, 16, 21, 23–25
Compressibility factor, 15, 16, 21, 192, 222, 442
Compression cycles, 21–23
Compression ratio:
    definition, 5
    per stage, 26, 27
    max. achievable in screw compressors, 160, 171
    vs. number of impellers, 223
Compressor control, 373–394
    objectives, 373, 374
Compressor fouling, 318–324
Compressor maps, 374–380
Compressor speeds, screw compressors, 171

Condensate recovery system, 148
Condensation, 13, 79, 187, 194
    evaluation by computer, 194
Connecting rod, 43, 44, 55
Connecting rod bolts, 43, 55
Constant life fatigue diagram, 356
Construction materials, 39, 40, 46, 50, 51, 57, 82, 86, 250, 306–312, 371
    for bearings, 40
    for crankshafts, 39
    for cylinders, 46, 50
    for piston rings, 52, 82, 86
    for piston rods, 57
    for rider bands, 52, 82
Continuous lubrication
    centrifuge effect in couplings, 354
    of gear couplings, 354
    viscosity optimization, 354
Contoured diaphragm coupling, 349–357
    configuration, 355
    cyclic stresses, 355, 356
    principal advantages, 353
    steels used in fabrication, 355, 356
Control of compressors, 373–394
    objectives, 373, 374
Control oil systems, 367–372
Control panel(s), 74, 95–97
Conversion table, 478, 479
Cooling:
    for centrifugal compressors, 226, 233, 239, 240
    of hyper compressors, 115, 117
Cooling provisions, 51, 60, 64, 79–82, 115, 117, 226, 233, 239, 240
Cooling water velocity, 80
Console, for jacket cooling, 81
Copper-filled PTFE, 86
Corrosion protection, for nonlubricated couplings, 357
Corrosive gas compression, 148
Counterweights, 39, 42
Couplings, 347–357
    function of a coupling, 352
    major requirements, 347
    rating parameters, 347–350
Crank-angle diagram, 40, 41, 56, 61
Crankcase pressurization, 106
Crankpin bearing, 40, 55
Crankshaft design, 37–39, 41, 42
Critical conditions, 15, 21
Critical pressure, 15, 21
Critical speed, 312, 314
    map, 314

Critical temperature, 15, 21
Cross-coupling effect, 259
Crosshead, 37, 43–45, 55
    bushings, 43, 54
    dummy type, 37, 41
    for hyper compressors, 121
Crossover (return bend), 236, 245–247
Cryogenic service, cooling omitted, 79
Cushioning (damping), 54, 56, 57
Cycle monitor, 84, 94
Cylinder bore, 195, 199
Cylinder clearance, 28, 31
    in hyper compressors, 112
Cylinder cooling, 51, 79–82
Cylinder heads, for hyper compressors,
    119
Cylinder liner(s), 46, 50, 113
    for hyper compressors, 113
Cylinder lubrication, 60, 79, 82–85, 196,
    197
    inadequate c.l., 79
Cylinder material, 46
Cylinder sealing, of hyper (secondary)
    compressors, 113

Dalton's law, 11, 13
Damped mode shape, 316
Damping, 259
Damping, pneumatic, 54, 56, 57
Deck-and-one-half valve, 58
Decomposition, 113, 125
Definition(s), of measurement units, 2, 3
Deflection analysis, 303
Delevitation, 335
Density, 18
Design point, 299
Dew point, 12
Diagnostic information, obtainable from
    magnetic bearing controls, 336
Diagnostic instruments, 88
Diagnostics, 97
Diaphragm cooling, 226
Diaphragm coupling, 349–354
    principal advantages, 353
Diaphragm valve, 163, 164
Diaphragms, 113, 127, 128, 135
Diester lubricants, 65, 196, 197
Diffuser, 248
Diffuser finish, effect on head and effi-
    ciency, 323
Diffuser return bend, 296
Diffuser width change, due to polymeriza-
    tion, 322

Diffusion losses, 304, 305
Digital phasemeter, 360
Digital pulsation studies, 125
Dimensionless numbers, 427, 428
Dimensions and weights, centrifugal com-
    pressors, 458–461, 480, 481
Discharge pressure(s):
    of diaphragm compressors, 136
    of hyper compressors, 111
    limit for labyrinth piston compressors,
        108
    of liquid ring compressor, 145, 146
    of sliding vane compressors, 143
Discharge temperature calculation, 442
Discharge temperature equation, 193
Discharge throttle unloading, 163
Discharge volumetric efficiency, 197
Discharge volute, 246–248
Disk pack coupling, 349
    principal advantages, 353
Displacement, 128, 136
    and clearance, 194, 195
Displacement-type switch, 91
Distance piece:
    closed and purged, 106
    single, 46, 65, 105
    two-compartment, 46, 47, 48, 65–67, 87
Divider block lubrication, 83–85, 93
    disadvantage of, 84
    instrumentation for d.b.l., 84, 93
Divider valve(s), 83
DMIX software, 193
Double-acting, 5, 50, 97, 99–109, 179
Double-acting mechanical seal (slide
    seal), for screw compressors, 166, 167
Double-acting thrust bearing effect, in
    magnetic bearings, 333
Double-deck valve, 56, 58
Double-flow compressors, 233, 238, 239
Double-helical screw, 152
Double-lobe construction, 145
Dowel pin, 55
Dowelling, 285
Drive mechanisms:
    for diaphragm compressors, 129–134
    for hyper compressors, 121–125
Dry gas seals, 325–345
    application range, 326
    operating experience, 330, 332
    operating principles, 326–330
    problems and solutions, 330, 331
    seal leakage, 332
Dry seal cartridge, 329

Dry seal upgrading, 331, 332
  configurations, 331
  materials, 331
Dry screw compressor(s), 154–156
Dummy crosshead, 37, 41
Dynamic compressors, 209–491
  driver availability, 216, 220
  performance, 217–226
Dynamic flow losses, 159, 161

Eccentric rotor mounting, 142
Eddy current devices, 94
Eddy current losses, in magnetic bear-
    ings, 332
Effective critical conditions, 441, 471
Efficiency, 23, 26, 31, 54, 56, 103, 104, 223,
    224, 474
Efficiency conversions, polytropic to adia-
    batic, 449
Efficiency optimization, 304
Efficiency and specific speed correlation,
    298, 304
Electron beam welding, 351
Electronic controls, for magnetic bearings,
    332
Electropneumatic transducer, 339
Encapsulated compressors, 106, 107
Energy, 9, 102, 103
  of a fluid, 374, 375
Energy savings, with magnetic bearings,
    336
Entropy (unavailability), 10
EP additives, 353, 354
Epoxy grout, 37
Equilibrium condition, 12
Equivalent gases, 17
Equivalent speed, 379, 407, 408
Error, control-related, 382
Eutectic devices, 93
Event sequencing, 374
Excess flow margin, effect on compressor
    performance, 415
Exothermic reaction, 113
Expander ring, 113
Expansion chamber, 54
Explosive decomposition, 113, 125
Explosive gas compression, 149
Extended distance piece, 65

Fabricated cylinder(s), 46, 48
Fan law deviation, 401, 407
Fan law effect, 400
Fan laws, 401, 448

Fault indication, of diaphragm compres-
    sors, 135, 136
Female rotor, 151, 156
Film seal, 263–265
Final discharge temperature, in liquid-
    injected screw compressors, 160
Finger-type unloaders, 71, 72
First law of thermodynamics, 9
Five-lobe rotors, 151
Fixed clearance pockets, 75, 76
Fixed pin design, 43
Fixed stator blading, 214
Flame arrestor effect, 145, 149
Flange welding, 290–293
Flapper valve, 93
Flare stack, venting to, 87
Flexible rotor, 312
Flexxor coupling, 348
Flexure pivot™ bearing, 271–274
Floating pin design, 43, 55
Floating ring seal(s), 166, 167, 261, 264
Flow blockage, in cooling water circuits, 80
Flow calculations, 448
Flow coefficient, 297, 409, 428–435, 474
  at high Mach numbers, 423
  of valves, 184
Flow instrumentation, 90
Flow measurement, 376
Flow path, 233, 244–250, 396
  rotating, 233, 244–246, 396, 405
  stationary, 247–250, 396
Flow range, of reciprocating compressors,
    33
Flush injection, 357
Flutter, 61
Forced cooling, 51, 52
Force-feed lubrication, 42, 43, 49
Forging:
  of crankshafts, 39
  of cylinders, 48
Forward pressure control, 368, 370
Fouling, 318–324
  of axial compressors, 358
  time rate of progress, 324
  (See also Polymerization)
Fouling deposits, sensed by torque moni-
    toring, 357
Foundation design, 35
Frame loads, 194–196, 200
Frame sizes:
  of centrifugal compressors, 450, 454,
    475–477
  of reciprocating compressors, 196

Frame(s), 43, 45
Fretting corrosion, in couplings, 353
Friction losses, 23, 102, 304, 403
Full-circulation cooling, 80, 81
Fusible elements, 93

Garter springs, 104, 105, 116
Gas, definition of, 12
Gas compression theory, 470
Gas constant, 12
Gas mixtures, 17
    determination of k-value, 448, 472
Gas properties, 437–444, 450, 454, 471
Gas pulsations, 125
Gas pumps, 137–141
    applications, 141
    capacities, 141
Gas recovery system, 87
Gas seals, 325–345
    application range, 326
    double arrangement, 327
    operating experience, 330, 332, 336
    seal leakage, 332
    single arrangement, 327, 328
    tandem arrangement, 327, 328
Gas temperature, in hyper compressors,
    112
Gas velocity, 54
Gas weight, effect on impeller perfor-
    mance, 403, 407
Gas whirl (swirl), 259
Gases, equivalent, 17
Gas-filled probe, 89, 90
Gas-shielded (TIG) welding, 308, 309,
    311
Gate rotor, 174
Gear couplings, 347, 348
    materials of construction, 353
Gear speed reducers, 77, 78
Geared compressors, 209, 210, 242
Glycol cooling, 51, 79
Graflon, 86
Graphical selection methods for centrifu-
    gal compressors, 462–469, 482–491
Graphite-filled PTFE, 82
Graphite rings, 104
Grease lubrication:
    of couplings, 354
    separation into oil and soap, 354
Groove pattern:
    in dry seals, 326, 327, 331, 332
    operating experience, 332
Grouting, 37

Guide vane angle, influence on head char-
    acteristic, 306, 390
Guide vane(s), 296
    adjustable, 218–221
    influence of position, 379
    stationary, 245, 246

H_2S service, 300
Half-frequency whirl, 268
Hardness:
    limitation in H_2S service, 300
    tests on impellers, 311
Head coefficient, 224, 225, 297, 299, 409,
    428–435
Head rise to surge (HRTS), 412, 413,
    416–418
    restricted options, 413
Head tank (overhead tank), 368, 370
Head-capacity relationship:
    of axial compressors, 214, 215
    of centrifugal compressors, 218–220
Head-flow curve shape, 396–410, 412–418
Heating effect on capacity, 187, 188
Heating elements, 367, 368
Heat of compression, effect on sealing
    liquid, 145
Heat removal, 78–82
    reasons for, 78
Heat treatment, 294, 295, 307
Helium leakage test, 135
High-capacity compression, 282
High-pressure cylinders, 115–121
High-pressure pistons, 114, 117
Hoerbiger controls, 73, 74
HOFIM (high-speed oil-free intelligent
    motor compressor), 341, 344, 345
    operating characteristics, 344, 345
Honeycomb seals, 259, 260, 262, 263
Hoop stress, 277
Horsepower, 190–192, 224
    of reciprocating compressors, 190–192
Horsepower adders (losses), 192
Horsepower deviation, 417
Horsepower loss (adders), 191, 192
Host communication, 374
Hot alignment monitoring, incorporated
    in torquemeter, 361
Hot spots, in compressor cylinders, 78
Hydraulic coupling, 221
Hydraulic fit runner disk, 273–275
Hydraulic fitup procedure, 274–276
Hydraulic injection pump, 128, 132
Hydraulic intensifier, 123

Hydraulic system, of diaphragm compressors, 129–134
Hydrodynamic oil film, 269
Hydrogen chloride compression, 149
Hydrostatic testing, 288, 296
Hyper compressors, 111–126
  capacities, 111
  power requirement, 112
  pressure developed, 111
  thermodynamics, 112
Hyperbolic contouring, 355

Ideal gas law, 10, 15, 16, 192
Impeller application range, 255, 256
Impeller blade shapes, 300
Impeller blade thickness, 300
Impeller clearance, 301
Impeller eye, 245, 246
Impeller exit width, restriction due to fouling, 321
Impeller inlet diagram, 299, 307
Impeller materials, 250, 306–312
Impeller size range, 255, 256
Impeller stage spacing, 293
Impeller technology, 250–257, 298–312
Impeller tip velocity, 402, 407, 474
Impeller welding, 306–312
Indicator card (pV diagram), 5–9, 29, 30, 56, 62, 70, 181–191
Indicator diagrams, for hydraulically driven hyper compressors, 124
Indikon torquemeter system, 360–365
  acceleration levels, 361
  inductive proximity probes, 363
  on-shaft calibration circuit, 361
  on-stream alignment monitoring, 361–364
  rotary transformer, 361–364
  torque pulsations, 362
Induced volume flow, 156, 160
Inducer impeller, 403, 404
Inducer section, 300, 309
Inductive bridge sensor, 334
Inhibitive coatings, 323
Injection rate, control of, 168
Inner casing cartridge, 229
Inner sleeve, in hyper compressors, 115
Instrumentation, for metallic diaphragms, 135, 136
Instruments, 88–96
Integral crossover, 236, 245
Integral feet, 284
Integral riveting, 252, 253

Integral (reset) windup, 385, 386
Integral step-up gears, 153
Intensifier, 123
Intercooler, 197
Intercooling, 26, 226, 233, 239, 240
Interference (Campbell) diagram, 318
Interlobe space, 151, 154
Intermediate pressure regulation, 168, 170
Interstage cooling, 226, 233, 239, 240
Invariant coordinates, 377
Involute profile, 137
Irreversibility, 23
Isenthalpic process, 183
Isentropic (adiabatic) compression, 22, 24
Isentropic exponent, 198
Isentropic horsepower, 192
Isentropic temperature, 192
Isocarbon® seal, 261, 264
Iso-cooling, 233, 239, 240
Isolation valves, 88, 89, 91
Isothermal compression, 21, 26
Isothermal law (Boyle's law), 10

Jacket cooling and heating, 78–81
Jacket heating, 78, 79
Jacket water console, 81
Jet compressor, for oil purifiers, 372
Journal bearings, 165, 268–274
Jumper piping, 80

K design, 106, 107
  of labyrinth piston reciprocating compressors, 106
k value, 20, 24, 25, 31, 33, 155
Kinematic similarity, 297
Knockout drum, instrumentation for, 91

Labyrinth piston compressors, 86, 99–109
Labyrinth seals, 259
Lateral critical speed, 314
LDPE (Low-density polyethylene), 111
Lead-filled PTFE, 86
Leak detection port, on metal diaphragms, 135
Leakage:
  of lobe-type blowers, 137
  of metal diaphragm compressors, 135
  of packing, 185, 186
  of piston rings, 184, 185
  of valves, 186, 187
Leakage, of gas:
  from labyrinth pistons, 103, 105–108
  from packing cases, 87, 88

Leakoff ports, in screw compressor seals, 165
Level instrumentation, 91, 92
Levitation, 335
Limiting control, 374, 392, 393
Liner(s), 46, 50, 113
    for hyper compressor cylinders, 113
Liquefaction, of gases, 13
Liquid injection (water washing), 357
Liquid level instrumentation, 91, 92
Liquid ring compressors, 145–149
Liquid-injected screw compressors, 154, 155, 160, 162, 167–170
    performance summary, 170–173
Load balancing, 374
Load reversals, 47, 75, 194
Lobe-type (blowers) compressors, 137–141
    applications, 141
    capacity range, 141
    operating principle, 140
    pressure ratings, 141
Lock ring, 128
Log decrement, 316, 317
Logarithmic decrement, 316, 317
Logic switches, 89, 96
Loop decoupling, 374, 393, 394
Loss of efficiency, due to fouling, 320, 321
Losses, 23, 62
    in reciprocating compressor sizing calculations, 197
Low moment coupling, 350
Low-density polyethylene, 111
Lube oil pumps, 368
Lube oil purifier, 372
Lubricants, 65, 117
    for hyper compressors, 117
Lubrication:
    of cylinder bore, 79, 82–85
    force-feed, 42, 43, 49, 60, 83, 92, 100
    of high-pressure compressors, 122
    limiting temperatures, 196
    of lobe-type blowers, 137
Lubrication systems, 367–372
Lubricator drive, 63
Lubricator flow rate, 63
Lubricator sump, 83

Mach number, 159, 161, 389, 402, 414, 421–423, 426
    consideration for multiple inlet compressors, 414
    effect of gas weight, 403
    measurement locations, 422

Magnetic bearing stator, 333, 339
Magnetic bearing systems, 325–345
    application range, 326
    controls, 332–336
    high-temperature performance, 326
    operating experience, 336
    operating principles, 332–335
    problems and solutions, 336, 337
Magnetic flux, 332
Magnetic particle inspection, 293, 295, 307, 311
Main nut, 128
Maintenance history, of screw compressors, 168, 170
Maintenance of vertically split compressors, 283–285
Makeup liquid, 145
Male rotor, 151, 156
Material stress, 227, 277, 304
    allowable in H_2S services, 300
Materials of construction, 39, 40, 46, 50, 52, 57, 82, 86, 250, 306–312
Mechanical compression cycle, in diaphragm compressors, 131–134
Mechanical contact seal, 106, 140, 142, 260, 264
Mechanical losses, 23, 159
    in centrifugal compressors, 431, 435, 453, 457
    in screw compressors, 159
Mechanical seal, 106, 140, 142, 260, 264
Mercury sealant, 113
Metal diaphragm compressors, 127–136
    drive mechanism, 129–134
    head design, 134–136
    terminology, 127, 128
Metal-arc (MIG) welding, 308
Metallic diaphragms, 113, 127, 128, 135
Metallic piston rings, for hyper compressors, 113
Mezzanine mounting, 283
Milled impeller blades, 302
Minicomputer, 97
Minimum wall thickness, 277
Mixed flow impeller(s), 258
Mixing area:
    effects of improper mixing, 421
    in sideload compressors, 419, 420, 421
Mode shape, 316
Modified Goodman diagram, for contoured diaphragm couplings, 355, 356
Molar heat capacity, 20
Molar specific heat, 20, 21

Mole(s), 17, 18
Mole fraction, 20
Molecular sieve, 107
Molecular weight, 17–19
  influence on shape of performance
    curve, 408
Mollier chart, for ethylene, 453, 492, 493
Mollier method of compressor performance prediction, 446, 452, 456
Monitoring instruments, 84, 88
Monitoring system, for magnetic bear-
  ings, 334
Monomer recovery system, 149
MOPICO motor pipeline compressor,
  341–344
  advantages, 342
  construction features, 341
  cost, 342
  dimensions and weights, 343
  operating characteristics, 342
  pipeline disadvantages, 343
Morganite Graflon, 86
Multiple fixed-volume clearance pockets,
  76
Multiple inlet compressors, 411–426
  critical selection criteria, 411–419
  design, 419–424
  testing, 424–426
Multiposition switch, 96
Multistage compression, 5, 8, 25–27, 108
Multistage overhung design, 242

N method, of compressor performance
    prediction, 447, 451
n value, 23
National Bureau of Standards, 193
Natural gas storage, applicability of
    HOFIM compressors, 344
Negative damping, 316
Niresist, 50
Nitrogen sparging, for lube oil purifica-
    tion, 372
Nodular iron, 46
No-flow switch, 84, 94
Noise suppression, 172
Nonconformist gases, 441
Noncontacting gear coupling, 350, 351
Noncontacting rod drop probes, 94
Noncooled compression, 79
Nonlubricated couplings, 348–350
Nonlubricated services, 49, 55, 65, 82, 85,
    86, 99–109
  limitations, 85

Nonoverloading, 217
Nonreversing loads, 47, 75
Normal clearance, 28
Nova, 325
Novacor, 321, 322
Nozzle fabrication, 288–291
Nozzle location (orientation), 228, 229,
    233, 234, 283, 284

Oil additives, not beneficial in couplings,
    353
Oil contamination, 177
Oil film seals, 263–265, 370
Oil passages, in crankshafts, 39
Oil scraper, 101
Oil separator(s), 85, 107
Oil slinger, 67, 101
Oil whirl, 268
Oil wiper ring, 65, 66
Oil-flooded single screw compressors,
    173–177
Oil-free compressor(s):
  helical screw type, 153–155
  nonlubricated recips, 49, 55, 65, 82, 85,
    86, 99–109
Oil-injected screw compressors, 154, 155,
    160, 162
  operating principle, 162
Oil-related fires, 336
On-line redundancy, 374
On-line shaft alignment monitoring,
    361–364
On-line torque measurement, 357–365
On-stream washing, 321
Open impellers, 250, 254, 300
Operating point, 376
Optimum efficiency, 299
O-rings, 227, 243, 277, 330
  explosive decompression, 331
Overhead tank, 264, 265
Overhung impeller design, 239–242
Overhung impeller turbomachinery,
    equipped with magnetic bearings, 335
Overload margin, 412, 413, 419, 423
Overpump valve, 131, 133
Overspeed testing, 303, 309
Over-the-impeller volute, 240, 246, 248
Oxygen compressors, 85, 86, 239

Packaged jacket cooling water system, 81
Packaged rotary screw compressor, 152,
    153
Packed plungers, 115, 117

Packing cases, 60, 63, 86–88
  for labyrinth compressors, 99
Packing cooling, 60, 63, 64, 86
Packing lubrication, 64
Packing pressurization, 88
Packing rings, 63, 86
  wear monitoring of, 94
Packing vents, 65–67, 86
Paraffin oils, 117
Parasitic flows, 414, 415
Parasitic shear loss, 336
Partial pressure, 11, 13, 14
Partial saturation, 14
Partition packing, 66
Passage width restriction, due to fouling,
  321
Passage-type unloader, 72–74
Peak torque, 37, 69
Peng-Robinson, 193
Perfect gas formula, 11, 12
Perfect gas law, 10
Performance:
  at new conditions, 427–435
  of reciprocating compressors, 69–97
Performance calculations, for centrifugal
  compressors, 445–491
Performance control, 374, 380–386
  response plots, 381, 386
Performance curve shape, 396–410,
  412–418
Performance deterioration, 357, 358
Peripheral velocity, 225, 396
Photoelasticity, 121
Pin load reversals, 47, 75
Pipeline compressor, 240
Piping connections, at compressor head,
  243
Piston materials, 52
Piston ring leakage, 184, 185
Piston ring seals, for lobe-type compres-
  sors, 137
Piston rings, 52, 55, 82, 86
  for hyper compressors, 113, 117
Piston rods, 55, 57, 63
Piston seals, for hyper compressors, 113
Piston speeds, 34, 49, 205
Piston(s), 46, 52, 55, 114
Plastic packing, 115
Plate springs, 121
Plate valve(s), 54, 56, 57
Plug-type unloader, 72, 73
Plungers, 117

Pneumatic elevator (unloading) applica-
  tions, 141
Pneumatic switches, 94, 95
Pneumatic transmitters, 95
Polymer deposits, sensed by torque moni-
  toring, 357
Polymerization, 319–321
  avoidance of, 155, 167, 168, 170
  factors influencing, 319
Polymers, in hyper compression services,
  117
Polytropic compression, 221–224, 375
Polytropic cycle, 22, 23
Polytropic efficiency, 375, 429–435
Polytropic head, 222–224, 374, 386, 397
  per stage, 224
Polytropic process, 127
Poppet valves, 56, 59, 121
  for hyper compressors, 121
Ported valve(s), 58, 97
Port-type unloader, 72–74
Position sensors, in magnetic bearings,
  332–334
Power amplifier, 334, 335
  loose connection experience, 337
Power calculation, for screw compressors,
  158, 159
Power consumption, of magnetic bearings,
  336
Power ratings:
  of labyrinth piston compressors, 109
  of reciprocating compressors, 33
Power requirement(s), 23, 27
  for dynamic compression, 224
Power tolerance, for screw compressors,
  159
Preliminary lubrication, 99
Preload, of compressor valves, 182
Pressure capability, 280, 281
Pressure coefficient, 224, 225, 297, 299,
  409, 428–435
Pressure and flow variations during
  surge, 388
Pressure head, 375
Pressure instrumentation, 88–94
Pressure limiter, 128, 130, 133
Pressure load, on rider bands, 86
Pressure pulsations, 125, 188, 189
  in hyper compressor piping, 125
Pressure ratings
  for centrifugal compressors, 244
  of reciprocating compressors, 33

Pressure regulation, on lube oil supply systems, 369
Pressure switches, 89
Pressure-balanced piston, 47
Pressure-dam bearing, 268, 269
Pressurization:
 of distance pieces, 67, 87
 of packing, 88
Pressurized crankcase, 106, 107
Pretensioning, 117, 125
Process variable (PV), 382, 384, 385, 392
Programmable controllers, 97
Properties of common gases, 437–444, 450, 454, 471
Proportional gain, 385
Proportional offset, 383, 385
Proportional-integral (PI) control algorithms, 382, 384, 385, 392
Proportional-integral-derivative (PID) control algorithms, 382, 384, 385, 392
Protective devices, 88–96
Pseudo critical conditions, 21, 441, 471
Pseudo molecular weight, 19
Pseudo reduced conditions, 16
Psychrometric work, 15
Pulsation effects, 37, 172, 188, 189
Pulsation snubbers, 125
Pulsation vessel, 197
Pump-to-point lubrication, 83–85
 disadvantage, 84
 instrumentation for, 84
Purge air, 335
PTFE (Teflon), 52, 55, 60, 82, 86, 285
PV diagram(s), 5–9, 29, 30, 56, 62, 70, 181–191
 of diaphragm compressors, 129, 132, 134
 of screw compressor, 158
PVC (polyvinyl chloride) compression, 149

Quality control, 293, 295, 307, 309, 311

Radial bearings, 268
Radial bladed impellers, 299
Radially split design (vertically split), 209, 211, 227
Radiography, 308
Range of impeller sizes, 303
Ratio of specific heats, 20, 21

Reciprocating components, 55
Reciprocating compressor:
 advantages, 179
 performance, 179–200
 sizing, 200–207
Recirculators (boosters), 111, 234
Recycle (bypass), 70
Recycle control, 373, 389, 390
Recycle valve, 389, 390, 393
Redlich-Kwong, 193
Reduced conditions, 16
Reduced flow rate, 377, 378
Reduced polytropic head, 377
Reduced pressure, 16
Reduced temperature, 16
Reevaluating performance, flow chart, 432
Reflex gauge, 91
Refrigeration process, 237, 413, 414
 restricted selection options, 414
Refrigeration service, 237, 282, 283, 413, 414
Reinjection service, 244
Relative humidity, 14
Reservoir tank, 367
 instrumentation associated with, 367–369
Reset (integral) windup, 385, 386
Residual unbalance, desired in couplings, 351, 352
Resistance curves, 376
Resistance temperature detectors (RTDs), 89, 93, 94
Retrofit, of axial inlet, 338, 339
Return bend (crossover), 245–247, 305, 396
Return channel, 296, 297, 305
Reversal of load, 194
Reversing flow (surge), 217–220, 276, 373, 374, 378, 386–393, 397, 404–407
Reynolds number, 428
Rider rings (rider bands), 52, 53, 55, 82, 86
 wear monitoring for, 94
Rise to surge, 299, 301
River water, 51, 81
Riveted impellers, 250–253, 307
Rocking beam, 123
Rod drop indicators, 94
Rod load limitations, 109
Rod load reversal, 47, 75

Rod loads, 71
Rod stuffing box (packing case), 60, 63,
    86–88
    for labyrinth compressors, 99, 104, 105
Rolled threads, 43, 57, 63
Room-temperature vulcanizing joint com-
    pound (RTV), 288
Ross Hill Controls, 342
Rotary piston blowers, 137–141
    applications for, 141
    capacities, 141
Rotary piston machines, 137–141
    applications, 141
    capacities, 141
Rotary positive blowers, 137–141
Rotary screw compressors, 151–177
    application areas, 154
    operating principles, 155, 156
    performance calculations, 156–159
Rotary transformer, 361–364
Rotating flowpath, 247–250
Rotary piston, 155
Rotating stall, 259
Rotative speed, of reciprocating compres-
    sors, 34
Rotor application range, 255, 256
Rotor balancing, 257
Rotor dynamics, 312–318
Rotor instability, 312, 315
Rotor laminations, 332
    failures and solutions, 336
Rotor nomenclature, 248, 249
Rotor technology, 247–250
RTDs, 89, 93, 94
RTV (room-temperature vulcanizing joint
    compound), 288
Running misalignment, 349

Sacrificial coating, 324
    payback for, 324
Safety coupling, 350, 351
Safety enhancement, with magnetic bear-
    ings, 336
Saturated gas, 14
Saturation, 15, 194
Saturation temperature, 12
Scavenging effect, 149
Scavenging gas, 106
SCFM (definition), 27
Screw compressors, 151–177
    advantages, 171
    cost, 171

Screw compressors (*Cont.*):
    disadvantages, 171–173
Scrubbers, 281
Seal balance, 282
Seal gas filter, 328, 330
Seal gas monitoring, 328–330
Seal oil console, 370–372
Seal oil supply systems, 367–372
    materials of construction, 371
Sealing:
    of hyper compressors, 113
    of lobe-type blowers, 137
Sealing arrangements, for screw compres-
    sors, 166
Sealing liquid, heat rise of, 145
Sealing problems, labyrinth compressors,
    103
Seals:
    for centrifugal compressors, 259–268
    labyrinth type, 259
Second law of thermodynamics, 9
Secondary compressors, 111–126
    capacities, 111
    power requirements, 112
    pressure developed, 111
    thermodynamics, 112
Section curves, not suitable for sideload
    compressors, 426
Section mismatching, 281
Sectional head rise, 419
Selection procedure, centrifugal compres-
    sors:
    English units, 451, 469–481
    metric units, 455, 469–481
Self-contained cooling, 51, 53, 64
Self-equalizing bearing, 271
Semiautomatic welding, 308
Semi-open impellers, 250
Separator (sour seal oil trap), 264, 368,
    370, 371
Separator(s), 145, 148
    in double-screw compressors, 169
    in single-screw compressors, 175
Separator sump instrumentation, 91
SermaLon coating, 322
Sermetel® W coating, 357
Sewage treatment plants, 141
Shaft sleeves, 249
Shaft vibration criteria, 313
Shear ring groove, 243, 279
Shear-ring enclosure, 228, 229, 241–243,
    296

Shockless entry, 406
Shop performance test:
  documentation used for predicting per-
    formance under new conditions,
    427–430
  of sideload compressors, 425
Shortcut calculations, for centrifugal com-
  pressor selection, 445–492
Shrinking techniques, 119
Shunt holes, 259, 261
Shutdown piston, 261, 264
Shutdown seal, 261, 264, 327
Sideload compressors, 411–426
Sideload pressure level, 419
Sidestream compression, 233, 235–238,
  293
Sight flow indicators, 90
Signal noise, 384
Silencing, 172
Single dry seals, 327, 328
Single-acting, 5, 50
Single-screw compression, 173–177
Single-stage compression, 5
  limitations, 25, 33
Sintered carbide, 113, 114, 116
Sleeve, for coast-down bearing, 334
Sleeve bearings, 268
Slide piston (slide valve), 163, 165, 175,
  176
Slide ring (mechanical face) seals, 166, 167
Slide valve (slide piston), 163, 165, 175,
  176
Sliding feet, 285
Sliding vane compressors, 142, 143
  capacities, 143
  operating principles, 142
  pressure capabilities, 143
Slip factor, 397
Slot welding, 309–311
Snubbers, 125
Soil condition(s), 35
Solid-state instruments, 91
Sonic velocity, 408, 420
Sound levels, in screw compressors, 172
Sour seal oil trap (drainer, separator),
  264, 368, 370, 371
Spacer sleeves, 248–250
Specific gas constant, 12
Specific gravity, of gases, 19
Specific heat, 20, 21
Specific humidity, 14
Specific mechanical energy of a fluid, 374
Specific speed, 254, 297, 298

Specific volume, 18, 112
  of ethylene at high pressures, 112
Speed, of diaphragm compressors, 136
Speed-based departure from fan laws, 408
Speeds, of pistons, 34, 49
Spillover volute, 246–248
Splash lubrication, 100
Splitter block lubrication, 83–85
Springs, 61, 104
Stability, 217–220, 258, 268, 312, 315,
  415–417
  improvement with honeycomb seals,
    262, 263
Stable range, 409, 415–417, 419
  at high Mach numbers, 423
Stage efficiency, 474
Stage design criteria, 296–306
Stage performance, 396–410
Stage pressure ratio limitations, 109
Stage spacing, 293, 312
Stage stability, 409
Stage stability improvement, 407
Staging, 196–198
Stagnation enthalpy, 375
Standard pressure and temperature
  (STP), 18
Standstill pressure, 106
Starting torque, 69
Static-filled cooling, 51, 79
Static levitation, 335
Stationary flowpath, 247–250, 296
Stator blades:
  adjustable, 215
  fixed, 214
Steady-flow process, 23
Step cylinder, 46, 49
Stepless capacity control, 73, 74
Stepless flow adjustment, for screw com-
  pressors, 163
Step-up gears, 153, 210, 242
Stonewall (choked flow), 220, 389, 396,
  397, 401, 402, 405, 406
  limited options with multiple inlet com-
    pressors, 413, 414
Straight-through compression, 233
Strain gauge, 121, 361
Strain gauge measurement, 279
Stress analysis, 303
Stress monitoring, 121
String (train) arrangement, 235–237, 323
Stroke:
  of labyrinth piston compressors, 109
  of reciprocating compressors, 34, 109

Suction throttling, 71, 163, 218–221
Suction valve throttling, 71, 163, 218–221
Suction valve unloading, 71–75
    drawbacks of, 74, 75
Suitability under uprated conditions, 429,
    430
    centrifugal compressors, 429, 430
Sulfuric acid, used as a sealing liquid, 149
Supercharging, of screw compressors, 168,
    170
Support head, 128
Surface finish:
    fouling risk, finish effects, 321, 322
Surge, 217–220, 276, 373, 374, 378,
        386–393, 396, 397, 404–407, 412, 413
    effect of surge, 389
Surge avoidance, in screw compressors,
    155
Surge drums, 125
Surge margin, 412, 413
Swirl, 306
Switch certification, 94
Switch logic, 89, 96
Switch set points, 95
Symbols, for units of measurement, 3
Synchronous excitation, 316
Synchronous motor applications, 342, 345
Synchronous vibration, 315
Synthetic lubricants, 117, 148, 174, 196,
    197

Tailrod, 47
Tandem cylinder(s), 46, 49, 50
Teflon (PTFE), 52, 55, 60, 82, 86, 285
Temperature, of gas in last stage of hyper
    compressors, 112
Temperature control, 81, 89, 93, 97
    on lube oil supply systems, 369
Temperature instruments, 89, 93
Temperature rise:
    in cooling water circuits, 80
    in dry screw compressors, 159, 171
    in liquid-injected compressors, 160
Temperature stratification, 424
Temperature-vapor pressure relationship,
    12
Test results:
    accuracy, 426
    documentation, 427
    evaluation, 426
Testing procedure, for centrifugal com-
    pressors, 425

T-groove seals, 331, 332
Theoretical compression cycles, 21–23
Theoretical power requirement, 26, 27,
    192
Theoretical volume flow, in screw com-
    pressors, 157
Theory of compression, 3–31
Thermocouples, 89, 93, 97
Thermodynamics, laws of, 9
Thermometers, 89
Thermostats, 99
Thermosyphon cooling, 51, 54
    application range, 79, 80
Three-dimensional blading, 301, 302
Three-piece impeller design, 302, 306
Throttling, of compressor suction,
    218–220
Throughbolts, 125
Thrust balance, 239, 256, 257
Thrust bearings, 165, 270–272, 274
    for screw compressors, 165
Thrust control, 339
Thrust disk, 248, 271, 273
Thrust-reducing seals, 325, 338–341
    controls for t-r seals, 340, 341
Tie-bar, 45
Tie-rod, 45
Tilting pad bearing, 268–270
    five-pad type, 269, 270
    load-between-pad, 269
    load-on-pad, 269
Timing gears, for helical screw compres-
    sors, 154
Torque, 37, 69
    fluctuations, 318
    peaks, 37, 69
Torque sensing, 357–365
Torque transmission, 347–357
Torquetronics torque sensing device,
    360–362
Torsional concerns, 78
Torsional natural frequencies, 316, 317
TorXimitor™ torque sensing device, 359,
    360
Total pressure, 11, 13
Train (string), 235–237
Transfer valve, 368
Transient torsional problem, 318
Transparent liquid gauge, 91
Trapped bushing seal, 266–268
Traps, 91, 92
Tridimensional fatigue strain, 121

Trimetal, 40
Triple diaphragm, 135
Truncated cylinder, 46, 49
Trunk piston compressor, 50
T-S diagrams, 16, 193
Tungsten carbide, 57, 113, 114, 116
Turndown map, 416–418
    required for documentation of test
       results, 426
Two-compartment distance piece, 46, 47,
    48, 65–67, 87

Ultrasonic inspection, 39, 308
Unavailability (entropy), 10
Unbalance response, 315, 316
Unbalance sensitivity, 312, 315
Unbalanced forces, 34, 40
Unit volume, in screw compressors, 157
Universal gas constant, 12
Unloading, 69, 71–75
    manual, 71
    pneumatic, 71
    stepless, 73, 74
Unstable control of compressors, 384, 385,
    394
U-tube seals, 113

Vacuum, limitation to attainment, 148
Vacuum dehydrator, 372
Valve area, 54
Valve crab assembly, 71
Valve dynamics, 62, 63
Valve flutter, 61
Valve guard, 57
Valve leakage, 186, 187
Valve lift, 54, 63
Valve motion, 61
Valve preload, 182
Valve reversal, 53
Valve seal, 57
Valve-in-piston compressor (VIP compres-
    sor), 96, 97
Valves:
    for hyper compressors, 119–121
    for low molecular weight gases, 73
    temperature monitoring, 5, 53, 55–60,
       97
    for VIP compressors, 97
Vaned diffusers, 406
    adjustable, 406
Vaneless diffuser, 248, 395, 406, 407
    narrowing to improve stability, 407

Vapor, definition of, 12
Vapor pressure, 12, 13
Vapor pressure curves, 438
Variable axial load, 340
Variable frequency drive, 342, 345
Variable inlet guide vanes, 379
Variable speed drivers, 77, 78, 342, 345
Variable speed performance, 218–220
Variable volume clearance pockets, 76–78
VDI Specification 2045, 159
Vector diagrams, 397, 398–406
Velocity distributions, 305
Velocity head, 375
Velocity relationships, 397–406
Vent cavities, in lobe-type blowers, 137
Venting:
    of distance pieces, 87
    of packing cases, 86, 87
Vertical orientation of cylinders, 36–39
Vibration instrumentation, 90
Vibration measurement, 312, 313
Vibration transducers, 312
Vinyl chloride recovery system, 148
VIP (valve-in-piston) compressor, 96, 97
Volume fraction(s), 20
Volume percent, of constituents, 18, 19
Volume ratio, 155
Volume references, 27
Volumetric compressor, 145
Volumetric efficiency, 28, 70, 75, 180, 182
    modifying factors, 29, 31, 70, 76
Volute, 240, 246, 248
Volute-type diffuser, 240

Water quality, 81
Water seal, 145–149
Water separation, in liquid-injected screw
    compressors, 168, 169
Water vapor, 13, 28
Water washing (flush injection), 357
Water-glycol cooling, 51, 79
Water-lubricated cylinders, 38
Water-sealed floating rings, 166, 167
Wear:
    accelerated due to cool cylinders, 79
    accelerated due to high forces, 86
Weight basis, of gas mixtures, 21
Weight reduction, achieved with magnetic
    bearings, 336
Welded cylinder(s), 46, 48
Welded impellers, 253–255, 301–312
Whispair jet, 140

Wiper rings, 65, 66
Wobble plate(s), 285, 288, 295
Working phases, of rotary screw compressors, 154–157
Wrap angle, in screw compressors, 160, 161
Wraparound flange, 137

Yoke, 123
Y-orientation of cylinders, 34

Z charts, 16
Z factor, 15, 16, 21, 23, 24
Zierau, S. M., 313
Z-type blading, 252, 253

ABOUT THE AUTHOR

Heinz P. Bloch is an internationally respected authority in
all areas of machinery operation, troubleshooting, and
repair. He was with the Exxon Corporation for over 20
years, and is now the president of Process Machinery Co.
Mr. Bloch is also the author or coauthor of eight other books,
including *Improving Machinery Reliability*, *Machinery
Failure Analysis*, *Machinery Component Maintenance and
Repair*, and *Major Process Equipment Maintenance*, as well
as more than 70 technical papers.